Ecological Studies, Vol. 176

Analysis and Synthesis

Ecological Studies

Volumes published since 1998 are listed at the end of this book.

M. Scherer-Lorenzen Ch. Körner E.-D. Schulze (Eds.)

Forest Diversity and Function

Temperate and Boreal Systems

With 102 Figures and 22 Tables

Springer

Dr. Michael Scherer-Lorenzen
Federal Institute of Technology Zurich (ETH)
Department of Agricultural and Food Sciences
Institute of Plant Sciences
Universitätsstr. 2
8092 Zürich, Switzerland

Prof. Dr. Christian Körner
University of Basel
Institute of Botany
Schönbeinstr. 6
4056 Basel, Switzerland

Prof. Dr. Ernst-Detlef Schulze
Max-Planck-Institute for Biogeochemistry
Hans-Knöll-Str. 10
07745 Jena, Germany

ISSN 0070-8356
ISBN 3-540-22191-3 Springer-Verlag Berlin Heidelberg New York

Library of Congress Control Number: 2004110616

Springer-Verlag is a part of Springer Science+Business Media
springeronline.com

Printed in Germany

Editor: Dr. Dieter Czeschlik, Heidelberg, Germany
Desk editor: Dr. Andrea Schlitzberger, Heidelberg, Germany
Cover design: *design & production* GmbH, Heidelberg, Germany
Typesetting and production: Friedmut Kröner, Heidelberg, Germany

31/3150 YK – 5 4 3 2 1 0 – Printed on acid free paper

Foreword

This past decade has witnessed a flurry of research activity focusing on the significance of the diversity of organisms regarding the capacity of ecosystems to capture resources and their role in regulating the stability and resilience of these systems. The results of this activity have been chronicled, debated and summarized, as noted in this volume. This discourse has been colored, of course, by the kinds of information available. Unfortunately, there are few results from explicit experiments on the diversity/functioning/stability relationships. We have had to turn to "experiments" that nature has provided, i.e. observing functioning in systems of varying diversity. However, as noted in this volume, it is difficult to draw firm conclusions from observational studies because of the complexity of natural environments in any given place and across gradients. It is just not possible to hold all habitat conditions constant, with only diversity varying, under natural conditions. This is not to say that such field observations are not of crucial importance in providing reasonable starting points for experimentation. It is just that these observations need to be augmented with experiments to clarify confounding factors, even though such experiments are difficult and costly.

During this past decade over a million hectares per year of natural forests have been converted to plantation forests, globally, as well as comparable amounts in afforestation projects on lands not previously forested. Since plantations are normally monocultures, with added fertilizers and pesticides, it becomes important to compare the full environmental and economic costs and benefits of varying species mixtures to see where savings can be made. Unfortunately, such full-cost accounting in agronomic practices is rarely done and hence society pays the price of off-site environmental impacts in the long run.

This volume is an important first step in launching the kinds of experiments and measurements that are needed to bring us much further along in our understanding of the role of diversity, at many levels, of the perennial woody systems that cover so much of the surface of the earth. The pitfalls of analysis of the information available are clearly demonstrated and yet the need for new and more comprehensive information is also made evident. In particular, a much greater investment must be made in long-term experiments that explicitly address the main questions posed in this book. This, in turn, will require that science-funding sources recognize that high-quality information on the ecosystem services provided by long-lived ecosystems, of varying complexities, which cover much of the earth's surface, will need a greater time commitment and a higher level of financial resources that are normally committed to ecological research. It has been clearly demonstrated that to unravel complex ecosystem processes in forest systems, such as those at Hubbard Brook in New Hampshire, there is no substitute for experimentation.

We have done the easy stuff, working experimentally with herbaceous communities, and have learned a great deal about the diversity/functioning/stability relationship. However, we now must move on to the next level and address those ecosystems that control a good portion of the carbon, nutrient and water balances of the earth – the forests. This volume will provide a important template for this next phase.

Stanford
September 2004

Harold Mooney

Preface

Evaluating the extent to which biological richness is necessary to sustain the earth's system and the functioning of individual ecosystems has emerged as one of the central research themes in ecology. Does biodiversity matter for ecosystem integrity, functioning, and the provision of goods and services?[1] This was the main question asked by the initial book summarizing this field of research (Schulze and Mooney 1993) that appeared as Volume 99 in the *Ecological Studies* Series. Since then, this field has largely been developed through the use of model systems, both mathematical and real (e.g., Hector et al. 1999; Loreau 2000; Tilman et al. 2001). For very practical reasons, the experimental/observational model systems were small in size, short-lived and even-aged, mainly herbaceous assemblages or microbial microcosms. Results obtained with these systems have stimulated the scientific debate enormously, and have been the basis and support for theory development. In essence, it was ascertained that, along with abiotic factors, biological richness is a major determinant of ecosystem functioning (Loreau et al. 2001). However, woody species were by and large excluded from this research, although forests (including plantations) cover over 30 % of the earth's land area (FAO 2001), produce 65 % of the annual carbon fixation (net primary production; Lieth 1975), and store almost 80 % of the carbon biomass of the planet (Watson et al. 2000). In addition, they are facing even larger changes in their biological diversity than herbaceous vegetation (WRI 2000), and the significance of forests for the vast majority of the world's terrestrial biodiversity has been acknowledged by the UN Convention on Biological Diversity (CBD) in its thematic work program on forest biological diversity (CBD 2001). It is thus a key ecological question whether links between diversity and functioning exist in forest ecosystems in a manner similar to those found in other ecosystems.

[1] By "ecosystem functioning" we mean the activities, processes, and properties of ecosystems that are influenced by their biota (see also Naeem et al. 2002). By "ecosystem goods and services" we mean the benefits people obtain from ecosystems (see also Daily 1997).

Besides possible effects of biodiversity on ecosystem processes, biodiversity-dependent ecosystem goods and services provided by forests are important economic factors in most countries and contribute to the welfare of millions of people (WRI 2000). The search for patterns and principles of the relation between biological richness and ecosystem functioning is therefore not only of scientific importance, but should provide guidance for the sustainable use of the world's forest resources.

This volume summarizes the results of a LINKECOL workshop held in Weimar, Germany, 13–15 June 2002. The basic idea behind the workshop was to extend the ongoing debate about the relationship between biodiversity and ecosystem processes to the forest realm. The meeting assembled experts from various fields of forestry, biology, and ecology who illustrated and synthesized existing knowledge on various aspects of the functional significance of forest diversity. The linkages between species and functional group diversity among various categories of biota were explored, asking the question, whether diversity in one trophic level (in this case trees) affects diversity in other functional groups (e.g., canopy insects). The workshop examined in particular the significance of the presence of a multitude of players for ecosystem processes, such as stand productivity or biogeochemical cycles. It further explored the significance of biological richness for the robustness/sensitivity of ecosystems to extreme events, stresses, and various forms of disturbances and interventions. Finally, the importance of diversity at different scales was considered, ranging from stand management to global issues.

We thank the European Science Foundation for their support under the "Linking Community and Ecosystem Ecology" Program (LINKECOL), and the Max-Planck Society. We would also like to express our thanks to all who contributed their efforts and ideas. We believe that the discussions and personal contacts made at the workshop and the contributions to this volume will act as a starting point for exploring the field of biodiversity and ecosystem functioning in boreal and temperate forests.

Zürich, Basel, and Jena
September 2004

M. Scherer-Lorenzen
Ch. Körner
E.-D. Schulze

References

CBD – Secretariat of the Convention on Biological Diversity (2001) Handbook of the convention on biological diversity. Earthscan Publications, London, Sterling

Daily GC (1997) Nature's services – societal dependence on natural ecosystems. Island Press, Washington, DC

FAO – Food and Agriculture Organization (2001) Global forest resources assessment 2000. Main report. FAO, Rome

Hector A, Schmid B, Beierkuhnlein C, Caldeira MC, Diemer M, Dimitrakopoulos PG, Finn J, Freitas H, Giller PS, Good J, Harris R, Högberg P, Huss-Danell K, Joshi J, Jumpponen A, Körner C, Leadley PW, Loreau M, Minns A, Mulder CPH, O'Donovan G, Otway SJ, Pereira JS, Prinz A, Read DJ, Scherer-Lorenzen M, Schulze E-D, Siamantziouras A-SD, Spehn E, Terry AC, Troumbis AY, Woodward FI, Yachi S, Lawton JH (1999) Plant diversity and productivity experiments in European grasslands. Science 286:1123–1127

Lieth H (1975) Primary production of the major vegetation units of the world. In: Lieth H, Whittaker RH (eds) Primary productivity of the biosphere. Springer, Berlin Heidelberg New York, pp 203–215

Loreau M (2000) Biodiversity and ecosystem functioning: recent theoretical advances. Oikos 91:3–17

Loreau M, Naeem S, Inchausti P, Bengtsson J, Grime JP, Hector A, Hooper DU, Huston MA, Raffaelli D, Schmid B, Tilman D, Wardle DA (2001) Biodiversity and ecosystem functioning: current knowledge and future challenges. Science 294:804–808

Naeem S, Loreau M, Inchausti P (2002) Biodiversity and ecosystem functioning: the emergence of a synthetic ecological framework. In: Loreau M, Naeem S, Inchausti P (eds) Biodiversity and ecosystem functioning: synthesis and perspectives. Oxford Univ Press, Oxford, pp 3–11

Schulze E-D, Mooney HA (1993) Biodiversity and ecosystem function. Ecological Studies, vol 99. Springer, Berlin Heidelberg New York

Tilman D, Reich PB, Knops J, Wedin D, Mielke T, Lehmann C (2001) Diversity and productivity in a long-term grassland experiment. Science 294:843–845

Watson RT, Noble IR, Bolin B, Ravindranath NH, Verardo DJ, Dokken DJ (2000) Land use, land-use change, and forestry. IPCC. Cambridge Univ Press, Cambridge

WRI – World Resources Institute (2000) World Resources 2000–2001: people and ecosystems: The fraying web of life. World Resources Institute, Washington, DC

Contents

Part C Biogeochemical Cycles

Contributors

BARRANTES, O.

Facultad de Veterinaria, Departamento de Agricultura y Economía Agraria, Miguel Servet 177, 50013 Zaragoza, Spain, e-mail: Olivia.Barrantes@unizar.es

BALDOCCHI, D.D.

University of California, Berkeley, Department of Environmental Science, Policy and Management & Berkeley Atmospheric Science Center, Ecosystem Science Division,151 Hilgard Hall, Berkeley, California 94720, USA, e-mail: baldocchi@nature.berkeley.edu

BORNIK, Z.

Green College, 6201 Cecil Green Park Road. Vancouver, BC V6T 1Z1, Canada, e-mail: zosiabornik@hotmail.com

BROCKERHOFF, E.

Forest Research, P.O. Box 29237, Christchurch, New Zealand, e-mail: Eckehard.Brockerhoff@ForestResearch.co.nz

DHÔTE, J.-F.

Laboratoire d'étude des Ressources Forêt-Bois, UMR INRA-ENGREF 1092, 14 rue Girardet, 54042 Nancy, France, e-mail: dhote@nancy-engref.inra.fr

DUELLI, P.

Swiss Federal Research Institute WSL, 8903 Birmensdorf, Switzerland, e-mail: peter.duelli@wsl.ch

GLEIXNER, G.

Max-Planck-Institute for Biogeochemistry, Hans-Knöll-Strasse 10, 07701 Jena, Germany, e-mail: ggleix@bgc-jena.mpg.de

Gracia, C.

Universitat de Barcelona, Departament d'Ecologia, Av. Diagonal 645, 08028 Barcelona, Spain, e-mail: cgracia@ub.edu

Hahn, V.

Max-Planck-Institute for Biogeochemistry, Hans-Knöll-Strasse 10, 07701 Jena, Germany, e-mail: vhahn@bgc-jena.mpg.de

Hättenschwiler, S.

University of Basel, Institute of Botany, Schönbeinstrasse 6, 4056 Basel, Switzerland. Present address: Center of Functional Ecology and Evolution, CEFE-CNRS, 1919 route de Mende, 34293 Montpellier, France, e-mail: Stephan.Haettenschwiler@unibas.ch

Hector, A.

Institute of Environmental Sciences, University of Zurich, Winterthurerstr. 190, 8057 Zürich, Switzerland, e-mail: ahector@uwinst.unizh.ch

Holdenrieder, O.

Federal Institute of Technology Zurich (ETH), Department of Environmental Sciences, Forest Pathology and Dendrology Group, Rämistr. 101, 8092 Zürich, Switzerland, e-mail: ottmar.holdenrieder@env.ethz.ch

Ibáñez, J.J.

Universitat Autònoma de Barcelona, Centre for Ecological Research and Forestry Applications, 08193 Bellaterra, Barcelona, Spain, e-mail: j.ibañez@creaf.uab.es

Inchausti, P.

Ecole Normale Supérieure, Laboratoire d'Ecologie, 46 rue d'Ulm, 75005 Paris, France. Present address: ECOBIO UMR 6553, Université de Rennes 1, 35042 Rennes, France, e-mail: Pablo.Inchausti@univ-rennes1.fr

Jactel, H.

Laboratory of Forest Entomology and Biodiverity, UMR BIOGECO, INRA, 69 Route d'Arcachon, 33612 Cestas Cedex, France, e-mail: herve.jactel@pierroton.inra.fr

Jones, H.E.

Centre for Ecology and Hydrology, Merlewood, Windermere Road, Grange-over-Sands, Cumbria LA11 6JU, UK. Present address: The Orchard, Oldfield Road, Windermere, Cumbria LA23 2BY, UK, e-mail: hej44@btinternet.com

Körner, Ch.

University of Basel, Institute of Botany. Schönbeinstr. 6, 4056 Basel, Switzerland, e-mail: Ch.Koerner@unibas.ch

Koricheva, J.

University of Turku, Department of Biology, 20014 Turku, Finland, e-mail: julkoric@utu.fi

Kramer, C.

Max-Planck-Institute for Biogeochemistry, Hans-Knöll-Strasse 10, 07701 Jena, Germany, e-mail: ckramer@bgc-jena.mpg.de

Mason, W.L.

Northern Research Station, Forest Research, Roslin, Midlothian EH25 9SY, Scotland, UK, e-mail: bill.mason@gsi.gov.uk

Mata, T.

Universitat Autònoma de Barcelona, Centre for Ecological Research and Forestry Applications, 08193 Bellaterra, Barcelona, Spain, e-mail: t.mata@creaf.uab.es

McNamara, N.

Centre for Ecology and Hydrology, Merlewood, Windermere Road, Grange-over-Sands, Cumbria LA11 6JU, UK. Present address: Centre for Ecology and Hydrology, Lancaster Environment Centre, Library Avenue, Bailrigg, Lancaster, LA1 4AP, UK, e-mail: NMCN@ceh.ac.uk

Müller-Starck, G.

Technische Universität München, Faculty of Forest Science and Resource Management, Section of Forest Genetics, Am Hochanger 13, 85354 Freising, Germany, e-mail: mueller-starck@forst.tu-muenchen.de

Mund, M.

Max-Planck-Institute for Biogeochemistry, Hans-Knöll-Strasse 10, 07745 Jena, Germany, e-mail: mmund@bgc-jena.mpg.de

Pautasso, M.

Federal Institute of Technology Zurich (ETH), Department of Environmental Sciences, Forest Pathology and Dendrology Group, Rämistr. 101, 8092 Zürich, Switzerland. Present address: Sheffield University, Animal & Plant Sciences, Biodiversity & Macroecology, Western Bank, S10 2TN Sheffield, UK, e-mail: m.pautasso@sheffield.ac.uk

Potvin, C.

McGill University Montreal, Stewart Biology Building, 1205 Docteur Penfield, W6/8 Canada, e-mail: catherine.potvin@mcgill.ca

Pretzsch, H.

Technische Universität München, Faculty of Forest Science and Resource Management, Chair for Forest Yield Science, Am Hochanger 13, 85354 Freising, Germany, e-mail: H.Pretzsch@lrz.tu-muenchen.de

Reynolds, G.

Imperial College London, Department of Agricultural Sciences, Wye Campus, Ashford, Kent TN25 5AH, UK, e-mail: glen@rsdanum.net

Sachse, D.

Max-Planck-Institute for Biogeochemistry, Hans-Knöll-Strasse 10, 07701 Jena, Germany, e-mail: dsachse@bgc-jena.mpg.de

Scherer-Lorenzen, M.

Max-Planck-Institute for Biogeochemistry, Hans-Knöll-Strasse 10, 07701 Jena, Germany. Present address: Federal Institute of Technology Zurich (ETH), Department of Agricultural and Food Sciences, Institute of Plant Sciences, Universitätstrasse 2, 8092 Zürich, Switzerland, e-mail: michael.scherer@ipw.agrl.ethz.ch

Scheu, S.

Darmstadt University of Technology, Institute of Zoology, Schnittspahnstrasse 3, 64287 Darmstadt, Germany, e-mail: scheu@bio.tu-darmstadt.de

Schmid, B.

Institute of Environmental Sciences, University of Zurich, Winterthurerstr. 190, 8057 Zürich, Switzerland, e-mail: bschmid@uwinst.unizh.ch

Schubert, R.

Technische Universität München, Faculty of Forest Science and Resource Management, Section of Forest Genetics, Am Hochanger 13, 85354 Freising, Germany. Present address: Hochschule Zittau/Görlitz (FH), Fachgruppe Biotechnologie, Külzufer 2, 02763 Zittau, Germany, e-mail: RSchubert@hs-zigr.de

Schulze, E.-D.

Max-Planck-Institute for Biogeochemistry, Hans-Knöll-Strasse 10, 07745 Jena, Germany, e-mail: dschulze@bgc-jena.mpg.de

Stenlid, J.

University of Agricultural Sciences, Department of Forest Mycology and Pathology, Box 7026–750 07 Uppsala, Sweden,
e-mail: Jan.Stenlid@mykopat.slu.se

Vayreda, J.

Universitat Autònoma de Barcelona, Centre for Ecological Research and Forestry Applications, 08193 Bellaterra (Barcelona), Spain,
e-mail: j.vayreda@creaf.uab.es

Vilà, M.

Universitat Autònoma de Barcelona, Centre for Ecological Research and Forestry Applications, 08193 Bellaterra (Barcelona), Spain,
e-mail: montse.vila@uab.es

Wirth, C.

Max-Planck-Institute for Biogeochemistry, Hans-Knöll-Strasse 10, 07745 Jena, Germany, e-mail: cwirth@bgc-jena.mpg.de

Ziehe, M.

Georg-August-University Göttingen, Institute of Forest Genetics and Forest Tree Breeding, Büsgenweg 2, 37077 Göttingen, Germany,
e-mail: mziehe@gwdg.de

Part A Introduction

1 The Functional Significance of Forest Diversity: the Starting Point

M. Scherer-Lorenzen, Ch. Körner, and E.-D. Schulze

1.1 Introduction

The dramatic and accelerating changes the earth's biota has undergone over the last decades have led to considerable research effort toward understanding the nature of biotic control over the processes within ecosystems. Predicting the consequences to the ecosystem of changes in species numbers, in distribution patterns of taxa, and in shifts of dominance that result in altered trophic interactions between organisms, has become a major challenge for community and ecosystem ecology. Does biodiversity matter for ecosystem integrity, functioning, and the provision of goods and services? This was the original question posed in a volume in *Ecological Studies* published in 1993 that started this field of research (Schulze and Mooney 1993). However, this question remained basically unanswered with respect to forests. It is the aim of the present book to summarize the state of knowledge with respect to forests, focusing on the temperate and boreal regions.

1.2 Applying a New Ecological Framework

The recent advances of research in the field of biodiversity and ecosystem functioning (Schulze and Mooney 1993; Kinzig et al. 2002b; Loreau et al. 2002) were accompanied by two remarkable features: first, a merging or increasing overlap of two disciplines in ecology that had followed separate ways in exploring the "nature" of ecosystems in the past, namely, population or community and ecosystem ecology (Likens 1992; Grimm 1995); second, and related to this first feature, the evolution of a new synthetic ecological framework that underlines the active role of the biota and its diversity in governing environmental conditions within ecosystems (Lawton 2000; Loreau et al. 2001; Naeem 2002) up to global processes (IPCC 2001).

Ecological Studies, Vol. 176
M. Scherer-Lorenzen, Ch. Körner, and E.-D. Schulze (Eds.)
Forest Diversity and Function: Temperate and Boreal Systems

In exploring biodiversity, community ecology has seen the distribution and abundance of species as a function of abiotic (physical and chemical) conditions and biotic (interactions among species such as competition or predation) factors. Examples for forests are: (1) the apparent increase in tree species richness along latitudinal gradients from boreal to tropical regions (Ricklefs 1977) or within continents (Silvertown 1985) reflects parallel gradients in physical conditions such as temperature and moisture, differences in time periods without major climatic changes, or many other factors varying in parallel with latitude (Pianka 1966; Stevens 1989; Iwasa et al. 1993); (2) differences between highly diverse early-successional woody communities and late-successional species-poor forests in central Europe have been explained by outcompetition of light-demanding species by shade-tolerant ones (Küppers 1984). In contrast, ecosystem ecology has looked at ecosystems independently of species diversity. It was the flow of energy and the fluxes and pools of elements that were important, although data were often taken on a species level and then aggregated to the whole ecosystem (Grimm 1995). The compilation of the results from the IBP (International Biological Program) study sites in deciduous forests may serve as an example here (Khanna and Ulrich 1991; Röhrig 1991). Similarly, biogeochemistry has treated ecosystems as series of linked compartments rather than as associations of species, although this always represented an operational convenience more than a hypothesis that species traits were irrelevant (Schimel 2001). However, the similarity among species in basic functional properties such as photosynthetic pathway, and the finding that plant productivity is dependent on the energy absorbed rather than on species identities, initially led to the use of earth system models that have little diversity content, but rather use only the color of the land surface (Mooney 2001).

Applying the new emerging framework, a specific ecosystem function is seen as a function of (1) biodiversity and the functional traits of the organisms involved, (2) associated biogeochemical processes, and (3) the abiotic environment. Thus, the active role of the biota and its diversity in governing environmental conditions is underlined. It is important to note, however, that even Tansley in his first definition of an ecosystem mentioned the influences of the organisms on the physical system, although not from a diversity perspective (Tansley 1935). The insight that biodiversity and the feedback of the biosphere on global processes cannot be neglected, and have a profound impact, has also been recognized by the modeling community: all but the most aggregated climate and ecosystem models incorporate the role of different functional types of plants defined by morphological and physiological traits (Schimel 2001; Schulze and Schimel 2001) – for instance being "broadleaf tree", "needleleaf tree", "C_3 grass", "C_4 grass", or "shrub" (Cox et al. 2000).

This volume explores the significance of tree diversity in temperate and boreal forests within this ecological framework, i.e., by exploring the relationship between forest biodiversity and ecosystem functioning.

1.3 The Road from Weidenberg to Weimar

More than 10 years of intensive research on biodiversity and ecosystem functioning has resulted in an exponentially growing number of publications, accentuated and synthesized by several important conferences and meetings. Although ecologists have been interested in effects of species and their numbers on ecological processes for a long time, the launch of the Scientific Committee of Problems of the Environment (SCOPE) program of 1991 entitled "Ecosystem Functioning of Biodiversity" definitively marked the start of the recent development of this scientific field. The start-up meeting held in Weidenberg/Bayreuth, Germany, in that year reviewed the state of knowledge (Schulze and Mooney 1993), which mostly consisted of a compilation of related studies from community and ecosystem ecology. It also marked the start of a hypothesis-based formulation of a comprehensive and articulated conceptual framework, graphically represented by a small number of hypothetical relationships between biodiversity and ecosystem processes, namely, that diversity shows (1) no effect on ecosystem function ("null hypothesis"), (2) a linear relationship, or (3) an asymptotic relationship wherein species loss initially has only a weak effect, but which accelerates as more species are lost (Vitousek and Hooper 1993). In the following period, an in-depth examination of the functional role of biodiversity in various ecosystems of the world was performed within the SCOPE program, later to be expanded as part of the Global Biodiversity Assessment (GBA; Heywood and Watson 1995; Mooney et al. 1996). It became clear that correlation studies looking at the impact of biodiversity on ecosystem processes could hardly detect any causal mechanisms of biodiversity effects and that covarying factors such as soil acidity or nitrogen could mask potential biodiversity-functioning relations. These ideas were originally formulated in a workshop at Mitwitz, Germany, in 1988, in which various experimental approaches of ecosystem studies were discussed (Mooney et al. 1991), ranging from natural catastrophes to designed layouts. Based on this knowledge and on results from earlier experiments on species interactions in multi-species communities, e.g., with algae (Tilman 1977) or with grasslands differing in species richness and composition (Tilman 1987), several experiments were initiated, manipulating biodiversity while keeping abiotic factors as constant as possible (e.g., Naeem et al. 1994; Tilman et al. 1996; Hector et al. 1999; for an overview see Schmid et al. 2002). Interestingly, the very first ecological experiment documented until now, which had also been analyzed by Darwin and mentioned in *On the Origin of Species* (Darwin 1872 p. 113), had a similar aim: to establish, on the basis of experiments, which species – both alone and in mixtures – make the most productive grasslands on different soil types (Hector and Hooper 2002). It is mainly these recent experiments that have spurred the tremendous debate and controversy among ecologists about the importance of biodiversity for ecosystem functioning, focusing on the validity of such experimental approaches, and on the relevance of several mechanisms responsible for the observed relations between diversity and function. In short, in the experiments that assemble communities differing in biodiversity by random draws of species from a fixed pool, it is difficult to separate effects due to the increasing probability that certain species with major impacts on ecosystem processes are present in higher diversity levels (the sampling effect) from effects due to niche complementarity (Aarssen 1997; Huston 1997; Wardle 1999; Scherer-Lorenzen, Chap. 17, this Vol.). Basis for the sampling-effect model is the notion that the functional characteristics of the dominant plants rather than their number largely control ecosystem processes (Grime 1997). Additionally, given the strong influence of extrinsic factors on both biodiversity and ecosystem processes, it has been questioned how relevant the patterns observed in biodiversity-functioning experiments are for interpreting species loss in natural communities (Grime 1997; Wardle et al. 1997; a review of this controversy is found in Kinzig et al. 2002a; Mooney 2002; Naeem et al. 2002). In 1999, a meeting held under the auspices of the International Geosphere-Biosphere Program–Global Change and Terrestrial Ecosystems (IGBP-GCTE focus 4) in Santa Barbara, California, USA, summarized the empirical findings and theoretical concepts that were published during the first 8 years since the first conference in Weidenberg. The resulting book documents the progress made in this field – in both conducting and interpreting experimental results and in developing sound ecological theory (Kinzig et al. 2002b). Another milestone in this series of important conferences was the "Synthesis Conference" held in Paris, France, in 2000, again organized under IGBP-GCTE and DIVERSITAS, which achieved a synthetic and balanced view of the knowledge and challenges in the fast growing area of research addressing biodiversity and ecosystem functioning (Loreau et al. 2001, 2002).

As one browses through the three important books that reviewed and summarized the knowledge about biodiversity-ecosystem-functioning research until now (Schulze and Mooney 1993; Kinzig et al. 2002b; Loreau et al. 2002), with the exception of the paper by Iwasa and colleagues (Iwasa et al. 1993) who modeled tree species diversity along latitudinal gradients (with a "traditional" community ecology perspective), no single contribution explicitly focuses on forest ecosystems. If forests are mentioned at all, it is only in relation to varying decomposer or litter diversities and their implications for soil processes such as decomposition (Mikola et al. 2002; Wardle and van der Putten 2002). Has the new field of research bypassed the forests? On the other hand, much work has been carried out in the forest sector on the ecological and socio-economic consequences of mixing (mostly commercially important) tree species, as compiled by Cannell et al. (1992), Kelty et al. (1992) and Olsthoorn et al. (1999). Further, the establishment of diverse forests is a legislative aim in European forest operations. But why have these findings been left almost unanalyzed within the biodiversity-ecosystem functioning framework (Scherer-Lorenzen et al., Chap. 17, this Vol., but see Bengtsson et al. 2000)? Among other reasons, it is this question that inspired the idea to organize a workshop in Weimar, Germany, in 2002 on the "Functional Significance of Tree Diversity in Temperate and Boreal Forests," experts from various fields of forest ecology invited to attend. This book summarizes the results of this workshop that was held under the auspices of the "Linking Community and Ecosystem Ecology" Program (LINKECOL) of the European Science Foundation.

1.4 Aims and Topics

The aim of our workshop was to check whether the statement made by von Cotta more than 175 years ago (1828) can be supported by re-analyzing the large amount of literature on mixed forests stands accumulated since then, and by compiling new data on this topic. In his "Instructions for silviculture" von Cotta noted: "Since not all tree species utilize resources in the same manner, growth is more lively in mixed stands and neither insects nor storms can do as much damage; also, a wider range of timber will be available everywhere to satisfy different demands..." (translation by H. Pretzsch). Productivity, resource use, pests, and disturbances: all these topics raised in this single sentence by von Cotta have been re-examined in the present volume. We only excluded socio-economic aspects – satisfaction of different demands – from our compilation, referring here to the work, for instance, of Olsthoorn et al. (1999).

To equally cover all forest biomes in one workshop and the volume at hand would clearly go beyond the scope of a concise review of existing knowledge and a focused discussion of diversity–function relationships. We therefore concentrate here on temperate and boreal forests, hoping that other forest types might be in the center of future discussions. Equally, a focus on a certain set of ecosystem traits and processes and functions is needed, and we selected three major groups that we think cover the most important aspects of ecosystem functioning: productivity and growth (Part B); biogeochemical cycles (Part C); and animals, pests, and disturbances (Part D).

The contribution by Körner (Chap. 2) introduces the concept of functional trait diversity, compiling a large amount of data on several traits of temperate tree species. The variation in those functional traits among species is enormous, and thus species richness and composition of forest communities could potentially have significant effects on ecosystem processes.

Part B covers a primary aspect of ecosystem functioning, namely, productivity and growth at the stand level, which integrate various processes in space and time, ranging from photosynthesis to mortality. Pretzsch (Chap. 3) first reviews theoretical considerations about consequences of mixing species for productivity. These hypotheses are then tested in long-term experiments that compare biomass production of species grown in monocultures and two-species mixtures. Following Pretzsch, who shows that generalizations about mixture effects are difficult to make even under standardized experimental conditions, the chapter by Vilá et al. (Chap. 4) explicitly focuses on factors that may confound diversity–productivity relationships resulting from observational studies. Exemplarily, they analyze a large data-set from the forest inventory of Catalonia, Spain, that supports a positive association between tree-species richness and stemwood production of trees (excluding the shrub layer). One aspect often ignored in studying biodiversity effects is the contribution of variation at the genetic level. Although the database for genetic diversity of tree species is relatively scarce, this aspect of biodiversity nevertheless has profound implications for an individual's ability to respond to stress, for its reproductive success, and for growth, as shown by Müller-Starck and colleagues for European tree species (Chap. 5).

Part C on biogeochemical cycles starts with a case study of a long-term mixture experiment at Gisburn, UK, where various aspects of nutrient cycling were studied in single-species stands and two-species mixtures (Jones et al., Chap. 6). How tree species diversity may or may not affect water use of plants is examined by Baldocchi (Chap. 7), theory and experimental data being reviewed across the scales of leaves, tree, and canopy. A crucial step in the cycling of nutrients within an ecosystem is decomposition of leaf litter. Because the chemical composition and structure of litter differs among tree species, diversity could also influence decomposition rates, as is discussed by Hättenschwiler (Chap. 8). The next two chapters explore potential effects of tree diversity for the carbon balance of forests, focusing on pools and fluxes within the soil compartment. While Gleixner et al. (Chap. 9) look at the processes and mechanisms responsible for the formation and the dynamics of soil organic matter at the molecular level, Mund and Schulze (Chap. 10) analyze the influence of silvicultural systems and their interactions with biodiversity for the soil carbon balance.

Part D on animals, pests, and disturbances deals with two widely accepted views, namely, that tree diversity should be positively correlated with the diversity of other trophic levels, and that it should lead to higher stability against biotic or abiotic disturbances. Scheu (Chap. 11) examines the plant diversity–animal diversity relationship by concentrating on the below-ground food web. He not only describes how tree diversity may determine the food web structure, but also how the below-ground community feeds back on the plant community. The link between plant diversity effects on higher trophic levels and a forest's response to disturbances is presented by Jactel et al. (Chap. 12) who performed a meta-analysis of tree diversity effects on insect pest infestations, testing the biodiversity–stability hypothesis. They explore the relevance of several responsible factors for such a relationship at both the stand and landscape levels, and conclude with implications for pest management. Pautasso and colleagues (Chap. 13) present a thorough review on the susceptibility of forest stands to fungal pathogens, which represents another important factor, besides pest insects, of economic risk in managed forests. They discuss how forest diversity may or may not influence this susceptibility, and how adopting the reverse view, i.e., how pathogens influence tree diversity, helps to understand the complex plant–pathogen dynamics. One of the most often cited examples for a positive diversity–stability relationship in forests is the presumed higher resistance of diverse tree communities to strong winds. However, as shown by Dhôte (Chap. 14), a closer look reveals a magnitude of interactions between factors contributing to the physical stability of a tree, including, besides wind itself, effects of location, developmental stage, and canopy characteristics. Finally, Wirth (Chap. 15) closes this section by relating the diversity of functional traits of boreal trees to the fire regime observed. He discusses implications for the biogeography of boreal forests and explores the significance of fire plant functional types for carbon cycling, linking the sections on disturbance and biogeochemistry.

What have we learned from analyzing the relationship between forest biodiversity and ecosystem functioning from different angles and what are the perspectives for future research? This is the subject of Part E. Scherer-Lorenzen et al. (Chap. 16) describe some experimental approaches for studying diversity–function relationships in forests that can overcome the big dilemma of observational studies, namely, that because of covarying factors it is nearly impossible to assign causality (see also Vilá et al., Chap. 4). Finally, in Chapter 17 we discuss the applicability of the main hypothesized mechanism for a positive relationship between plant diversity and ecosystem processes derived from grassland studies, namely, resource use complementarity, to forest ecosystems. The final synthesis reveals both the foresight of von Cotta (1828), and the progress made since then. But as in other prospering fields of science, a variety of questions about the significance of forest biodiversity remains to be answered.

References

Aarssen LW (1997) High productivity in grassland ecosystems: effects by species diversity or productive species? Oikos 80:183–184

Bengtsson J, Nilsson SG, Franc A, Menozzi P (2000) Biodiversity, disturbances, ecosystem function and management of European forests. For Ecol Manage 132:39–50

Cannell MGR, Malcolm DC, Robertson PA (eds) (1992) The ecology of mixed-species stands of trees. Blackwell, Oxford

Cox PM, Betts RA, Jones CD, Spall SA, Totterdell IJ (2000) Acceleration of global warming due to carbon-cycling feedbacks in a coupled climate model. Nature 408:1084–1087

Darwin C (1872) On the origin of species by means of natural selection. Murray, London

Grime JP (1997) Biodiversity and ecosystem function: the debate deepens. Science 277:1260–1261

Grimm NB (1995) Why link species and ecosystems? A perspective from ecosystem ecology. In: Jones CG, Lawton JH (eds) Linking species and ecosystems. Chapman and Hall, New York, pp 5–15

Hector A, Hooper R (2002) Darwin and the first ecological experiment. Science 295:639–640

Hector A, Schmid B, Beierkuhnlein C, Caldeira MC, Diemer M, Dimitrakopoulos PG, Finn J, Freitas H, Giller PS, Good J, Harris R, Högberg P, Huss-Danell K, Joshi J, Jumpponen A, Körner C, Leadley PW, Loreau M, Minns A, Mulder CPH, O'Donovan G, Otway SJ, Pereira JS, Prinz A, Read DJ, Scherer-Lorenzen M, Schulze E-D, Siamantziouras A-SD, Spehn E, Terry AC, Troumbis AY, Woodward FI, Yachi S, Lawton JH (1999) Plant diversity and productivity experiments in European grasslands. Science 286:1123–1127

Heywood VH, Watson RT (eds) (1995) Global biodiversity assessment. Cambridge Univ Press, Cambridge

Huston MA (1997) Hidden treatments in ecological experiments: re-evaluating the ecosystem function of biodiversity. Oecologia 110:449–460

IPCC – Intergovernmental Panel on Climate Change (2001) Climate change 2001: the scientific basis. Cambridge Univ Press, Cambridge, New York

Iwasa Y, Sato K, Kakita M, Kubo T (1993) Modelling biodiversity: latitudinal gradient of forest species diversity. In: Schulze E-D, Mooney HA (eds) Biodiversity and ecosystem function. Springer, Berlin Heidelberg New York, pp 433–451

Kelty MJ, Larson BC, Oliver CD (eds) (1992) The ecology and silviculture of mixed-species forests. Kluwer Academic Publishers, Dordrecht, Boston

Khanna PK, Ulrich B (1991) Ecochemistry of temperate deciduous forests. In: Röhrig E, Ulrich B (eds) Ecosystems of the world 7: temperate deciduous forests. Elsevier, Amsterdam, pp 121–163

Kinzig AP, Pacala SW, Tilman D (2002a) Preface. In: Kinzig AP, Pacala SW, Tilman D (eds) The functional consequences of biodiversity: empirical progress and theoretical extensions. Princeton Univ Press, Princeton, pp xiii–xvii

Kinzig AP, Pacala SW, Tilman D (eds) (2002b) The functional consequences of biodiversity: empirical progress and theoretical extensions. Princeton Univ Press, Princeton

Küppers M (1984) Carbon relations and competition between woody species in a central European hedgerow. 1. Photosynthetic characteristics. Oecologia 64:332–343

Lawton JH (2000) Biodiversity and ecosystem processes: theory, achievements and future directions. In: Kato M (eds) The biology of biodiversity. Springer, Berlin Heidelberg New York, pp 119–131

Likens GE (1992) The ecosystem approach: its use and abuse. Ecology Institute, Oldendorf

Loreau M, Naeem S, Inchausti P, Bengtsson J, Grime JP, Hector A, Hooper DU, Huston MA, Raffaelli D, Schmid B, Tilman D, Wardle DA (2001) Biodiversity and ecosystem functioning: current knowledge and future challenges. Science 294:804–808

Loreau M, Naeem S, Inchausti P (eds) (2002) Biodiversity and ecosystem functioning: synthesis and perspectives. Oxford Univ Press, Oxford

Mikola J, Bardgett RD, Hedlund K (2002) Biodiversity, ecosystem functioning and soil decomposer food webs. In: Loreau M, Naeem S, Inchausti P (eds) Biodiversity and ecosystem functioning: synthesis and perspectives. Oxford Univ Press, Oxford, pp 169–180

Mooney HA (2001) Biological diversity, evolution and biogeochemistry. In: Schulze E-D, Heimann M, Harrison S, Holland E, Lloyd J, Prentice IC, Schimel D (eds) Global biogeochemical cycles in the climate system. Academic Press, San Diego, pp 279–284

Mooney HA (2002) The debate on the role of biodiversity in ecosystem functioning. In: Loreau M, Naeem S, Inchausti P (eds) Biodiversity and ecosystem functioning: Synthesis and perspectives. Oxford Univ Press, Oxford, pp 12–17

Mooney HA, Medina E, Schindler D, Schulze E-D, Walker BH (eds) (1991) Ecosystem experiments. Wiley, Chichester

Mooney HA, Cushman JH, Medina E, Sala OE, Schulze E-D (eds) (1996) Functional roles of biodiversity – a global perspective. Wiley, Chichester

Naeem S (2002) Ecosystem consequences of biodiversity loss: the evolution of a paradigm. Ecology 83:1537–1552

Naeem S, Thompson LJ, Lawler SP, Lawton JH , Woodfin RM (1994) Declining biodiversity can alter the performance of ecosystems. Nature 368:734–737

Naeem S, Loreau M, Inchausti P (2002) Biodiversity and ecosystem functioning: the emergence of a synthetic ecological framework. In: Loreau M, Naeem S, Inchausti P (eds) Biodiversity and ecosystem functioning: synthesis and perspectives. Oxford Univ Press, Oxford, pp 3–11

Olsthoorn AFM, Bartelink HH, Gardiner JJ, Pretzsch H, Hekhuis HJ, Franc A (eds) (1999) Management of mixed-species forests: silviculture and economics. IBN-DLO, Wageningen

Pianka ER (1966) Latitudinal gradients in species diversity: a review of concepts. Am Nat 100:33–46

Ricklefs RE (1977) Environmental heterogeneity and plant species diversity: an hypothesis. Am Nat 111:376–381

Röhrig E (1991) Biomass and productivity. In: Röhrig E, Ulrich B (eds) Ecosystems of the world 7: temperate deciduous forests. Elsevier, Amsterdam

Schimel D (2001) Preface. In: Schulze E-D, Heimann M, Harrison S, Holland E, Lloyd J, Prentice IC, Schimel D (eds) Global biogeochemical cycles in the climate system. Academic Press, San Diego, pp xvii–xxi

Schmid B, Joshi J, Schläpfer F (2002) Empirical evidence for biodiversity-ecosystem functioning relationships. In: Kinzig A, Tilman D, Pacala S (eds) Functional consequences of biodiversity: experimental progress and theoretical extensions. Princeton Univ Press, Princeton, pp 120–150

Schulze E-D, Mooney HA (eds) (1993) Biodiversity and ecosystem function. Springer, Berlin Heidelberg New York

Schulze E-D, Schimel D (2001) Uncertainties of global biogeochemical predictions. In: Schulze E-D, Heimann M, Harrison S, Holland E, Lloyd J, Prentice IC, Schimel D (eds) Global biogeochemical cycles in the climate system. Academic Press, San Diego, pp 3–14

Silvertown J (1985) History of a latitudinal diversity gradient: woody plants in Europe 13,000–1000 Years B.P. J Biogeography 12:519–525

Stevens GC (1989) The latitudinal gradient in geographical range: how so many species coexist in the tropics. Am Nat 133:240–256

Tansley AG (1935) The use and abuse of vegetational concepts and terms. Ecology 16:284–307

Tilman D (1977) Resource competition between planktonic algae: an experimental and theoretical approach. Ecology 58:338–348

Tilman D (1987) Secondary succession and the pattern of plant dominance along experimental nitrogen gradients. Ecol Monogr 57:189–214

Tilman D, Wedin D, Knops J (1996) Productivity and sustainability influenced by biodiversity in grassland ecosystems. Nature 379:718–720

Vitousek PM, Hooper DU (1993) Biological diversity and terrestrial ecosystem biogeochemistry. In: Schulze E-D, Mooney HA (eds) Biodiversity and ecosystem function. Springer, Berlin Heidelberg New York, pp 3–14

Von Cotta H (1828) Anweisung zum Waldbau. Arnoldische Buchhandlung, Dresden

Wardle DA (1999) Is "sampling effect" a problem for experiments investigating biodiversity-ecosystem function relationships? Oikos 87:403–407

Wardle DA, Zackrisson O, Hörnberg G, Gallet C (1997) The influence of island area on ecosystem properties. Science 277:1296–1299

Wardle DA, van der Putten WH (2002) Biodiversity, ecosystem functioning and above-ground-below-ground linkages. In: Loreau M, Naeem S, Inchausti P (eds) Biodiversity and ecosystem functioning: synthesis and perspectives. Oxford Univ Press, Oxford, pp 155–168

2 An Introduction to the Functional Diversity of Temperate Forest Trees

CH. KÖRNER

2.1 Introduction

Because of the ease of handling, the study of the functional significance of plant diversity became in large a domain of grassland or old field research. Research with simple, synchronized herbaceous/grass communities of various composition helped in establishing a conceptual framework, in essence explaining effects of plant species diversity on ecosystem functioning as either being related to some sort of complementarity of resource use, to species interactions (facilitation and other mutualisms), or to insure against system failure in cases of severe disturbance or stress that affect different species to different degree (e.g., Loreau 2000; Scherer-Lorenzen et al., Chap. 17, this Vol.; see also the Preface to this volume). While not yet fully understood in these simpler systems, these plausible concepts still await their test in longer-lived plant stands, natural ecosystems and forests in particular. Short-lived plants can replace each other in one to few years and may spread clonally with associated ecosystem processes following rapidly. It may take more than 100 years to change the composition of forests and even longer for the associated ecosystem (soil) adjustments. Given the longevity of trees, year by year compound-interest effects of small trait differences may accumulate and become effective within the life of an individual. This chapter explores some of these trait differences among temperate forest tree species and presents a few comparisons with species in other climatic zones.

Does it matter which and how many tree species compose the forests that cover a third of Europe's land area? This question is not as new as one might think. When wood was the sole source of heating energy, the overall productivity of forests was more important than the production of a certain type of straight timber. Early forest research of the 19th century had explored the yield of monospecific vs. two- or multi-species stands and produced clear evidence that the yield of mixed forests (e.g., evergreen conifers with deciduous species) can, on average, be 10–20 % (some extreme cases up to 50 %) higher

Ecological Studies, Vol. 176
M. Scherer-Lorenzen, Ch. Körner, and E.-D. Schulze (Eds.)
Forest Diversity and Function: Temperate and Boreal Systems

than that of any monoculture in the same area (e.g., Gayer 1886; Table 2.1; see also Pretzsch, Chap. 3, this Vol.). However, the positive effects were commonly found on good soils, whereas on poor soils, mixtures may yield even less than certain monocultures (Kenk 1992). According to Wiedemann (1951), the range may go from +18 to −17 % of volumetric yield for mixed stands of spruce/beech and oak/beech, depending on location (soil).

Primarily mass-oriented considerations (productivity) became less significant when the dominant economic role of forests shifted from fuel to timber and energy was supplied from fossil sources. A massive rebuilding of the forested landscape followed, leading to forest types and species abundance strongly in contrast to the potential (Fig. 2.1). Today, 70 % of the European forests are conifer forests and 30 % broadleaved deciduous forests, the reverse of the pattern one would see in a natural landscape (Table 2.2). Most of these stands are close to monospecific.

In this volume, several authors are synthesizing current knowledge about the effects of forest species diversity on forest functioning and integrity. This introductory chapter will explore the extent to which common temperate zone forest tree species differ in their functional traits. As will be shown, these differences are substantial, and thus the presence or absence of a certain species could potentially influence ecosystem processes significantly. However, some of these traits may become functional only when different species interact and be less significant in monospecific stands. Other traits may emerge more strongly in monospecific stands and become insignificant in mixed stands. With the current evidence it seems nearly impossible to differentiate the expression and function of traits by species neighborhood. This assessment will thus consider variations in traits irrespective of the assemblage type in which sample trees have grown.

Although the temperate forest tree diversity in Europe is only about 10 % of that of temperate North America and less than 5 % of that found in temperate East Asia (largely because of past glaciation cycles and the east-west orienta-

Table 2.1. Examples of tree-species-mixture effects in the older forestry literature

Forest type	Increase in timber yield (%)	Reference
Fagus + *Pinus* vs. *Pinus* only	+ 20 to + 50 depending on site	Ertfeld (1953)[a]
Fagus + *Larix* vs. *Larix* only	+ 18	Burger (1941)[a]
Fagus + other deciduous *vs. Fagus* only	+ 26	Ertfeld (1953)[a]
Picea + *Abies* vs. *Picea* or *Abies* only	+ 37	Vanselow (1937)[a]
Fagus + *Picea* vs. *Fagus* only	+ 26	Ertfeld (1953)[a]
Fagus + *Picea* vs. sum of monocultures	+ 12 to + 18	Wiedemann (1943)

[a] Assmann (1961)

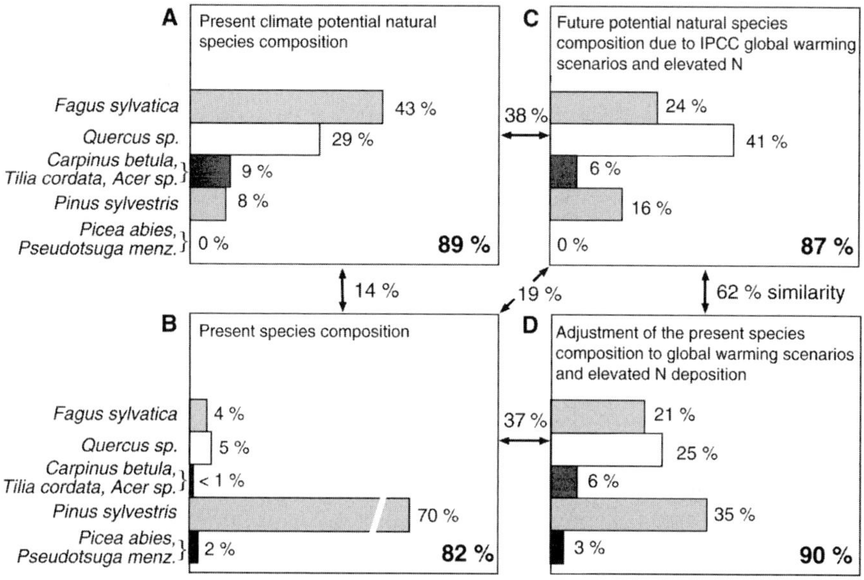

Fig. 2.1. Current potential natural forest ecosystems in the NE German lowlands (**A**) have been converted to type **B** forests over the last two centuries. A warmer climate with associated drier summers and high N input would induce succession toward a different type of unmanaged forest (**C**), but would also need adjustments in plantation forests (type **D**) to cope well with the new situation. *Bottom right percentages* refer to the contribution of these species to total forest area. (Anders and Hofmann 1996)

Table 2.2. Actual percentage of relevant tree species in relation to the total forest cover in various European countries. (Ellenberg 1996)

	Netherlands	Germany, western part	Germany, eastern part	Poland	Czech Republic and Slovakia	Austria	Switzerland
Forest cover (%)	7	28	27	24	34	46	27
Coniferous trees	86	69	79	88	69	85	70
Picea abies	+	42	25	9	49	58	40
Pinus sylvestris, Larix decidua	60	27	54	76	15	20	10
Abies alba	+	+	+	3	5	7	20
Exotic species, etc.	26	+	+	+	+	+	+
Deciduous trees	14	31	21	12	31	15	30
Fagus sylvatica	2	23	12	3	16	10	25
Quercus robur/ petraea	9	8	5	4	6	2	+
Soft wood	2	+	4	3	2	2	+
Others	1	+	+	2	7	1	5

tion of the continental divide), the ca. 20 native key tree species composing European temperate forests (out of a total of ca. 60 European temperate tree species) cover a wide range of functional traits, possibly as wide as their American and Asian counterparts. Some of this variation will be reviewed here.

Forest diversity can be defined in many ways by accounting for taxonomic composition (species diversity), diversity in tree age, functional diversity among taxa, genotype differences within taxa, and spatial diversity introduced by canopy stratification and different clustering of taxa/functional types (forest mosaics). In a given forest, trees may be grouped by a number of traits into so-called functional groups or functional types. The criteria for such groupings are nearly endless (Körner 1993; Mooney et al. 1995; Smith et al. 1997) and the transitions among groups are often gradual. Here is a selection of such criteria, some of which will be discussed later:

- successional stage of forest (Fig. 2.2)
- native vs. exotic (neophytes)
- light/shade tolerance (Fig. 2.3)
- photosynthetic capacity (Fig. 2.4)
- maximum leaf diffusive conductance for water vapor and stomatal density (Figs. 2.5, 2.6, 2.7)
- crown architecture (Fig. 2.8) and its change with tree age
- soil moisture preference
- mycorrhizal association
- maximum rooting depth (Fig. 2.9)
- maximum height of trees (dominant, subdominant, and minor trees)
- biometric relationships (Fig. 2.10)
- tree vigor, i.e., typical relative growth rate (Figs. 2.11, 2.12)
- ability to fix nitrogen with symbionts
- nitrogen recovery during leaf senescence
- phylogenetic origin (e.g., conifers vs. broad-leaf angiosperms)
- pollination (wind vs. insects)

	Early successional	Late successional
Evergreen conifers	*certain Pinus* sp.	*Abies, Picea*
Deciduous non conifers	*Populus, Alnus, Betula, Acer*	*Quercus, Tilia, Fagus, Carpinus*

Fig. 2.2. An example for grouping common temperate forest tree genera into functional types, here by their predominant position in a successional sequence and by phylogenetic origin. The latter corresponds to a set of other functional traits such as evergreen vs. deciduous, monopodial vs. sympodial (architecture), slow vs. rapid litter recycling, etc. Each of the four groups can be subdivided into ecological groups (also at the species level) depending on preferences for soil moisture, nutrients, and light

- age of first fruiting (Fig. 2.13)
- masting behavior (none vs. pronounced)
- fruit size (large vs. small)
- litter quality (see Hättenschwiler, Chap. 8, this Vol.)
- fire resistance (see Wirth, Chap. 15, this Vol.), etc.

2.2 Successional Stage

This category does not refer to a specific character, but rather to a syndrome, i.e., a typical combination of traits. Successional position of a species is one of the particularly important characteristics (Fig. 2.2). Any functional assessment of a diverse forest would have to account for the presence or absence of these cardinal functional types of trees.

There is a rich literature, impossible to exhaustively review here, that documents characteristic differences between early and late successional tree species. Among the common traits of early successional species are small seed size, high relative growth rate, low shade tolerance, high rates of photosynthesis, high water consumption, high tissue nutrient concentration, low tissue density, high specific leaf area, early reproduction, shorter life span. Early successional species tend to produce deeper roots early in their life, late successional ones tend to spread most of their fine roots beneath the litter layer, and these differences are inherent (Gale and Grigal 1987). Early succession tree species also produce seeds more regularly (no masting behavior). A most significant trait of earlier successional species is less efficient nutrient recovery from leaves at senescence and softer, more rapidly decomposing litter (Reich et al. 1997; Eckstein et al. 1999; Hättenschwiler, Chap. 8, this Vol.). The faster growth of early successional species is clearly facilitated by the mentioned tissue properties and a greater leaf area fraction (Walters et al. 1993). However, these traits change during ontogeny, so that seedlings or saplings of some late successional species may exhibit early successional characteristics (Gleeson and Tilman 1994), and some early successional species may develop late successional characteristics as they age (as for instance in pine species). Some of these traits even influence the responsiveness to environmental changes such as atmospheric CO_2 enrichment, with early successional species for instance being more responsive under high nutrient supply and late successional being more responsive in low light (Bazzaz and Miao 1993).

2.3 Variation in Gas Exchange Capacity and Associated Leaf Traits

The maximum rate of leaf net photosynthesis under ambient CO_2 concentration (photosynthetic capacity) varies from 2 to 22 μmol m^{-2} s^{-1} among temperate forest tree species (Figs. 2.3 and 2.4; Larcher 1969; Jarvis and Stanford 1986; Ceulemans and Saugier 1991, who summarize a global range of even 3 to 30 μmol m^{-2} s^{-1}). These very large differences are not associated with tree taxonomy, as is illustrated by the fact that both conifer and deciduous tree species jointly cover this range (Fig. 2.3). There is a weak negative correlation with shade tolerance of a species (see for instance Koike et al. 2001), but even within one of these categories photosynthetic capacity varies by a factor of 5. Because many other traits such as tissue density, tissue duration, dry matter allocation, and nutrient use influence growth, these differences in leaf photosynthesis do only weakly translate into relative growth rate. In other words, a similarly high photosynthetic capacity may be found in fast-growing willow and slow-growing oak, with the latter simply producing fewer leaves per unit of axial wood. This compensation at the canopy level is one of the reasons why

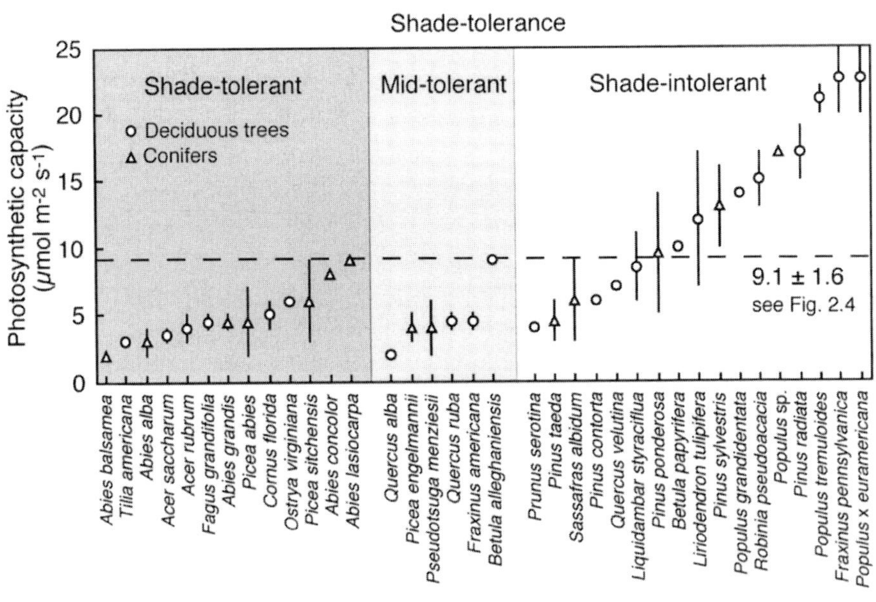

Fig. 2.3. Photosynthetic capacity of three groups of temperate tree species known for their preference for low, medium, and high light requirement/tolerance. Note the variation within each group. (Barnes et al. 1998)

the global variation of closed-canopy photosynthesis is much smaller, 15–40 μmol m^{-2} s^{-1}, according to Ceulemans and Saugier (1991).

Temperate forest trees do not represent a group distinctively different in terms of gas exchange capacity from tree species belonging to other climatic zones. There is also no systematic trend that would separate conifers from broad leaf evergreen and temperate deciduous species (on a projected leaf area basis), all varying around the global mean of ca. 9 μmol m^{-2} s^{-1} (Körner 1994; for deciduous–conifer comparisons, see also Schulze 1982). The variation is thus not clearly related to climate or leaf morphology, but rather seems to reflect leaf longevity (amortization) and canopy density (light climate). The more open tree crowns are (low LAI) the higher the leaf-specific photosynthetic capacity. As the net result of such balancing traits, the mean annual productivity of natural forests across the humid parts of the globe is almost completely explained by the duration of the growing period (Körner 1998). There is no general correlation between photosynthetic capacity and wood productivity in forests (Ceulemans and Saugier 1991). This does of course not hold for fertilized and yield-optimized plantation forests, conditions under which certain genera (*Populus*, *Eucalyptus*) with high photosynthetic capacity are clearly superior in terms of productivity to other genera. Hence, the diversity of photosynthetic capacity in tree species is not a trait to be seen in isolation. Such diversity generally follows from life history characteristics of whole

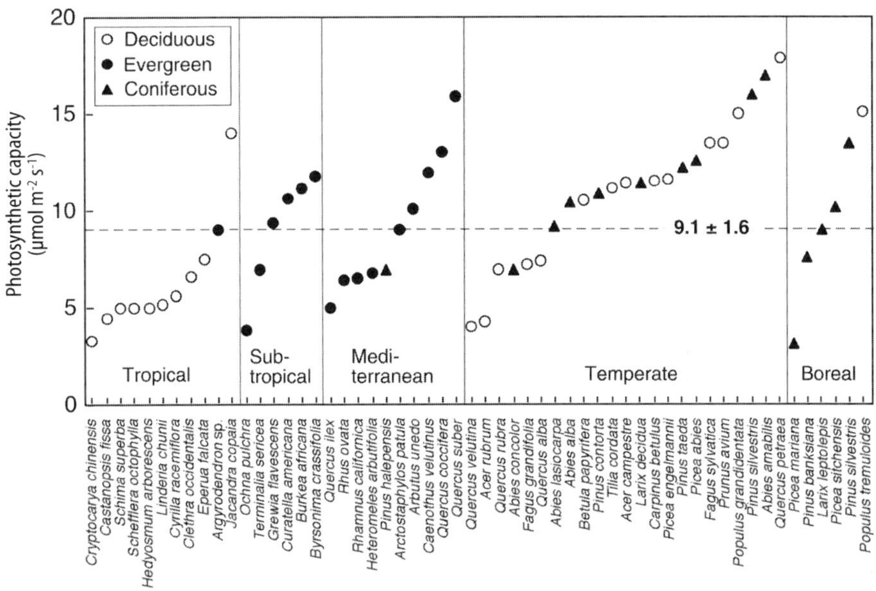

Fig. 2.4. The variation of photosynthetic capacity in forest tree species across climatic zones (from various sources compiled by Körner 1994)

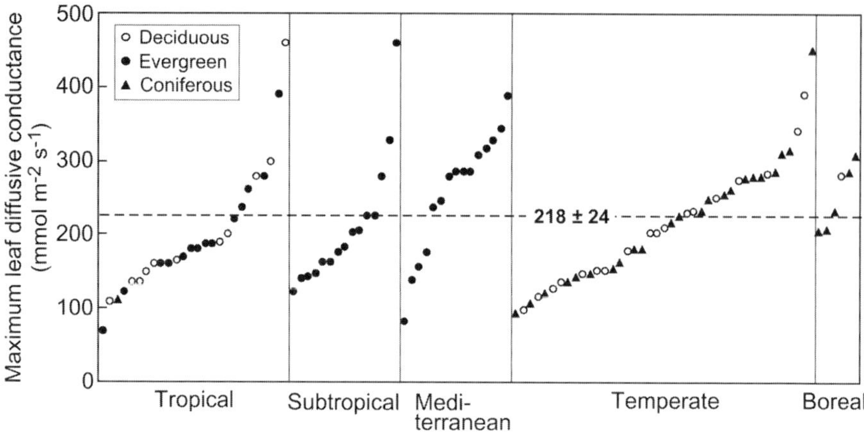

Fig. 2.5. Maximum leaf diffusive conductance in trees: temperate zone forest trees match the rest of the world examples. (Körner 1998)

plants and leaves, and thus is a consequence of these, rather than of a primary driver per se. Nevertheless, photosynthetic capacity has indicator value for life history.

It is well established that photosynthetic capacity is correlated with potential vapor loss, best represented by the maximum leaf diffusive conductance, in essence representing the highest stomatal conductance (Körner et al. 1979; Wong et al. 1979; Schulze et al. 1994). Hence, similarly to Fig. 2.4, Fig. 2.5 demonstrates a lack of any systematic trend in maximum leaf conductance across the globe's climatic zones. In other words, temperate zone forest trees cover the same range of maximum conductance from ca. 90 to 450 mmol m^{-2} s^{-1}, i.e., a fivefold range as photosynthetic capacity. When using a projected leaf area (rather than total leaf surface area), there is also no systematic difference in this respect between conifers and deciduous trees (Körner 1994). In part, this is due to the three-dimensional shape of conifer needles, accumulating a large amount of photosynthetic machinery when projected to the needle's silhouette. Late successional deciduous trees also have stomata on the lower leaf side only, whereas prominent conifer taxa such as *Picea* have stomata on all needle sides. The difference between deciduous trees and evergreen conifers highlighted in earlier works stems primarily in this reference area problem. Species that may occur in the same mixed forest may exhibit a nearly threefold difference in maximum leaf conductance from fully sunlit (top of canopy) as well as from shaded (below canopy) leaves (Fig. 2.6).

These gas exchange properties are associated with mesophyll thickness and stomatal density. Temperate zone forest trees vary in stomatal frequency per mm^2 between 100 and 600, with an almost twofold mean difference between shade- and sun-exposed leaves within the same tree crown (Fig. 2.7;

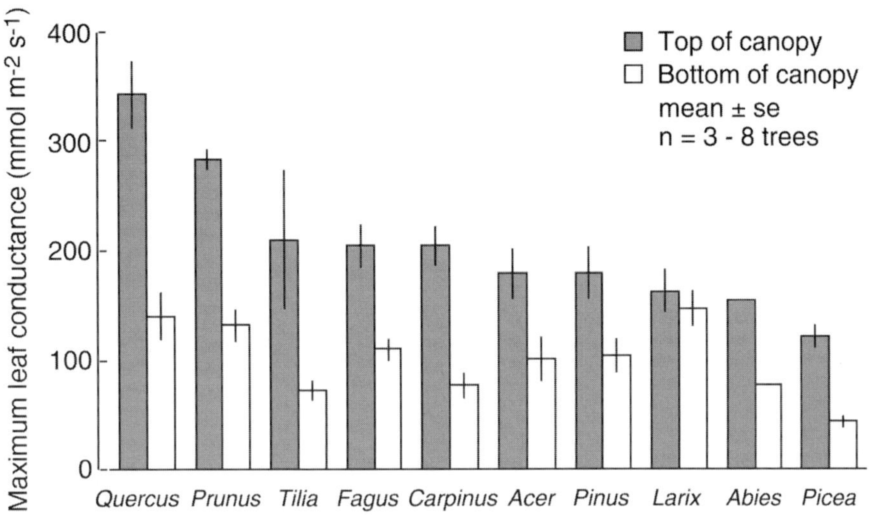

Fig. 2.6. An example of how maximum leaf diffusive conductance may vary within a single temperate forest . Species from *left* to *right*: *Q. petraea, P. avium, T. platyphyllos, F. sylvestris, C. betulus, A. campestre, P. sylvestris, L. decidua, A. alba, P. abies* (data from the Swiss canopy crane near Basel; S. Guillon, S. Keel, S. Pepin, Ch. Körner, unpubl. data)

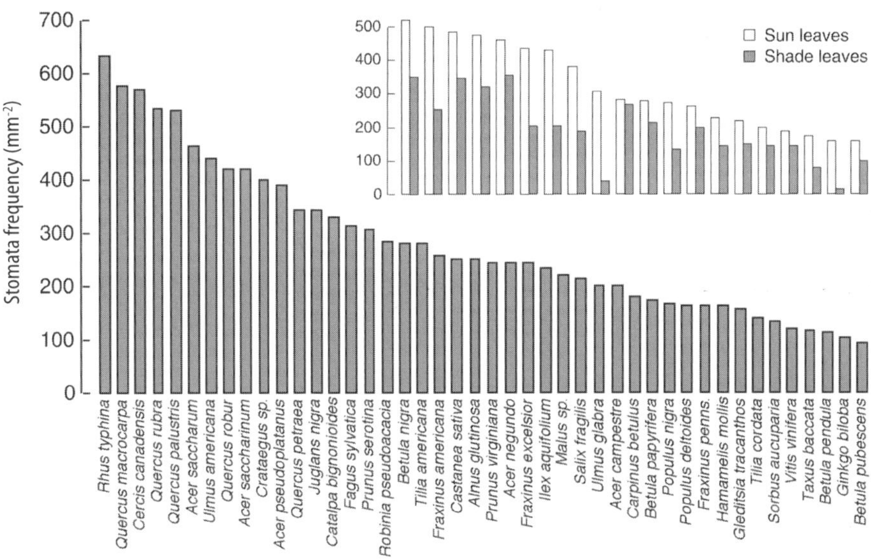

Fig. 2.7. Stomatal frequency in temperate zone forest trees. The *inset* shows the sun vs. shade leaf difference for a subsample of these species

see also Elias 1988). As a rule, sun-exposed leaves have more stomata per unit leaf area than shaded leaves, which has to do with the greater photosynthetic capacity of sun leaves, which in turn is auto-correlated with greater mesophyll thickness and amount of leaf nitrogen per unit leaf area (e.g., Bassow and Bazzaz 1997; Mitchell et al. 1999). Note that a high stomatal density is not necessarily indicative of a high leaf diffusive conductance and vice versa. A plot of maximum leaf conductance and stomatal density across species reveals no correlation (not shown), which reflects the additional significance of pore size and pore geometry for gas diffusion. This makes it rather difficult if not impossible to draw meaningful conclusions from stomatal density on leaf functioning (Poole et al. 1996).

2.4 Tree Stature, Crowns and Roots

The way trees intercept solar radiation and interact with wind, temperature, and precipitation is strongly modified by the shape and density of their crowns (e.g., Smith et al. 1997). New forest canopy access facilities such as cranes now permit the close exploration of canopy surfaces in situ. Those who had the vision of a uniform green carpet, with crowns of different species lining up to form a relatively smooth canopy, intercepting direct solar radiation and interacting with wind in a plane, will rapidly be convinced that this is the wrong picture. The vertical topography within the outer canopy surfaces of even an even-aged deciduous temperate forest may easily range over 15 m in height, include broad domes, sharp cones, a dense and narrow versus an open and deeply structured leaf display, "hills" and "canyons", reflecting tree crown shapes as they develop under competitive conditions (Fig. 2.8). *Fagus sylvatica*, for instance, develops crowns which are relatively open at the top permitting light to penetrate deep into a leafy subcanopy. In contrast, canopy trees of *Quercus petraea* form dense umbrellas with a very sharp light gradient. *Carpinus betulus* holds an intermediate position between *Fagus* and *Quercus*. Crowns of *Tilia platyphyllos* form impenetrable, sharply pointed cones, with almost complete light extinction within a 50-cm leaf coat. In sharp contrast, *Prunus avium* (not shown in Fig. 2.8) presents a widely open (spaced) crown. Conifers, on the other hand, often exhibit clumped leaf area distribution (cluster effect, Whitehead et al. 1984), permitting exceptionally high LAI in some species. All these differences in canopy structure and leaf arrangement translate into different degrees of aerodynamic coupling to the atmosphere (the "omega factor", Jarvis and MacNaughton 1986). These architectural traits vary a lot with age. Genera like *Acer* and *Pinus*, for instance, exhibit monopodial (acrotone) growth early in their life, but form umbrella-shaped canopies in their mature age.

At their edges, the crowns in the canopy as seen in Fig. 2.8 show heavy mechanical damage from wind, when branches beat each other. An early

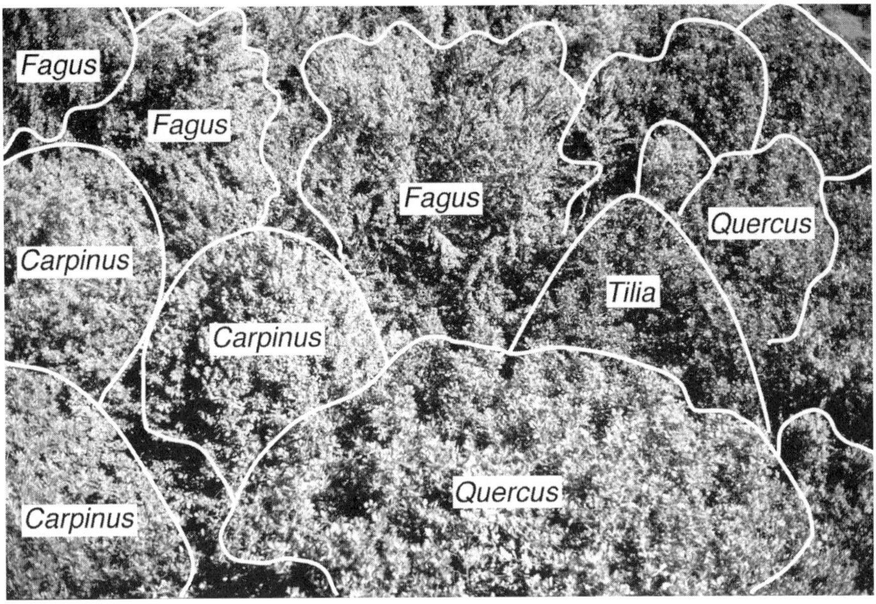

Fig. 2.8. Crown architecture creates a highly diverse forest surface with humps and canyons, gullies and ridges, brushes, and umbrellas. Here, a view on the surface of a 120-year-old forest seen from the gondola of the Swiss canopy crane near Basel. For further explanations see the text

snowfall before leaf shedding caused massive branch breakage in oak, little damage in *Fagus*, and none at all in the flexible *Carpinus* crowns. Crown shape and associated mechanical properties are significant components of the functional significance of forest species diversity (see Dhôte, Chap. 14, this Vol.).

Structural canopy diversity gets particularly high when conifers are mixed with deciduous species. Besides crown shape, the flexibility of branches and the summer versus evergreen leaf types have significant influence on forest canopy functioning. Heavy winter storms, such as the one in December 1999 in western Europe, caused much less damage in leafless deciduous trees than in species such as spruce. While monospecific spruce stands had been flattened in places, nearby mixed stands had a high percentage of surviving deciduous trees (ensuring continued forest functioning) and only "selectively" broken or felled conifers.

A key consequence of crown architecture is light penetration. Tree species differ substantially in how they intercept light in their crowns (e.g., Parker et al. 1989; Simpson et al. 1990; Sumida 1993) and which LAI they accumulate at steady-state canopy closure. As a rule of thumb, species with high photosynthetic capacity build canopies with lower LAI, with a large fraction of leaves intercepting direct solar radiation. Often such species belong to the early successional series and are in danger of being outcompeted by species that produce denser or multilayer crowns and are more shade tolerant (e.g., Küppers 1989). Differences in crown architecture (and light penetration) contribute to positive mixture effects on productivity ("overyielding"; i.e., mixtures yielding more biomass than the most productive monocultures under the same growth conditions; Kelty 1992; Pretzsch, Chap. 3, this Vol.). Crown development of individual species has a strong influence on forest succession, and hence plays an important role in models of forest dynamics (e.g., Shugart et al. 1992; Smith et al. 1997; Bugmann and Solomon 2000; Pretzsch 2001; Grote and Pretzsch 2002).

Among the many morphological features of trees, maximum rooting depth has particularly great functional significance. Most temperate trees reach an average depth of ca. 3 m, but the range is 1.5 to 7.5 m, i.e., fivefold (Fig. 2.9). Among conifers, pines grow deepest (7.5 m), among deciduous species, oaks hold the record (ca. 4 m). Much greater rooting depths occur in semiarid woodlands, where depths exceeding 50 m have been reported (Canadell et al. 1996; Jackson et al. 1996). Among the most selective factors for tree species are periodic droughts. While deep roots may have a higher cost and compete with investments in shallow roots for nutrient acquisition, they ensure a moisture supply during periods of topsoil desiccation. They may also contribute to nutrient accessibility in upper soil horizons during such dry periods through

Fig. 2.9. Diversity of rooting depth in temperate forest tree species. (Canadell et al. 1996)

"hydraulic lift" (plant-driven, nighttime moisture relocation from deep to shallow soil strata; e.g., Dawson 1993). The functional significance of a differentiation of rooting strategies among taxa had been clearly documented for Mediterranean woody species, with drought deciduous species being less deeply rooted than evergreens (Davis and Mooney 1986). In temperate deciduous forests, oaks are clearly in advantage over other species (as are pines among conifers) during dry periods. The centennial heat with associated drought in the summer 2003 documented the advantage of oak over beech at the Swiss canopy crane experimental site. While almost no rain reached the rooted soil over 3 months, oak exhibited continuous high stem sap flux rates, whereas more shallowly rooted beech had to drastically reduce water loss and showed leaf senescence at exposed places (S. Leuzinger and Ch. Körner, unpubl.).

While the significance of tree height, crown shape, and rooting depth is obvious, the functionality of other morphological features is often unclear. There are remarkable biometric relationships among structural traits, which means that these traits follow certain common trends across taxa. For instance, maximum tree size correlates with maximum leaf size, at least for small species. However, the double logarithmic representation of such data fogs the fact that there is no such correlation in very tall species. A closer correlation exists between leaf and fruit size (Fig. 2.10), but again the logarithmic scale visually diminishes the large actual variation among species. Covariation of important plant traits have been explored by several authors (e.g., Westoby et al. 2002) and may best be handled as syndromes rather than as individual factors associated with plant success in a given environment.

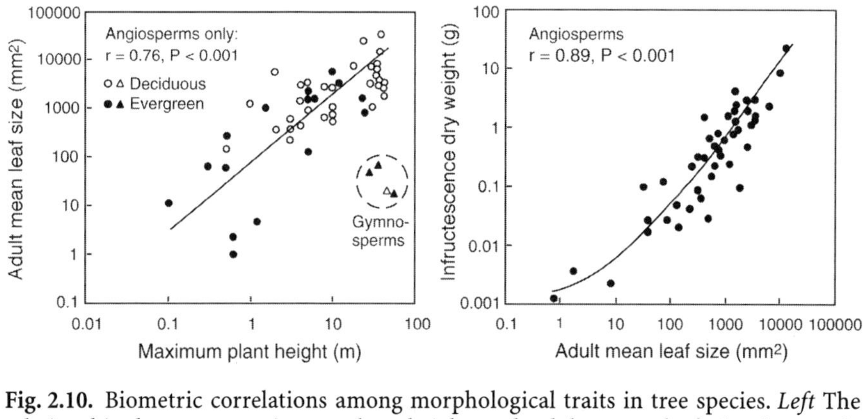

Fig. 2.10. Biometric correlations among morphological traits in tree species. *Left* The relationship between maximum plant height and adult mean leaf size in woody angiosperms and gymnosperms; *right* the relationship between leaf size of adult plants and infructescence dry weight among 47 woody species (excluding conifers). The correlation analysis refers to $\log_{10}$-transformed data. (Cornelissen 1999)

2.5 Growth and Reproduction

The ultimate combined outcome of species-specific characteristics associated with successional stage, physiology, and morphology is growth and reproduction of trees. Tree species may be inherently slow or inherently fast growing either in their seedling or later growth stages, and as a consequence productivity varies among forest taxa, as is well documented in the relevant forest literature (so-called growth tables).

As an example I refer to the variation in relative growth rate among temperate tree taxa in their seedling stage under standardized (optimal) growth conditions (Fig. 2.11). These close-to-maximum rates vary from around 0.01 g g^{-1} day^{-1} in *Ilex* and *Taxus* to ca. 0.12 g g^{-1} day^{-1} in some fast-growing deciduous species. For evergreens *Prunus lauracerasus* holds a 0.08 g g^{-1} day^{-1}

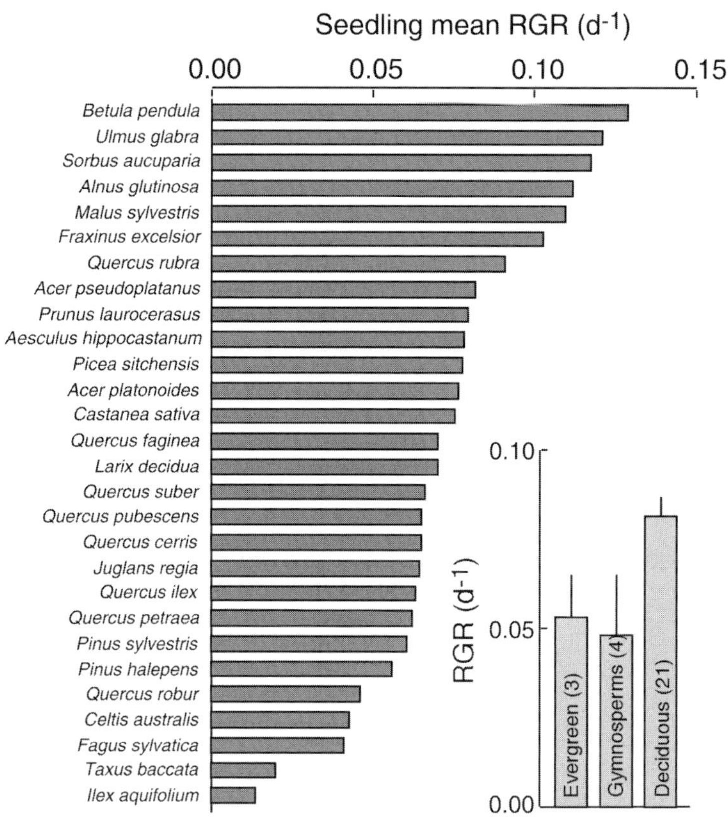

Fig. 2.11. The relative growth rate (*RGR*) among seedlings of temperate zone forest species grown under optimal growth conditions in a controlled environment (from Cornelissen et al. 1996). *Inset* Cornelissen et al. (1998) arrived at similar relationships between RGR in evergreen and deciduous tree species in a larger sample of species

record. Remarkably, these differences in relative (!) growth rate do not reflect seed size. Small-seeded, light-demanding species are found among the fastest growing, whereas large-seeded, shade-tolerant species such as *Fagus sylvatica* and *Quercus robur* are found at the lower end. A negative correlation between seed size and initial growth rate has been documented widely (e.g., Moore 1993; Gleeson and Tilman 1994) and reflects two, not causally linked trends, namely, that early successional species tend to have small seeds and their leaf physiology is adjusted for use of high light intensities. The larger sized seeds of late successional species ensure long-term survival in low light environments and resistance to periodic leaf litter burial; they do not represent reserves for rapid growth. However there are many other environmental drivers, including predation, which may affect seed size (Grubb and Metcalfe 1996).

It is not quite clear why evergreen conifers, but also broad leaf evergreens, tend to grow slower in their seedling stage (even under optimized growth conditions) than deciduous angiosperms (Fig. 2.11, inset). In part, this may reflect the standardized common test conditions, which do not account for the actual differences in seed bed conditions these groups of species may be selected for. More likely this relates to inherently smaller specific leaf area (SLA) in evergreens, which is associated with needle longevity (Schulze 1982; Veneklaas and Poorter 1998).

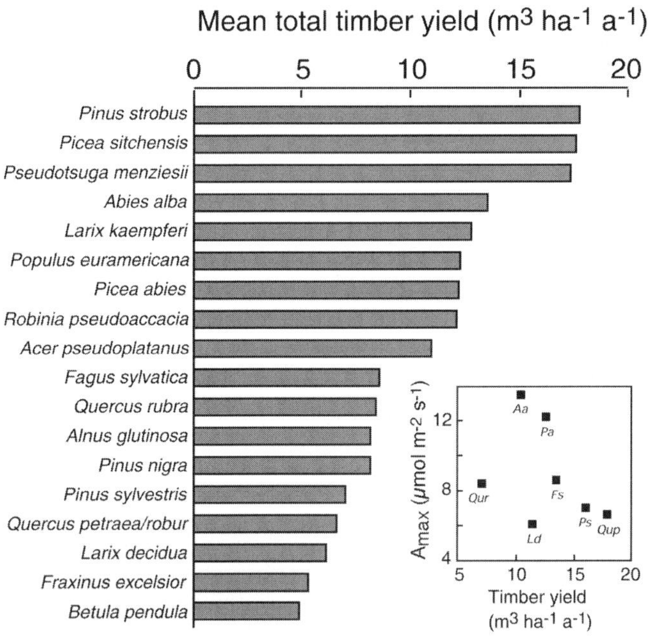

Fig. 2.12. The variation in timber yield of temperate zone forest tree species during their "productive" periods, grown in monoculture. (Kramer 1988)

As can be seen from Fig. 2.12, the maximum relative growth rate of seedlings has no predictive value for the productivity of adult trees. *Betula pendula* shows the highest seeding RGR, but the lowest timber yield. Rank lists like this can and have been used to predict forest yield in mixed stands but also to parameterize forest dynamics models (see Pretzsch, Chap. 3, this Vol.; Shugart et al. 1992). These monoculture data may not simply apply to growth under interspecific competition, but provide a first approximation of the potential growth rate of various taxa during canopy closure and thereafter.

The diversity of reproductive characteristics of temperate forest trees is as large as that in growth traits. In addition to the comments on seed size above, a few general comments on reproductive age and fruiting behavior should suffice here. Some species reach maturity (first fruiting) after 15 years of growth, while others may take 50 years (Fig. 2.13). Good seed crops are produced annually in early succession species, but may occur only every 3 to 7 years in late successional species (masting behavior). Masting (mass fruiting) has not yet found a conclusive explanation. A common one is avoidance of herbivory. However, if this were true, all or most masting species (e.g., *Fagus* and *Quercus*) in a forest would have to mast in synchrony to be effective against generalist herbivores. An alternative explanation is exhaustion of

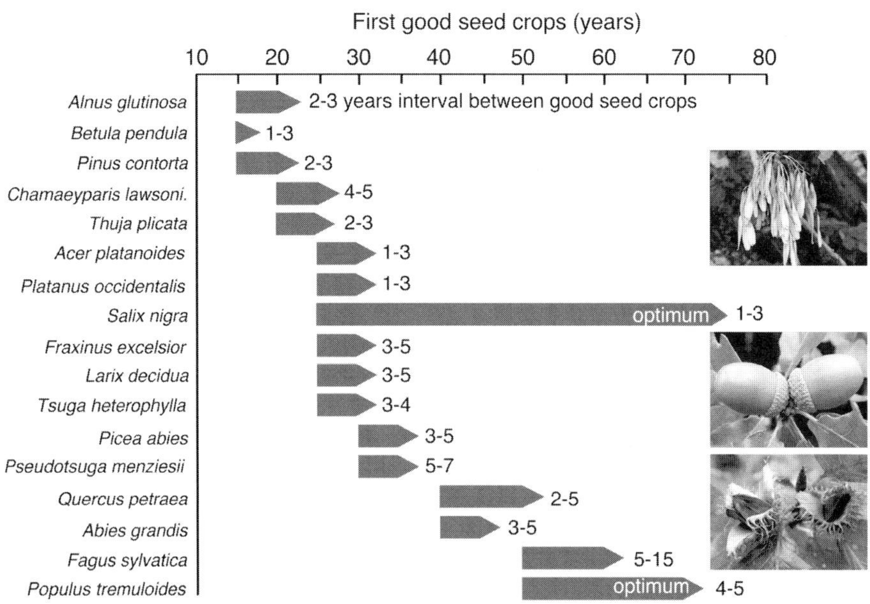

Fig. 2.13. Reproductive traits of temperate forest trees: time of first good seed crop and recurrence of fruiting. Note that the earlier tree species fruit, the more regularly they fruit (no masting behavior).

reserves, as indirectly evidenced by narrower tree rings observed in some of the earlier works (cf. Schulze 1982). However, recent data do not indicate any significant demand in reserves for mass fruiting in *Fagus* and *Quercus* (Hoch et al. 2003), but perhaps this is already a consequence of elevated CO_2 in combination with nitrogen deposition (Körner 2003). Masting intervals had become shorter in recent years in central Europe (from discussion with foresters and personal observation). In a pine plantation in North Carolina, elevated CO_2 had accelerated the fruiting both in time and mass, suggesting a faster tree development (LaDeau and Clark 2001).

2.6 Species-Specific Responses to Global Environmental Change

Several hundred publications have explored the differential responses of species and genotypes to global environmental changes such as CO_2 enrichment and enhanced soluble nitrogen deposition (e.g., Körner and Bazzaz 1996). In forest trees, for obvious practical reasons, most of this screening for species-specific responses (in this case to elevated CO_2) was done in seedlings under more or less controlled growth conditions, yielding broad patterns of responsiveness, in part reflecting the growth conditions under which traits were studied (for reviews see, e.g., Eamus and Jarvis 1989; Saxe et al. 1998; Norby et al. 1999; Körner 2000; Gruber 2003; pers. observ.). The larger the experimental facilities (and hence the studied trees) became, the less was tree species diversity a central theme. Instead of re-visiting this large literature, I will present three cases from works with elevated CO_2 which focused on native temperate forest tree species diversity and which used a minimum interference approach, i.e., as natural as feasible growth conditions were applied. The aim of this section is to illustrate variation in responses of tree species which could influence future forest composition. I will start with seedlings, followed by saplings and adult trees.

In the seedling stage, temperate forest trees are particularly sensitive to in situ CO_2 enrichment, because a higher CO_2 concentration permits growth in deeper shade (a shift of the photosynthetic light compensation point). However, this interaction between light and CO_2 at the seedling stage is extremely species specific, as has been shown by a field test under natural light and soil conditions in an old growth forest (Hättenschwiler and Körner 2000). The growth stimulation of seedlings ranged from zero to more than 50 %. When light was very dim (1 % of sunlight) *Fagus sylvatica* championed over the four other species tested (zero response in *Abies*). When the understory light was 4 % of full sunlight, *Abies alba* showed the strongest response (near zero response in *Fagus*). Hence, a shift in the understory light regime from 1 to 4 % of above canopy light caused a complete reversal of the ranking of species with respect to their growth response to a high CO_2 regime. Such species-specific responses in conjunction with atmospheric change may alter the sequence of RGR dramatically, as shown in Fig. 2.11, and affect future forest composition.

The differential growth responses of tree species to CO_2 enrichment are largely not understood. For instance, it is still unclear whether fast-growing tree species are taking greater advantage of elevated CO_2 concentrations than slow-growing species, as was an initial impression derived from studies with young plants grown with unlimited nutrient supply (Poorter 1993). Would the ranking shown in Figs. 2.11 and 2.12 become more pronounced in a high CO_2 world (cf. Tangley 2001)? Are evergreen conifers more or less responsive than deciduous trees? Figure 2.14 illustrates a case where this was tested with communities of saplings of spruce and beech, growing naturally together on

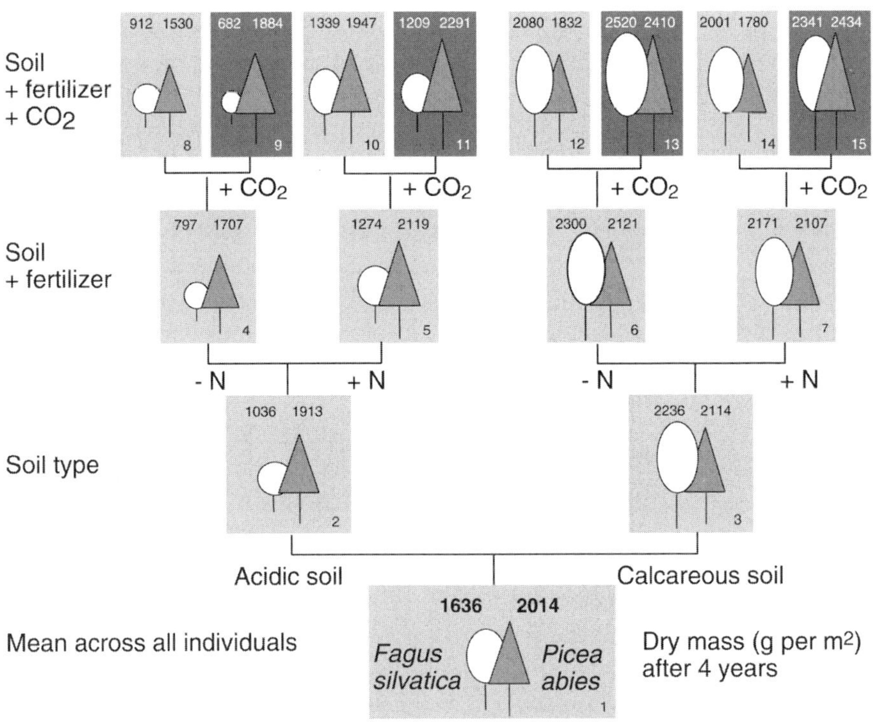

Fig. 2.14. The influence of soil type on the species-specific responses of beech and spruce to elevated CO_2 and wet nitrogen deposition. The diagram shows the cascade of treatment effects on total aboveground tree biomass after 4 years of growth in mixed stands of standardized genetic diversity. What is possibly the most complex experimental design ever used in CO_2-enrichment research with trees still does not account for age structure and spatial patchiness as it may occur under field conditions, illustrating the challenge of experimental biodiversity research

either acidic or calcareous soils. Mixed communities of these two species were exposed to factorial treatment of increased soluble nitrogen deposition and CO_2 enrichment on these two substrate types for 4 years. By the time the largest trees were 2 m in height and LAI had reached a steady state of close to 6, significant species effects became apparent (Spinnler et al. 2002). Yet, the threefold interaction of species with soils, N deposition, and CO_2 enrichment revealed quite unexpected results. It depended on the soil type whether beech or spruce took greatest advantage. Within a given soil type, N deposition was decisive for the CO_2 effect. Fields 9 and 13 in Fig. 2.14 illustrate the extreme cases, with *Fagus* either becoming a massive "loser" under elevated CO_2 or a likely "winner". It seems the decisive tree trait here was the species' preference for calcareous soil and, presumably, the respective mycorrhization (Wiemken et al. 2001). This example illustrates that it depends on soil conditions, whether, in which direction, and to what extent the presence of certain species will influence system behavior.

The third example is from an ongoing CO_2 enrichment experiment with ca. 100-year-old forest trees (Pepin and Körner 2002). After three seasons of CO_2 enrichment, these 30–35 m trees exhibited a number of different responses, hardly any of which could have been predicted from previous research. Insofar as can be judged to date, adult *Fagus* does not save any water under high CO_2 as is commonly predicted from theory, but *Carpinus* does. *Quercus* holds an intermediate position. Leaf nitrogen depletion, a common phenomenon under elevated CO_2, is found in *Quercus* and *Fagus* but not in *Carpinus*. Figure 2.15 illustrates the possible consequences. One of the major forest pests,

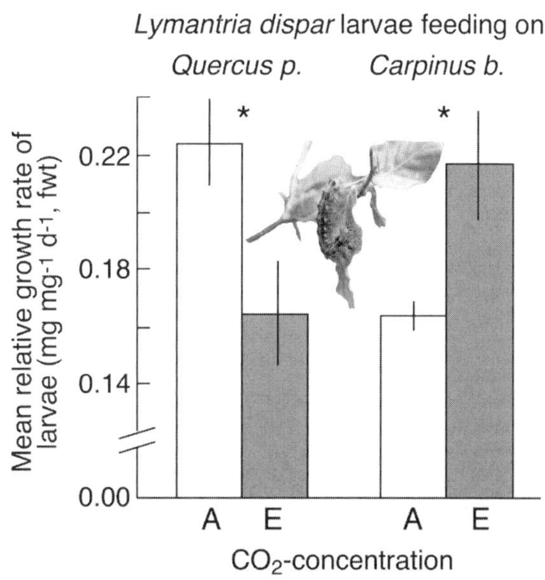

Fig. 2.15. The tree-species-specific response of larval growth of the moth *Lymantria dispar* feeding on adult trees of *Quercus petraea* and *Carpinus betulus* exposed to ambient and elevated CO_2 concentrations in a forest near Basel. (Hättenschwiler and Schafellner 2004)

the moth *Lymantria dispar* enhances its growth rate when feeding in situ on high-CO_2-exposed *Carpinus*, but significantly declines in vigor when forced to feed on high-CO_2 *Quercus*. Altered abundance of tree species is a potential long-term outcome of the contrasting water and pest responses and their interaction (among many other influences). The growth responses to elevated CO_2 of these large trees are not yet available, but it seems a drought as experienced in 2003 makes *Fagus* more receptive to CO_2 enrichment (more growth) than the other species. Hence, patterns of drought response for different tree species as documented for current CO_2 concentrations (e.g., Leuschner et al. 2001) may be altered by changes in CO_2.

2.7 Outlook

The traits of forest trees reviewed here illustrate the range and types of key characteristics which may alone or in combination with others determine the presence or absence of certain taxa in a given habitat. In turn, these characteristics may contribute in various ways to the functioning of the whole forest. There are many other tree traits not accounted for here, which may be crucial for biodiversity effects on forest functioning, as for instance bark resistance to fire (e.g., Schiller 2002 for pines; other examples in Wirth, Chap. 15, this Vol.), pathogen and herbivory resistance (references in Pautasso et al., Chap. 13, this Vol., and Jactel et al., Chap. 12, this Vol.), resistance to waterlogging, certain soil preferences, mycorrhization (e.g., Read 1993; Smith and Read 1997), and the ability to form symbiosis with N_2-fixing bacteria (e.g., Jones et al., Chap. 6, this Vol.; Binkley 1992; Binkley et al. 2003).

While it is obvious that the abundance of certain taxa with certain traits will influence forest functioning (e.g., deciduous trees will do better during winter storms than evergreen trees), it is very difficult to scale from individual species' traits to stand properties. The more a single species will dominate, the more likely its traits will translate in a predictable way into stand characteristics. The more even and rich the representation of taxa in a given stand, the more will individual traits be diluted among the traits of others and predictions of species' effects on forest functioning will become near to impossible.

Given the multitude of interactions of traits across taxa, an empirical assessment seems the safest avenue toward identifying the significance of species identity to forest functioning. Such a post hoc identification of the significance of traits may be the needed shortcut and simplification to overcome the sheer endless number of traits that may theoretically feed into the functioning of a forest. Sensitivity to elevated CO_2 is a good example where a priori tree typologies were not really helpful in predicting responses. Complex interactions of different tree taxa as illustrated in Fig. 14 need to be taken into account in any attempt toward drawing projections of forest responses from tree species' responses. Two rather different conclusions would have been published had this experiment been conducted on one of the two soil types only.

Besides experimental manipulations of forest diversity (e.g., mixed plantations, selective removal or addition of species; see Scherer-Lorenzen et al., Chap. 16, this Vol.; Jones et al., Chap. 6, this Vol.; Cannell et al. 1992; Ewel et al. 1999), future forest biodiversity research should capitalize on the existing spectrum of forest diversities in the landscape that have been the result of forest management. Different management types not only created different tree species mixtures, but also altered age distribution, another facet of forest diversity to be explored (Mund and Schulze, Chap. 10, this Vol.; Smith et al. 1997). Given the longevity of trees, their natural or artificial abundance in mature mixtures represent established test situations which can never be created within the time frame of conventional experiments. These "test sites" do exist, though the "experimental design" may be noisy and loaded with a multitude of environmental and historical covariables that can confound results. A careful site characterization and large site numbers are needed to retain a reasonable signal-to-noise ratio. There is little alternative with adult trees than to utilize existing patterns in order to obtain a realistic picture of responses to help in evidencing and understanding the significance of forest diversity for ecosystem functioning. The range and the spectrum of functional traits of tree species as presented here may help in selecting promising mixtures of species, formulate hypothesis about the impact of species traits, and assist in model parameterization.

References

Anders S, Hofmann G (1996) Erforschung des ökologischen Wandels – Voraussetzung für künftige Waldbewirtschaftung. Mitt Bundesforschungsanst Forst-Holzwirtsch Hamburg 185:129–139

Assmann E (1961) Waldertragskunde. BLV, Munich

Barnes BV, Zak DR, Danton SR, Spurr SH (1998) Forest ecology. Wiley, New York

Bassow SL, Bazzaz FA (1997) Intra- and inter-specific variation in canopy photosynthesis in a mixed deciduous forest. Oecologia 109:507–515

Bazzaz FA, Miao SL (1993) Successional status, seed size, and responses of tree seedlings to CO_2, light, and nutrients. Ecology 74:104–112

Binkley D (1992) Mixtures of nitrogen$_2$-fixing and non-nitrogen$_2$-fixing tree species. In: Cannell MGR, Malcolm DC, Robertson PA (eds) The ecology of mixed-species stands of trees. Blackwell, Oxford, pp 99–123

Binkley D, Senock R, Bird S, Cole TG (2003) Twenty years of stand development in pure and mixed stands of *Eucalyptus saligna* and nitrogen-fixing *Facaltaria moluccana*. For Ecol Manage 6226:1–10

Bugmann HKM, Solomon AM (2000) Explaining forest composition and biomass across multiple biogeographical regions. Ecol Appl 10:95–114

Cannell MGR, Malcolm DC, Robertson PA (eds) (1992) The ecology of mixed-species stands of trees. Blackwell, Oxford

Canadell J, Jackson RB, Ehleringer JR, Mooney HA, Sala OE, Schulze ED (1996) Maximum rooting depth of vegetation types at the global scale. Oecologia 108:583–595

Ceulemans RJ, Saugier B (1991) Photosynthesis. In: Raghavendra AS (ed) Physiology of trees. Wiley, London, pp 21–50

Cornelissen JHC (1999) A triangular relationship between leaf size and seed size among woody species: allometry, ontogeny, ecology and taxonomy. Oecologia 118:248–255

Cornelissen JHC, Castro Diaz P, Hunt R (1996) Seedling growth, allocation and leaf attributes in a wide range of woody plant species and types. J. Ecol 84:755–765

Cornelissen JHC, Castro-Diez P, Carnelli AL (1998) Variation in relative growth rate among woody species. In: Lambers H, Poorter H, van Vuuren MMI (eds) Inherent variation in plant growth. Physiological mechanisms and ecological consequences. Backhuys, Leiden, pp 363–392

Davis SD, Mooney HA (1986) Water use patterns of four co-occurring chaparral shrubs. Oecologia 70:172–177

Dawson TE (1993) Hydraulic lift and water use by plants: implications for water balance, performance and plant-plant interactions. Oecologia 95:565–574

Eamus D, Jarvis PG (1989) The direct effects of increase in the global atmospheric CO_2 concentration on natural and commercial temperate trees and forests. Adv Ecol Res 19:1–55

Eckstein RL, Karlsson PS, Weih M (1999) Leaf life span and nutrient resorption as determinants of plant nutrient conservation in temperate-arctic regions. New Phytol 143:177–189

Elias P (1988) Stomata in forest communities: density, size and conductance. Acta Uni Carolinae Biol 31:27–41

Ellenberg H (1996) Vegetation Mitteleuropas mit den Alpen, 5th edn. Ulmer, Stuttgart

Ewel JJ, ODowd DJ, Bergelson J, Daehler CC, DAntonio CM, Gomez LD, Gordon DR, Hobbs RJ, Holt A, Hopper K.R, Hughes CE, LaHart M, Leakey RRB, Lee WG, Loope LL, Lorence DH, Louda SM, Lugo AE, McEvoy PB, Richardson DM, Vitousek PM (1999) Deliberate introductions of species: Research needs – Benefits can be reaped, but risks are high. Bioscience 49:619–630

Gale MR, Gigal DF (1987) Vertical root distribution of northern tree species in relation to successional status. Can J For Res 17:829–834

Gayer K (1886) Der gemischtes Wald. Seine Begründung und Pflege, insbesondere durch Horst- und Gruppenwirtschaft. Parey, Berlin

Gleeson SK, Tilman D (1994) Plant allocation, growth rate and successional status. Funct Ecol 8:543–550

Grote R, Pretzsch H (2002) A model for individual tree development based on physiological processes. Plant Biol 4:167–180

Grubb PJ, Metcalfe DJ (1996) Adaption and inertia in the Australian tropical lowland rain-forest flora: contradictory trends in intergeneric and intrageneric comparisons of seed size in relation to light demand. Funct Ecol 10:512–520

Gruber F (2003) Welche Witterung bestimmt die Fruchtbildung bei der Rotbuche? AFZ – Der Wald 5:246–250

Hättenschwiler S, Körner Ch (2000) Tree seedling responses to in situ CO_2-enrichment differ among species and depend on understorey light availability. Global Change Biol 6:213–226

Hättenschwiler S, Schafellner C (2004) Gypsy moth feeding in the canopy of a CO_2-enriched mature forest. Glob Change Biol 10, in press

Hoch G, Richter A, Körner Ch (2003) Non-structural carbohydrates in temperate forest trees. Plant Cell Environ 26:1067–1081

Jackson RB, Canadell J, Ehleringer JR, Mooney HA, Sala OE, Schulze ED (1996) A global analysis of root distributions for terrestrial biomes. Oecologia 108:389–411

Jarvis PG, McNaughton KG (1986) Stomatal control of transpiration: scaling up from leaf to region. Adv Ecol Res 15:1–49

Jarvis PG, Sandford AP (1986) Temperate forests. In: Baker NR, Long SP (eds) Photosynthesis in contrasting environments. Elsevier, Biomed Div, pp 199–236

Kelty MJ (1992) Comparative productivity of monocultures and mixed-species stands. In: Kelty MJ, Larson BC, Oliver CD (eds) The ecology and silviculture of mixed-species forests. Kluwer, Dordrecht, pp 125–141

Kenk GK (1992) Silviculture of mixed-species stands in Germany. In: Cannell MGR, Malcolm DC, Robertson PA (eds) The ecology of mixed-species stands of trees. Special publication of the British Ecological Society 11. Blackwell, Oxford, pp 53–63

Koike T, Kitao M, Maruyama Y, Mori S, Lei TT (2001) Leaf morphology and photosynthetic adjustments among deciduous broad-leaved trees within the vertical canopy profile. Tree Physiol 21:951–958

Körner Ch, Scheel JA, Bauer H (1979) Maximum leaf diffusive conductance in vascular plants. Photosynthetica 13:45–82

Körner Ch (1993) Scaling from species to vegetation: the usefulness of functional groups. In: Schulze ED, Mooney HA (eds) Biodiversity and ecosystem function. Ecological Studies, vol 99. Springer, Berlin Heidelberg New York, pp 117–140

Körner Ch (1994) Leaf diffusive conductances in the major vegetation types of the globe. In: Schulze ED, Caldwell MM (eds) Ecophysiology of photosynthesis. Ecological Studies, vol 100. Springer, Berlin Heidelberg New York, pp 463–490

Körner Ch (1998) Alpine plants: stressed or adapted? In: Press MC, Scholes JD, Barker MG (eds) Physiological plant ecology. Blackwell, Oxford, pp 297–311

Körner Ch (2000) Biosphere responses to CO_2 enrichment. Ecol Appl 10:1590–1619

Körner Ch (2003) Carbon limitation in trees. J Ecol 91:4–17

Körner Ch, Bazzaz FA (1996) Carbon dioxide, populations, and communities. Academic Press, San Diego

Kramer H (1988) Waldwachstumslehre. Paul Parey, Hamburg

Küppers M (1989) Ecological significance of above-ground architectural patterns in woody plants: a question of cost-benefit relationships. Tree 4:375–379

LaDeau SL, Clark JS (2001) Rising CO_2 levels and the fecundity of forest trees. Science 292:95–98

Larcher W (1969) The effect of environmental and physiological variables on the carbon dioxide gas exchange of trees. Photosynthetica 3:167–198

Leuschner C, Backes K, Hertel D, Schipka F, Schmitt U, Terborg O, Runge M (2001) Drought responses at leaf, stem and fine root levels of competitive *Fagus sylvatica* L. and *Quercus petraea* (Matt.) Liebl. trees in dry and wet years. For Ecol Manage 149:33–46

Loreau M (2000) Biodiversity and ecosystem functioning: recent theoretical advances. Oikos 91:3–17

Mitchell KA, Bolstad PV, Vose JM (1999) Interspecific and environmentally induced variation in foliar dark respiration among eighteen southeastern deciduous tree species. Tree Physiol 19:861–870

Mooney HA, Lubchenco J, Dirzo R, Sala OE (1995) Biodiversity and ecosystem functioning: ecosystem analysis. In: Heywood VH, Watson RT (eds) Global biodiversity assessment. Cambridge Univ Press, Cambridge, pp 328–452

Moore PD (1993) For richer, for poorer. Nature 366:515

Norby RJ, Wullschleger SD, Gunderson CA, Johnson DW, Ceulemans R (1999) Tree responses to rising CO_2 in field experiments: implications for the future forest. Plant Cell Environ 22:683–714

Norman JM, Jarvis PG (1974) Photosynthesis in Sitka spruce (*Picea sitchensis* (Bong.) Carr.). III. Measurements of canopy structure and interception of radiation. J Appl Ecol 11:375–398

Parker GG, O'Neill JP, Higman D (1989) Vertical profile and canopy organization in a mixed deciduous forest. Vegetatio 85:1–11

Pepin S, Körner Ch (2002) Web-FACE: a new canopy free-air CO_2 enrichment system for tall trees in mature forests. Oecologia 133:1–9

Poole I, Weyers JDB, Lawson T, Raven JA (1996) Variations in stomatal density and index: implications for palaeoclimatic reconstructions. Plant Cell Environ 19:705–712

Poorter H (1993) Interspecific variation in the growth response of plants to an elevated ambient CO_2 concentration. Vegetatio 104/105:77–97

Pretzsch H (2001) Modellierung des Waldwachstums. Parey, Berlin

Read DJ (1993) Plant-microbe mutualisms and community structure. In: Schulze ED, Mooney HA (eds) Biodiversity and ecosystem function. Ecological Studies, vol 99, Springer, Berlin Heidelberg New York, pp 181–209

Reich PB, Grigal DF, Aber JD, Gower ST (1997) Nitrogen mineralization and productivity in 50 hardwood and conifer stands on diverse soils. Ecology 78:335–347

Sampson DA, Smith FW, Long JN (1990) Use of canopy light interception to estimate forest leaf area index. Bull Ecol Soc Am [Suppl] 71:313

Saxe H, Ellsworth DS Heath J (1998) Tansley review no 98 Tree and forest functioning in an enriched CO_2 atmosphere. New Phytol 139:395–436

Schiller G (2000) Ecophysiology of *Pinus halepensis* Mill. and *P. brutia* Ten. In: Ne'eman G, Trabaud L (eds) Ecology, biogeography and management of *Pinus halepensis* and *P. brutia* forest ecosystems in the Mediterranean basin. Backhuys, Leiden, pp 51–65

Schmid B (2002) The species richness-productivity controversy. Trend Ecol Evolut 17:113–114

Schulze ED (1982) Plant life forms and their carbon, water and nutrient relations. Encycl Plant Physiol New Ser 12B. Springer, Berlin Heidelberg New York, pp 616–676

Schulze ED, Kelliher FM, Körner Ch, Lloyd J, Leuning R (1994) Relationships among maximum stomatal conductance, ecosystem surface conductance, carbon assimilation rate, and plant nitrogen nutrition: a global ecology scaling exercise. Annu Rev Ecol Syst 25:629–660

Shugart HH, Leemans R, Bonan GB (1992) A systems analysis of the global boreal forest. Cambridge Univ Press, Cambridge

Smith SE, Read DJ (1997) Mycorrhizal symbiosis, 2nd edn. Academic Press, London, 605 pp

Smith TM, Shugart HH, Woodward FI (1997) Plant functional types. IGBP book series 1. Cambridge Univ Press, Cambridge

Spinnler D, Egli P, Körner Ch (2002) Four-year growth dynamics of beech-spruce model ecosystems under CO_2 enrichment on two different forest soils. Trees 16:423–436

Sumida A (1993) Growth of tree species in a broadleaved secondary forest as related to the light environments of crowns. J Jpn For Soc 75:278–286

Tangley L (2001) High CO_2 levels may give fast-growing trees an edge. Science 292:36–37

Veneklaas EJ, Poorter L (1998) Growth and carbon partitioning of tropical tree seedlings in contrasting light environments. In: Lambers H, Poorter H, van Vuuren MMI (eds) Inherent variation in plant growth. Physiological mechanisms and ecological consequences. Backhuys, Leiden, pp 337–361

Walters MB, Kruger EL, Reich PB (1993) Growth, biomass distribution and CO_2 exchange of northern hardwood seedlings in high and low light: relationships with successional status and shade tolerance. Oecologia 94:7–16

Westoby M, Falster DS, Moles AT, Vesk PA, Wright IJ (2002) Plant ecological strategies: some leading dimensions of variation between species. Annu Rev Ecol Syst 33:125–159

Whitehead D, Edwards WRN, Jarvis PG (1984) Conducting sapwood area, foliage area, and permeability in mature trees of *Picea sitchensis* and *Pinus contorta*. Can J For Res 14:940–947

Wiedemann E (1943) Der Vergleich der Massenleistung des Mischbestandes mit dem Reinbestand. Allg Forst Jagdztg 119:123–132

Wiedemann E (1955) Ertragskunde und waldbauliche Grundlagen der Forstwirtschaft, 2nd edn. Sauerländer, Frankfurt

Wiemken V, Laczko E, Ineichen K, Boller T (2001) Effects of elevated carbon dioxide and nitrogen fertilization on mycorrhizal fine roots and the soil microbial community in beech-spruce ecosystems on siliceous and calcareous soil. Microb Ecol 42:126–135

Wong SC, Cowan IR, Farquhar GD (1979) Stomatal conductance correlates with photosynthetic capacity. Nature 282:424–426

Part B Productivity and Growth

3 Diversity and Productivity in Forests: Evidence from Long-Term Experimental Plots

H. PRETZSCH

3.1 Introduction: The Mixed Stands Issue – A Central European Perspective

At the beginning of regular forest management and systematic forestry sciences the primary objective was sustainable timber production. The sustainability principle, originally conceived by von Carlowitz (1713), served both Hartig (1791, 1804) and von Cotta (1828) as the basis from which they proceeded to develop methods for the sustainable management of forest enterprises. The main focus at the time was the sustainable production of timber. Attention was also called to diversity, protective, and recreational functions, inter alia, by von Hagen (1867), but none of these elements were included in the rules and regulations on sustainable planning. There was a widespread belief that, in the wake of sustainable timber production, all other forest functions would be automatically fulfilled. It was not until Dieterich's forest-functions theory was publicized (1957) that forests began to be discussed as habitats and recreational areas, with functions such as the protection of climate, soil, and water. Today, there is international consensus that the multiple function of forests includes protection of forest resources, health and vitality of forest ecosystems, production of wood and other forest products, biological diversity, and protective and socioeconomic functions (MCPFE 2000). The sustainability of biodiversity is becoming an imperative, similar to the sustainability of timber production in the past. Yet, what is the relationship between biodiversity and productivity? In the following we shall concentrate on how tree species' diversity and forest productivity are interrelated, a question of particular relevance to forestry practice.

Hartig, considered the forefather of forestry science, commented on the mixed stands issue as follows (Hartig 1791, p. 134): "...the mixing of deciduous and coniferous species is not advantageous, as the coniferous trees generally tend to supplant deciduous ones and because one type of tree impedes the growth of the other; so that no mixed deciduous and coniferous forests

Ecological Studies, Vol. 176
M. Scherer-Lorenzen, Ch. Körner, and E.-D. Schulze (Eds.)
Forest Diversity and Function: Temperate and Boreal Systems

should be established with intent" (translation by the author = t. by a.). Concerned about serious production losses in mixed stands, Hartig (1804, p. 40) recommended: "All mixed stands with coniferous and deciduous species should be converted into pure stands of the constituent species, as soon as circumstances permit" (t. by a.). This was contradicted by von Cotta (1828, p. 115): "Endeavours to establish pure stands everywhere is based on an old and highly detrimental prejudice.... Since not all tree species utilize resources in the same manner, growth is more lively in mixed stands and neither insects nor storms can do as much damage; also, a wider range of timber will be available everywhere to satisfy different demands..." (t. by a.). This opinion was supported by Gayer (1886, p. 31): "The mixed forest does not only produce more, but also more valuable commercial timber than that grown in pure stands" (t. by a.). Statements by Möller (1922, pp. 41–42) are even more optimistic: "...if we design stands of shade-intolerant and shade-tolerant tree species, ...the potential for timber production is raised even more; the reason being that it is now possible to go considerably farther in the stratification of age classes than in the design of pure stands with only a single layer" (t. by a.). Wiedemann, a professional yield scientist, dampens the optimism voiced by the above silviculturists (Wiedemann 1951, p. 341) saying "...even in silviculture, room must be given to hard facts next to emotions" (t. by a.). It was not until data was evaluated from long-term experiments, under observation in many European countries since the founding of the Forestry Research Stations in 1870 to 1880, that a clearer picture was conveyed of the productivity in pure and mixed stands that differentiated between species and sites.

First evaluations of long-term experimental plots put a damper on hopes for increased yield through mixture. The reason for this was their revelation of far greater productivity in monocultures of Norway spruce (*Picea abies*) and Douglas fir (*Pseudotsuga menziesii*) on many sites in temperate and boreal zones than in any mixture (Schwappach 1912; Wiedemann 1949; Assmann 1970; Schober 1975). Whenever the primary objective is to maximize dry biomass production, then in many places there is no alternative to pure stands of these species. Similar advantages of monospecific stands were noted for *Pinus* species in Mediterranean and for *Eucalyptus* as well as *Albizia* species in subtropical and tropical climate zones (Weck 1955; Kelty 1992).

By contrast, in grassland ecosystems, most studies found evidence for a logarithmic rise in productivity with increasing numbers of species (e.g., Hector et al. 1999; Loreau et al. 2001), as also reported for North American forests (Caspersen and Pacala 2001). According to these authors, production gains are most obvious when monocultures and two-species mixtures are compared. In this chapter, we therefore concentrate on pure stands and mixed stands composed of two species. Such two-species mixtures, especially those mixed by groups, are predominant in mixed forests (Bartelink and Olsthoorn

1999) and have been scientifically studied more intensively than any others (Kelty 1992).

I summarize some results from mixed-stand experimental plots in my network of permanent plots, and outcomes of other mixed-stand experiments reported in the literature in order to quantify the extent of increased or reduced productivity in mixed stands and to identify corresponding causes. Going beyond the studies of Cannell et al. (1992), Kelty (1992), and Olsthoorn et al. (1999), I present the following new aspects: first, suitable approaches and measures for productivity comparisons will be introduced; second, benchmarks for productivity increases or decreases for commercially important tree species in temperate and boreal zones will be derived. Then I will explore the relationships between productivity in pure and mixed stands as a function of species and site conditions. The decisive factor is to include stand management and risk in the yield comparison.

The chapter incorporates results of the long-term experimental network, surveyed by the Chair of Forest Yield Science of the Technical University of Munich. This network involves experiments in pure and mixed stands that are unique as far as observation time, sampling volume, and spectrum of silvicultural treatment methods are concerned (Pretzsch 2002, pp. 133–138).

3.2 Theoretical Considerations

3.2.1 Ecological Niche, Site–Growth Relationships

Considerations on the productivity $p_{1,2}$ of a mixed stand composed of two species usually refer to the productivity p_1 or p_2 of corresponding pure stands on the same site. Given species interaction without synergistic effects on growth the productivity $p_{1,2}$ of a mixed stand is represented as pure stand's growth weighted with the mixture proportions m_1 and m_2, i.e., $p_{1,2}=m_1\,p_1+m_2\,p_2$. The crucial factors for results from mixtures are the ecological niches of the species and their compatibilities, since this is what determines productivity on any given site. Beneficial effects from species interactions that enhance yield of a mixed stand are of particular interest in this context. The chances for an enhancement of biomass production by mixture depend on the relationship between p_1 and p_2 and on potential beneficial mixture effects.

The relationships will be explained using two model examples composed of species occupying similar and different ecological niches (Fig. 3.1a, b). The unimodal dose–response curves represent the different niches inasmuch as they reflect the dependence of productivity on growth conditions typical of the species. For simplicity's sake growth conditions on this graph are plotted

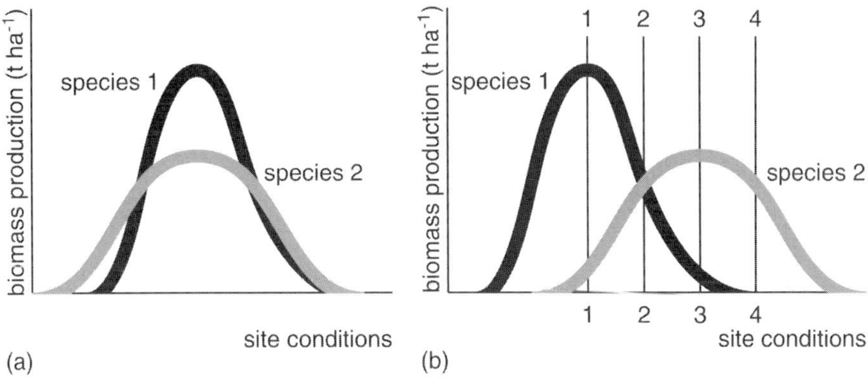

Fig. 3.1a, b. Dry biomass production of various tree species in relation to site conditions. **a** Productivity of two tree species with similar ecological amplitude but different levels of production. **b** Productivity of two tree species with different ecological amplitudes. *Numbers 1* to *4* below the abscissa represent different site conditions to which species 1 and 2 show different growth responses (cf. Fig. 3.2)

using one dimension only and are composed of the n-dimensional vector of environmental factors (e.g., temperature, pH of the soil, storm–snow load) and availability of resources (e.g., irradiation, water, nutrient supply, atmospheric carbon dioxide, etc.).

In the first case (Fig. 3.1a), the tree species 1 and 2 occupy similar ecological niches, but they differ clearly in growth yield on the given site. Productivities p_1 and p_2 for the superior and inferior tree species, respectively, vary to the extent that in these instances the addition of the inferior tree species will usually cause a reduction in stand productivity. Examples of this are the superiority of Douglas fir (*Pseudotsuga menziesii*) over Scots pine (*Pinus sylvestris*) or of red oak (*Quercus rubra*) over common oak (*Quercus petraea*) through a wide range in ecological amplitude for these species. Both examples compare an indigenous with a foreign species; corresponding examples for two indigenous species can be hardly found.

Relationships become more complicated if the species in the mixture occupy distinctly different niches (Fig. 3.1b). Let us assume species 1 and 2 are mixed in stands on four different sites (site conditions 1 to 4). Depending on site conditions yield relationships will consequently vary considerably. On site 1, optimal for species 1, the addition of species 2, inferior on this site, will become a burden. On site 2, well suited to both species, productivity is in balance. On site 3, optimal for species 2, the inferior species 1 will have a slowing-down effect on growth. Examples from practice demonstrating this kind of inferiority are a mix of common beech (*Fagus sylvatica*) and common oak (*Quercus petraea*) stands on fresh calcareous sites, or of Norway spruce (*Picea abies*) and common beech (*Fagus sylvatica*) stands on acidic, cool, and moist

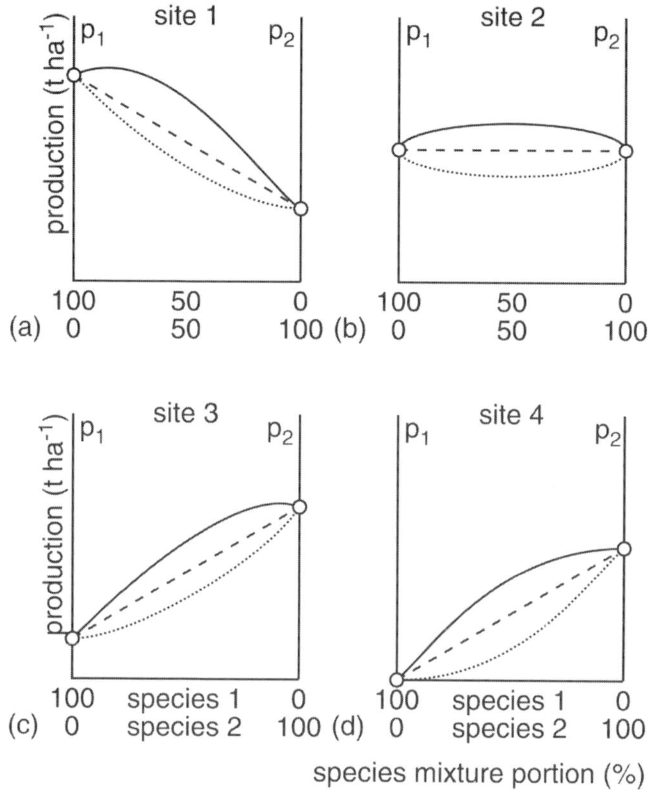

Fig. 3.2a–d. Dry biomass production for two species in pure and mixed stands under different site conditions 1 to 4 (cf. Fig. 3.1). Dry biomass productivities p_1 and p_2 of species 1 and 2 are shown for pure stands (*left* and *right ordinate*, respectively). The *connecting lines* show expected values for productivity $p_{1,2}$ in the mixture for different mixture proportions. The *linear connecting lines* (*dashed*) represent no interaction effects between species 1 and 2, the *continuous* and *dotted lines* reflect positive and negative effects from the species mixture, respectively

sites. In the first case, it is common oak, and in the second common beech that can only be sustained by silvicultural treatment. On site 4 species 1 will disappear sooner or later, while species 2 will achieve good productivity. Figure 3.2a–d depicts productivities p_1 and p_2 of species 1 and 2, respectively, on sites 1 to 4. The right-hand and left-hand ordinates plot productivities for species 1 and 2 in pure stands, the abscissa the mixture proportion. For sites 1 to 4 the resultant relationships are $p_1>p_2$, $p_1=p_2$, $p_1<p_2$, and $p_1<p_2$ with $p_1=0$, respectively. The example explains why the site-related productivity relationships diverge and serves to warn against generalizing results obtained from a limited spectrum of site conditions. If the mixed species don't interact at all or if mixture effects on growth cancel each other, productivity $p_{1,2}$ in the mixed

stand will lie on the straight reference line between p_1 and p_2 (dashed lines). In this case, an increase in the mixture portion would be reflected in a proportional increase of $p_{1,2}$.

3.2.2 Quantification of Effects from Species Interactions

If growth deviates positively or negatively from this straight reference line (convex and concave curves in Fig. 3.2, respectively), this is indicative of species interaction that will either increase or reduce productivity. Of special interest are positive deviations and their causes. Kelty (1992) distinguishes between "competitive reduction" (i.e., two or more species have reduced competition in mixture compared to pure stands, and they use the resources more efficiently) and "facilitation" (i.e., in mixture one species affects positively the growth of another species).

Misunderstandings frequently arise from the fact that *relative* superiority of productivity is frequently confused with *absolute* superiority. We adapt the term "overyielding" from agricultural science and population biology, and define relative and absolute superiority of mixed stands over the monoculture "non-transgressive" and "transgressive" overyielding, respectively (Hector et al. 2002).

Productivity superiority of the mixture versus the pure stand is *relative* if species 1 and 2 of the mixture together produce more than each constituent species on comparable pure stands of identical size, such that $p_{1,2}>m_1\ p_1+m_2\ p_2$ (= non-transgressive overyielding). The percent relative superiority or inferiority in productivity is calculated from $\Delta p_{rel}=[p_{1,2}/(m_1\ p_1+m_2\ p_2)-1]\cdot 100$, where p_1 and p_2 equal productivity of species 1 and 2 in the pure stand, respectively, $p_{1,2}$ that of the two species in mixture, m_1 and m_2 are the proportions of species 1 and 2 in the mixed stand, calculated for instance from the proportions of dry biomass for both species (w_1 and w_2) in the mixed stand: $m_1=w_1/(w_1+w_2)$ and $m_2=w_2/(w_1+w_2)$, respectively. In Fig. 3.2 the solid convex (as seen from below) lines represent beneficial interaction effects, while the dotted concave lines stand for negative interaction effects through mixture. Which of the component species is favored or suppressed can be determined by analogy. Growth of species p_1 measured in the pure stand is compared with that in the mixture $p_{1,(2)}$. If, as in the case of the European larch and Norway spruce mixture (cf. Fig. 3.6 c), the result is $p_{1,(2)}>m_1\ p_1$ and $p_{(1),2}>m_2\ p_2$, this would indicate favorable mixture effects from which both tree species benefit.

We use the term *absolute* superiority of the mixed stand over the pure stand (transgressive overyielding) where $p_{1,2}>\max\ (p_1, p_2)$. We are dealing with absolute inferiority where $p_{1,2}<\min\ (p_1, p_2)$. Stand production thus lies above or below that for pure stands for species 1 and 2. For absolute superiority and inferiority we therefore use the expressions $\Delta p_{abs}=[p_{1,2}/\max(p_1, p_2)-1]\cdot 100$ and $\Delta p_{abs}=[p_{1,2}/\min(p_1, p_2)-1]\cdot 100$, respectively. For a better understanding of these relationships, we shall return one more time to our model example. If the convex, solid lines in Fig. 3.2 exceed p_1 as well as p_2, this would indicate a case of absolute superiority. On site 2 this is true for all mixture proportions and for sites 1 and 3 whenever mixture proportions approach that of the more productive pure stand.

3.2.3 Yield Gains Through Risk Distribution

If one species in a mixture is more sensitive to disturbances, the more robust species may then, on account of its better adaptation, profit from the weakening or mortality of the inferior species. Assuming, e.g., a shift in environmental factors and resource availability with unfavorable effects on species 1 (Fig. 3.3, arrow 1→3), a significant decrease in production in pure stands would be the result. However, if a second species is added to the stand that is better adapted to the new growth conditions, this would enable that species to have a stabilizing effect on growth. Species 2 would then make better use of available resources and consequently improve its productivity and space sequestration. The same would happen if one species would disappear completely because of biotic calamities or natural mortality. In this case, the remaining species in the mixture would be able to recover the loss in production through accelerated growth, as suggested by the so-called insurance hypothesis (Yachi and Loreau 1999). This buffering by the remaining species would improve with the regularity of its distribution over the stand area. In both cases, the advantage of mixtures lies in risk distribution as a consequence of silvicultural diversification.

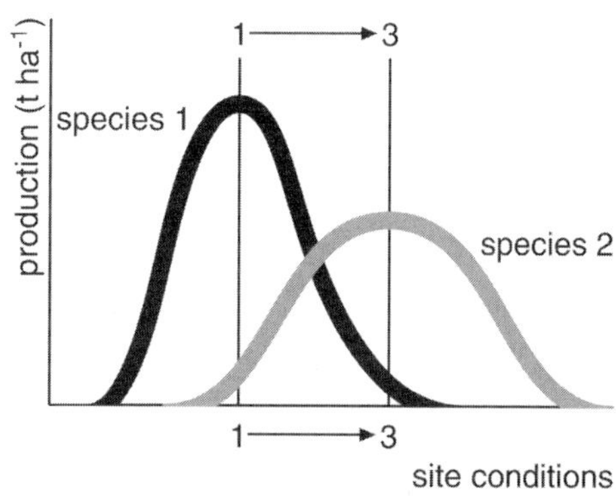

Fig. 3.3. Dry biomass production for species 1 and 2 (*black* and *grey lines*, respectively) in relation to site conditions. The shift in site conditions from 1 to 3 (*arrows*) results in a considerable increment loss for species 1. If species 1 and 2 are mixed, species 2 is capable of compensating for a loss in biomass production

3.3 Empirical Considerations

In order to keep the following empirical considerations as sound and valid as possible, they are based on findings in long-term experimental plots and avoid less reliable inventory data. For yield comparisons, only rarely are data on dry biomass available; information is usually confined to stem volume, which is considered to be of greater relevance in forestry practice. However, the yield comparisons below are mainly based on total dry biomass production (t ha^{-1}), since interspecific differences in wood density have to be eliminated in order to get meaningful results (Assmann 1970). Volume data were converted into dry biomass using specific wood densities reported by Trendelenburg and Mayer-Wegelin (1955) and Knigge and Schulz (1966). Specific wood densities for the individual species are as follows: common beech (*Fagus sylvatica*) 0.554 t m^{-3}, Norway spruce (*Picea abies*) 0.377, Scots pine (*Pinus sylvestris*) 0.431, common oak (*Quercus petraea*) 0.561, Douglas fir (*Pseudotsuga menziesii*) 0.412, European larch (*Larix decidua*) 0.487, common maple (*Acer pseudoplatanus*) 0.522, and common ash (*Fraxinus excelsior*) 0.564 t m^{-3}.

3.3.1 Productivity in Mixtures

Among the dominant tree species of a growth region, usually some are superior in total growth to all others, e.g., *Pinus pinaster* on the Iberian Peninsula, *Picea sitchensis* in Atlantic Western Europe. In central Europe, Norway spruce (*Picea abies*) and Douglas fir (*Pseudotsuga menziesii*) are often superior in productivity. There is hardly any reduction in this preeminence, which is related to stem volume, if the biomass from branches, leaves, and roots as well as fruit is included (Assmann 1970). On many sites Norway spruce and Douglas fir therefore have the role of species 1 in the mixture constellation shown in Fig. 3.1a.

Figure 3.4 shows the hierarchy of *important* tree species in terms of total dry biomass production for ages 50 (Fig. 3.4a) and 100 (Fig. 3.4b) in pure stands. For the purpose of comparison, the volume yields (m^3 ha^{-1}) shown on conventional yield tables for the best and poorest sites were converted into dry biomass production (t ha^{-1}). Up to age 50 Douglas fir and Norway spruce yield two to four times as much as other commercial tree species, if grown in monocultures. At 100 years of age Douglas fir and Norway spruce are still the most productive species, although the difference from the other species has become smaller. Common beech and silver fir (*Abies alba*), at age 50 among the lower and intermediate third, have caught up considerably at age 100. Conversely, early culminating species such as Scots pine lose their superior positions. The change in productivity hierarchy between the

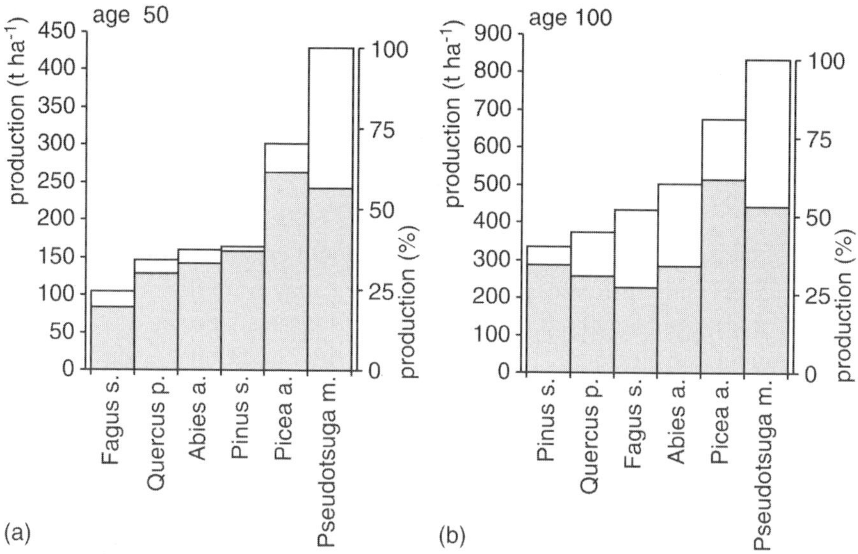

Fig. 3.4. Total yield in stem dry biomass production for selected commercial tree species at age 50 and 100 years (**a** and **b**, respectively). Production for best (total *bar* height) and most unfavorable yield classes (*gray bars*) are shown, calculated from yield tables for common beech (*Fagus sylvatica*; Schober 1967), common oak (*Quercus petraea*; Jüttner 1955), silver fir (*Abies alba*; Hausser 1956), Scots pine (*Pinus sylvestris*; Wiedemann 1948), Norway spruce (*Picea abies*; Assmann and Franz 1963), and Douglas fir (*Pseudotsuga menziesii*; Bergel 1985)

ages 50 to 100 underscores the effect of age on the results of yield comparisons.

Given the higher productivity in Norway spruce and Douglas fir stands, the admixture of other species usually causes production losses, since losses through the substitution of superior species by inferior species cannot be compensated for by the beneficial interactions between species in mixture. Let us assume the dry biomass production of Norway spruce in a pure stand comes to 800 and that of common beech to 480t ha^{-1} on a specific site. Let us further assume that 50 % common beech are added to the Norway spruce stand and that neutral interactions between the species prevail. This would reduce the production in the pure Norway spruce stand to $p_{1,2} = 0.5 \times 800$ t ha^{-1} $+ 0.5 \times 480$ t ha^{-1} $= 640$t ha^{-1}, i.e., to 80 % of the pure Norway spruce stand, a loss of 20 %. In the case of overyielding, the beneficial interactions from the mixture would have to compensate for this deficit. However, there are no examples of a mutual facilitation of Norway spruce and common beech to this extent. This implies that almost any admixture to the more productive pure Norway spruce stand would lower yield. The great superiority of pure Norway spruce and Douglas fir stands explains the outstanding progress of their cultivation in the past. Still, in view of the objective to achieve multi-criteria sustainability outlined above, the unilateral and short-term optimization of productivity is increasingly coming under criticism.

3.3.2 Dependence of Yield on Site Conditions

Our model example shows how greatly the yield relationship between two tree species with different ecological niches may vary from site to site (Fig. 3.2a–d). Contradictory results involving identical species mixtures are better understood through the inclusion of site conditions. I elucidate the influence of site conditions on yield relationships between pure and mixed stands using the most important and best studied mixture in Europe, i.e., Norway spruce/common beech. For this mixture a unique database involving 9 experimental areas with 33 experimental plots under permanent observation is available, with data going back to the beginning of the last century. Most of the experiments have been surveyed by the Chair of Forest Yield Science of the Technical University of Munich. Due to the unique length of time involved, growth and yield of Norway spruce and common beech in pure and mixed stands have been the subject of several investigations (e.g., Kennel 1965; Assmann 1970; Pretzsch 1992, 2003). Here, I report on only the relationship between site condition and growth. The site spectrum ranges from calcareous, warm, and dry sites in central and northern Germany, with a natural dominance of common beech, to acidic, cool sites with heavy precipitation in southern Bavaria, typical for natural Norway spruce stands. The experiments involve plots in pure and mixed stands with different mixture proportions. As the stands all have been established by natural regeneration the species are not exactly even in age. Nevertheless, apart from minor site-dependent variations in height growth the stands are mono-layered. From the start of the experiments, both pure and mixed plots have been subjected to moderate thinning from above, i.e., the closed canopy was maintained. The common beech component ranges from 30–50 %. Reference age for the results is 100 years.

At this age stem dry biomass stock ranges from 500–1,300 t ha^{-1} and 400–1,100 ha^{-1} for Norway spruce and common beech in pure stands, respectively. In mixtures the stock values lie between those for the pure stand plots of the constituent species. Depending on whether the site has favorable or unfavorable effects on either Norway spruce or common beech, these values will approach those in the pure spruce or beech stand. The same is true for total dry biomass production, for which data exist from the entire long observation period. Figure 3.5 represents total volume growth from mixed stands in relation to that from pure Norway spruce stands (100 % line) on adjacent sites with equal site conditions. The experiments are ranked such that those optimal for growth of Norway spruce sites are plotted on the left-hand side of the

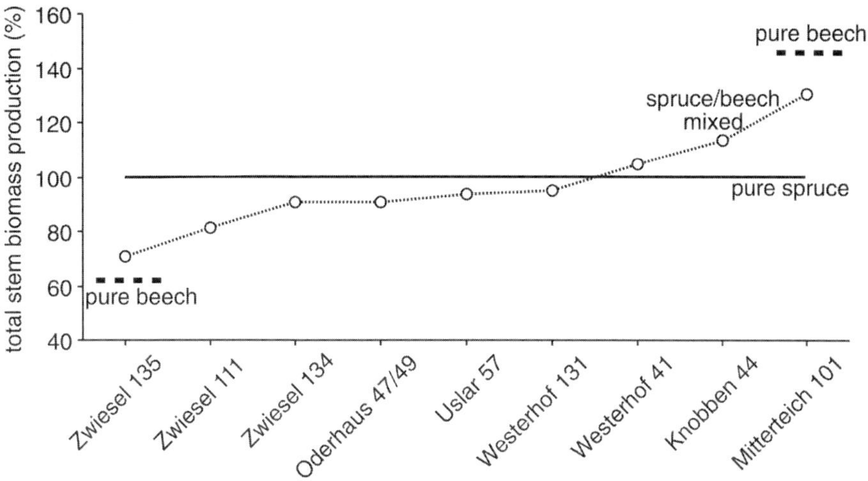

Fig. 3.5. Total stem biomass production of pure Norway spruce stands (*100% line*) and Norway spruce/common beech mixed stands in dependence on site conditions on long-term experimental plots at age 100. Since replications are not available for those long-term experiments, standard errors for the observations are lacking. See text for details

graph. From left to right, growth conditions for Norway spruce decrease while those for common beech improve. Mixture results on sites with optimal growth conditions for common beech are shown at far right. The yield for common beech, also related to the pure Norway spruce stand, is plotted for experiments "Zwiesel 135" and "Mitterteich 101," which mark the lower and upper yield spectra, respectively, for common beech (broken line). As can be seen, the admixture of common beech on typical Norway spruce sites reduces productivity to 70 % of the pure Norway spruce stand. On sites with optimal growth conditions for common beech, the inclusion of common beech to Norway spruce leads to a production increase of up to 130 % of pure Norway spruce. In the first case, the replacement of Norway spruce by common beech decreases yield, whereas in the second every admixture of common beech achieves a gain over productivity in the pure Norway spruce stand. The effects from species interactions Δp_{rel} between Norway spruce and common beech range between +10 and –20 %. In all available studies, however, the mixed stand of Norway spruce and common beech occupies a position between the corresponding pure stands, as far as productivity is concerned.

In a study by Jensen (1983), too, pure stands form the walls of the corridor, so to speak, in which the mixtures are positioned. Along a west–east transect through Jutland/Denmark he gives a model example of site-condition effects on the growth relationship between Norway spruce and silver fir. In the coastal dune belt, silver fir is superior to Norway spruce, the adjacent

Riss–glacial landscape leads to equivalent growth in silver fir and Norway spruce, whereas on the old inland moraines of the Würm glacial period silver fir is inferior to Norway spruce in dry biomass production. The decisive factor for the inland superiority of Norway spruce is its adaptability to low water supplies on acidic sites. By contrast, silver fir profits from better water availability and the more favorable nutrient supply in the coastal region. Even here, the evident beneficial effects from species interaction in mixtures are not sufficiently powerful to cause transgressive overyielding.

Yield limitations for mixtures composed of shade-tolerant trees such as Norway spruce/common beech and Norway spruce/silver fir are not transferable to mixtures consisting of shade-intolerant and shade-tolerant trees. Frivold and Kolström (1999) studied silver birch (*Betula pendula*), Scots pine, and Norway spruce growth in Finland, Sweden, and Norway. They emphasize that the potential superiority or inferiority of these species in mixtures is related to site conditions. Depending on site conditions, the effects of species interaction may be unfavorable, neutral, or beneficial, and in the latter case even lead to overyielding of mixed stands over the more productive pure stands. In southern and central Finland, Scots pine/silver birch mixtures surpass pure Scots pine and pure silver birch stands by 10 and 14 %, respectively (Mielikäinen 1980). For Norway spruce/silver birch mixtures a 10 to 15 % increase in production may occur compared with corresponding pure stands of these species, depending on the site (Mielikäinen 1985). In the oceanic regions of Norway and Sweden silver birch loses some of its increment capacity compared with coniferous species. There, Scots pine/silver birch mixtures do not achieve greater yield than the pure stands while Norway spruce/silver birch mixtures show a beneficial effect from mixtures only during the juvenile growth period (Frivold and Frank 2002).

3.3.3 Typical Mixture Effects on Yield

The examples shown in Fig. 3.6 illustrating antagonistic, neutral, and beneficial effects from species interaction (Fig. 3.6a–c) represent the frame for the mixture effects in a replacement series experiment from two-species mixtures to be expected in temperate and boreal zones (see Sect. 3.2.1 for theoretical background).

The Norway spruce/common beech experimental area “Freising 813” (Fig. 3.6a) represents negative effects from species interaction in mixtures (recognizable in the U shape of the connecting line of the total yield). With a proportion of 40 to 50 % common beech, total productivity decreases by about 30 % compared with expected values given neutral effects in mixture. The reaction of $p_{1,(2)}$ and $p_{(1),2}$ reveal that Norway spruce's productivity increases in proportion to the Norway spruce portion in the mixture, while that of common beech increases sub-proportionally. Common beech can

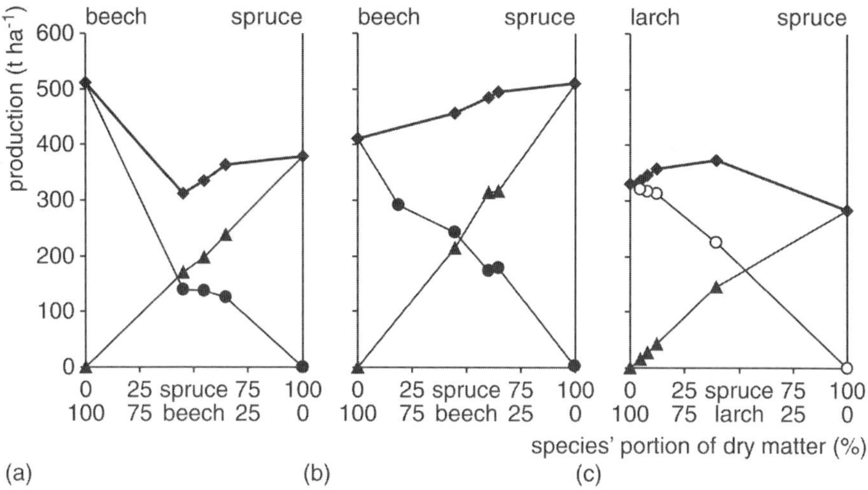

Fig. 3.6a–c. Typical response pattern of dry biomass production in two-species mixtures. Mixtures of Norway spruce and common beech in flat land areas with unfavorable mixture effects (**a**), Norway spruce and common beech in submontane areas with neutral effects (**b**), and European larch and Norway spruce in subalpine areas with beneficial mixture reactions (**c**). The dry biomass production from pure stands of common beech or European larch are plotted on the *left-hand ordinates*, while that for pure Norway spruce stands is plotted on the *right-hand ordinates*. The *upper connecting lines* (accentuated by *rhombuses*) represent total productivity $p_{1,2}$ of the mixed stands. The productivity for each mixture species $p_{1,(2)}$ and $p_{(1),2}$ is shown (*triangles* and *circles*, respectively). Data for are from **a** Pretzsch et al. (1998), **b** Pretzsch (1992), and **c** Zöhrer (1969)

thus be identified as the losing species that reduces the mixed stand's increment.

The Norway spruce/common beech experimental area "Zwiesel 111" (Fig. 3.6b) represents neutral effects from species interaction. On all plots of this experimental area, total dry biomass production increases in proportion to the portion of the constituent species in the mixture. Thus, total productivities of mixtures are between those of pure-stand plots. Accordingly, productivity gains are solely attributable to common beech being replaced by Norway spruce, which grows faster on this site, and not to beneficial interactions between the two species.

In both mixed Norway spruce/common beech stands (Fig. 3.6a, b) Norway spruce productivity (triangles) increases in proportion to its portion in the mixture. The reaction of common beech (circles), though, is negative or neutral. In the former case, beech is inferior to the more productive Norway spruce. In the latter case, a balance of competition is achieved. Wiedemann (1942, 1943, 1951) was already able to differentiate between the two reaction types. In northern Germany, Norway spruce/common beech mixtures with highly productive beech, approximately the same dry biomass as corresponding pure stands are achieved. For common beech of moderate growth the mixed stand biomass production is about 19% less than in pure common beech stands on identical sites. Kennel (1965) studied mixtures of Norway spruce and common beech in the Bavarian alpine foothills, the Bavarian Forest, and the Harz in Lower Saxony, as did Burger (1941) in Switzerland, with similar results. We note that in mixtures composed of two shade-tolerant species, dry biomass production in mixed stands never significantly exceeds that of comparable pure stands, and is often considerably lower.

On the other hand, beneficial effects from species interaction with over-yielding are often achieved in mixtures of shade-intolerant and shade-tolerant trees. Zöhrer (1969) provided evidence that the biomass production of European larch/Norway spruce mixtures in the Salzburger Land is superior to that of pure stands on identical sites (Fig. 3.6 c). With increase in the Norway spruce portion, total yield rises over-proportionately, reaches a peak at 40%, and then declines to the value of the pure Norway spruce stand. The position of the resultant data above the connecting line between the dry biomass production of both pure stands is indicative of the beneficial effects of species' interactions in the mixture. The European larch/Norway spruce mixture therefore surpasses the pure Norway spruce stand by 22–28% and the pure European larch stand by 2–13%. For mixed stands, composed of shade-intolerant and shade-tolerant species such as common oak/common beech, Scots pine/Norway spruce and Scots pine/common beech, Bonnemann (1939) and Wiedemann (1943, 1951) found similar beneficial effects from species interactions after 50 years of observation. For long-term Scots pine/common beech experimental areas in the Dübener Heide, Dittmar et al. (1986) noted beneficial interaction effects compared with the pure stand of 7–25%, depending on the age and structure of the mixture. Burger (1941) and Wimmenauer (1941) found the same strong superiority in European larch/common beech mixtures.

A considerably higher superiority of 50% is claimed by DeBell et al. (1989) for mixtures of *Eucalyptus saligna* and the leguminous and nitrogen-fixing tree species *Albizia falcataria* in Hawaii. Other examples are summarized by Kelty (1992). Compared with these yield relationships in the subtropics, mixture effects of about $\pm$30% for commercial tree species in temperate and boreal zones appear rather moderate.

3.3.4 Disturbances and Silvicultural Treatment

The greater the niche variations among the constituent species of a mixed stand, the more elastic will be its response in the face of disturbances (Sect. 3.2.3). An example of this are the annual increment values from the Norway spruce/common beech experimental area "Schongau 814" in the period 1960–1995 (Fig. 3.7). In contrast to the component common beech,

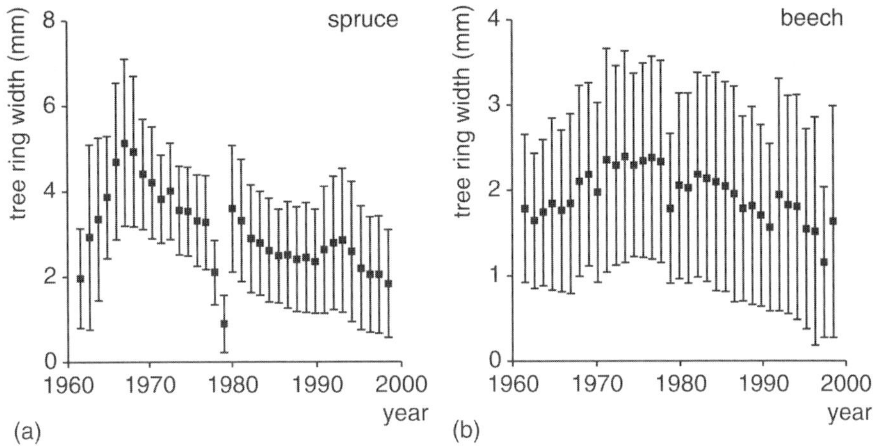

Fig. 3.7. Mean ± standard deviation of annual ring width for Norway spruce (**a**) and common beech (**b**) on the mixed stand experimental area Schongau 814 (n=193 for Norway spruce and n=87 for common beech). On that site Norway spruce shows much more sensitive reactions to the drought of the year 1976 than common beech

Norway spruce reacted to the drought of the year 1976 with a strong decrease in increment. In pure Norway spruce stands this would cause serious production losses. In Norway spruce/common beech mixtures, disturbances of this kind can be mitigated by compensatory growth of common beech (Fig. 3.3), as suggested by the "insurance hypothesis" (Yachi and Loreau 1999; Pretzsch 2003). Unfortunately, yield comparisons between pure and mixed stands usually refer to more or less undisturbed stands. Affected plots are abandoned after calamities or unplanned use and only undisturbed plots are kept under continuous observation. Statements about inferiority or superiority derived from these experiments therefore also apply merely to "normal" circumstances. If response patterns after disturbances were also considered yield comparisons would become more realistic. In the following, the example of Norway spruce and common beech will serve to illustrate that pure stands composed of these species respond in radically different ways to disturbances in the form of thinnings from mixed stands.

The ensuing analysis of thinning–growth relationships in pure stands is based on 19 thinning experiments with 26 Norway spruce and 30 common beech plots located in Bavaria and Lower Saxony. The oldest plots have been under regular observation since 1870–1880. With few exceptions they are composed of three plots each that are identical in site and age but were consistently managed according to the specifications for A, B, and C grade (slight, moderate, and heavy thinning from below) and thus cover a wide spectrum of stand densities (Pretzsch 2002, 2003). The analysis of density–growth relationships in mixed stands relies on data from a total of 23 experimental areas with Norway spruce/common beech mixtures in south Germany involving a total of 78 plots under observation since 1954. With a basal area spectrum ranging from 20–80 m^2 ha^{-1} at age of about 100, densities tend to vary even more here than in the pure stands. Removals from thinnings, remaining stand response, and hence total growth are quantifiable based on data from up to 20 routine inventories of removed, remaining, and total stand biomass.

Figure 3.8 represents the stem dry biomass yield for pure Norway spruce and common beech stands and mixed Norway spruce/common beech stands (Fig. 3.8a–c) on plots subjected to various grades of thinning in comparison with the untreated plots at age 100. If we first consider the response pattern in pure stands (Fig. 3.8a, b), it appears remarkable that dry biomass production in the transition from A grade to B grade rises by 5 to 10 % in either case. In the transition from B grade to C grade, we note a decrease in total growth. In common beech stands total growth yield for C grade is even higher than for A grade. After 130 years of experimental research we are therefore able to state for Norway spruce and common beech a significant increase ($p<0.05$) in growth from A grade to B grade and for Norway spruce a significant decrease ($p<0.05$) in growth in the transition from B grade to C grade. The relationship between density and growth hence represents an optimum curve. Any approach to maximum stand density is concomitant with growth reductions of 5 to 10 %. By contrast, the biomass production in Norway spruce/common

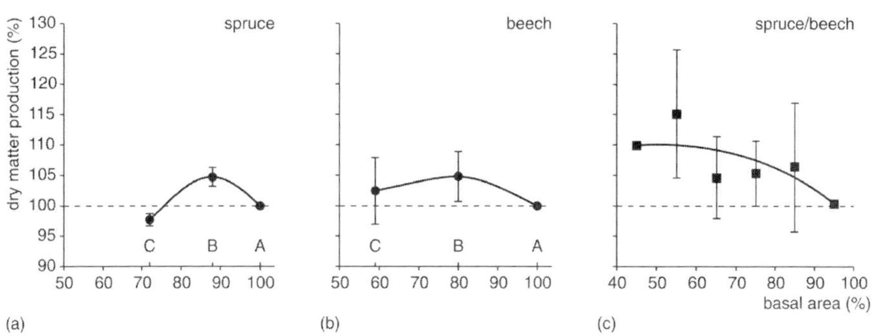

Fig. 3.8a–c. Density-growth relationships in pure and mixed stands from Norway spruce and common beech at age 100. *A*, *B*, and *C* grade, respectively, refers to slight, moderate, and heavy thinning from below in the pure stands. In the mixed stands, the experimental design included untreated plots, slight, moderate, strong, and accretion thinning. Means±SE. As the production of the untreated plots with maximal density were used as reference for the thinned plots and set to 100 %, they have no standard error. **a** Results from 9 Norway spruce thinning experiments with 26 plots, consistently slight, moderate, and heavy thinning since 1870. **b** Results from 10 common beech thinning experiments with 30 plots, consistently subjected to slight, moderate, and heavy thinning from below since 1870. **c** Results from 23 mixed stand experiments with 78 plots under observation since 1954. Dry biomass production in heavily thinned mixed stands is depicted without standard error, since replications were lacking

beech mixtures (Fig. 3.8 c) achieves similar values over a wide range of densities. Their is no significant reduction ($p>0.05$) of biomass production even when stand density is reduced to 50 %. In comparison with corresponding pure stands, Norway spruce/common beech mixtures are able to compensate for disturbances caused by thinnings through accelerated growth in the remaining stand.

The cause for this response pattern is the space sequestration by dominant Norway spruce and dominant but also subdominant common beech. This leads to an increase, from pure to mixed stands, in the total crown shading area and also in the frequency of multiple crown shading. In mixed stands removals from or losses in the upper crown layer can be compensated for by stronger growth in the lower layer. This buffer effect through a vertical stratification of the canopy becomes particularly effective with increasing age. In the mixed stand the remaining trees are able to close any gaps that may form by mortality, to slow down age-related breakup of the crown layer and to have a stabilizing effect on stand biomass production.

3.4 Discussion and Conclusions

Lack of data during the founding period of forestry sciences led to statements on yield relationships between pure and mixed stands that were based on faith rather than knowledge (Möller 1922; Wiedemann 1951). Today, thanks to long-term experiments, the vastly increased knowledge on two-species mixtures can be consolidated. Mixture effects may vary considerably depending on species mixture, site, silvicultural treatment, and risks. Compared with pure stands resource utilization can be improved by almost 30 % by combining early and late successional species, ontogenetically early and late culminating species, shade-intolerant and shade-tolerant tree species. However, where ecological niches and functional characteristics are similar, species may compete for the same resources in crown and root systems. The consequent effects from species interactions may be negative, with a reduction in productivity up to 30 %.

There is special potential for increased productivity in mixtures of about equally productive species on a given site which complement each other in the spatial-temporal utilization of space, leading to a reduction of competition (Kelty 1992). This can be achieved by joint growing space occupation with shade-intolerant species (e.g., European larch, Scots pine), semi-shade-tolerant species (e.g., Norway spruce, Douglas fir) and shade-tolerant trees (e.g., common beech, silver fir). This kind of stratification using species of different shade tolerance will allow light transmitted through the upper canopy to be used by the layers underneath. Gains in productivity are also achieved in tree mixtures where the temporal courses in seasonal growth period and in the aging process complement one another. Let us assume growth of species 1 in a two-species mixture culminates early and then declines rapidly. The decrease in total increment can then be made up for by addition of species 2, the development of which is anti-cyclic to that of the other. Assmann (1970) shows that species-specific periodicity is reflected in different time scales. Species which culminate early in the season also exhibit the same characteristic as regards lifespan. Mixtures of species with anti-cyclic seasonal growth characteristics also often complement one another in the aging process.

Temporal and spatial resource use complementarity, e.g., in Norway spruce/common beech mixtures, may also occur in combined form. In spring, before the leafing of common beech, more intense light can penetrate the stand and curtail the winter dormancy of Norway spruce, thus prolonging its seasonal growth period which, per se, is longer than that of common beech (Schober 1950/1951). From this kind of "job-sharing," e.g., Norway spruce and common beech, mixtures may draw advantages in resource utilization (Mitscherlich 1952). Beneficial interaction effects will be stronger the closer and more intensive the mixing of Norway spruce and common beech (Ellenberg et al. 1986).

Systematic yield gains by up to 30 and 50 % for grasslands and natural forest ecosystems in the transition from pure stands to two-species mixtures (Hector et al. 1999; Caspersen and Pacala 2001; Loreau et al. 2001; Pfisterer and Schmid 2002) can be transferred to managed forests only to a very limited extent. Presumably, in boreal and temperate forests niche differentiation is comparatively low due to species reduction in the course of the ice ages and due to the much slower evolutionary and co-evolutionary processes of long-lived trees. This may be a reason why increased efficiency in resource use and productivity of mixed stands compared with pure stands is much lower in long-lived woody ecosystems than in short-lived herbaceous stands. Many of the European forest stands are "artifacts" designed with very productive species such as Norway spruce and Douglas fir cultivated outside their natural habitats. Often, genetic variation in these species no longer reflects natural selection but a choice controlled by mankind's commercial criteria. These forests are therefore not designed for optimum niche utilization by the mixture species. Niche overlapping and risks may occur that are reflected in unfavorable effects from species interactions in the mixture.

The combination of several species is synonymous with a distribution of risks. As a rule mixed stands are more elastic in their response to changing site conditions and show greater resilience in the face of natural losses or calamities. Let us assume a pure Norway spruce stand on a site in the Bavarian alpine foothills with good water supply and acidic soil, where Norway spruce growth far surpasses that of common beech, but to which common beech is added to raise stand biodiversity and aesthetic value. As a consequence the replacement of Norway spruce by the slower-growing common beech under undisturbed development causes considerable yield loss because of opportunity costs. However, if the greater elasticity against disturbances is included in the calculation, opportunity costs may be considerably lower. This has been shown in simulation studies on the effect of climate change on the growth of pure and mixed Norway spruce and common beech stands in Germany (Pretzsch and Dursky 2002). Assuming temperatures within the growth period rise by 2 °C for the above site, precipitation in the vegetation period drops by 10%, and the growing season is prolonged by 10 days, a decrease would occur in Norway spruce productivity on that particular site by over 10%. The substitution of 30% of the Norway spruce by common beech, which is better adapted to the assumed changing climate conditions, could compensate for the climate-related increment losses in Norway spruce. In view of the increasing disturbances to which our forest ecosystems are being subjected through chemical emissions and climate change, the chances for advantages being realized through this kind of risk distribution by species mixture will probably rise in the future (Lindner and Cramer 2002).

Variations in stand density, too, are more easily compensated for in mixed stands. This property of mixed stands keeps their growth rate stable under lack of treatment and maximum density as well as under density reductions due to silvicultural treatment or calamity. In pure stands an optimum relationship exists between density and growth. The overlapping of different response patterns for Norway spruce and common beech leads to a considerably wider plateau in the density–growth relationship of mixed stands than for corresponding pure stands. The broad saddle in the resultant curve looks similar to the much discussed curve by Langsaeter (1941, p. 173; Fig. 3.3). In contrast to the pure stands under study, the approach to maximum density causes merely a slight and statistically insignificant decrease in growth. This important relationship is schematically represented in Fig. 3.9. The mixture (black) is compared with two pure stands (grey lines). The first case (upper line) assumes the production superiority of pure stands versus mixtures at average density. Whenever density is reduced due to some kind of disturbance, pure stands will respond with considerable increment loss and become inferior. By contrast, growth in the mixed stand remains stable over a wide range of densities. Even though mixed stands may be inferior under "stable" conditions they may develop superiority on account of their greater resilience in the face of perturbation or non-treatment. In the second instance (lower line) the pure stand is less productive than the mixture, even under "normal" conditions. In this case, given positive or negative deviations from average density, e.g., lack of treatment or unplanned disruption of stand canopy, the stand becomes even more inferior.

In short, the productivity relationship between pure and mixed stands under "normal" conditions may shift considerably once risks are included (Pretzsch 2003). The decisive factor here, in essence, is the probability of the occurrence of disturbances and damage. The temporarily dazzling productiv-

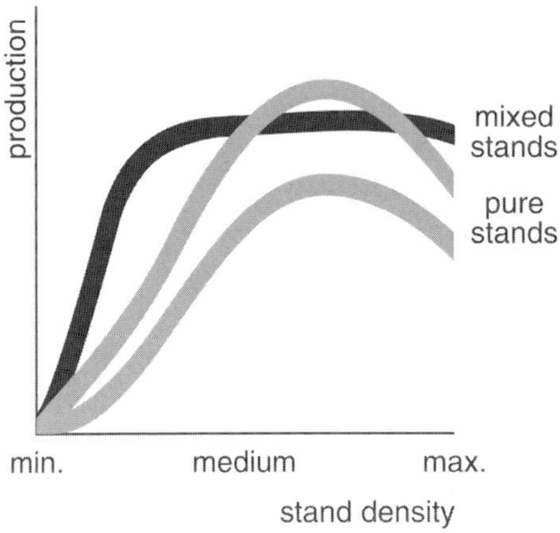

Fig. 3.9. Schematic representation of relationships between stand density and biomass production in pure and mixed stands. The overlapping of different response patterns for Norway spruce and common beech leads to a considerably wider plateau in the density–growth relationship of mixed stands than for corresponding pure stands. This reflects the greater growth resilience in mixed stands (*black line*) as compared with that in pure stands (*grey lines*)

ity superiority of artificial pure stands is often reversed and turns into inferiority if risks are included in the calculation (Pretzsch und Dursky 2002). Thus, while Norway spruce is overwhelmingly superior to common beech on many sites under normal conditions, storm damage in Norway spruce stands is four times as high as in common beech stands (von Lüpke and Spellmann 1999). To a considerable extent mixtures can overcome this kind of perturbation, equivalent to an abrupt reduction in stand density from medium to lower levels, without greater growth reductions. Of course, the above relationships between species number and biomass production do not alone justify decisions in favor of pure or mixed stand establishment or a certain stocking density. The outcome of such decisions may be quite different, depending on frame conditions and specific objectives of forest management – for instance if the major aim is quantity (e.g., pulp, fuelwood, C storage), quality (e.g., structural wood or veneer), or safety and risk prevention (stabilization against storms or erosion control). However, if we recall that a suitable mixture may raise dry biomass production by up to 30 % and, moreover, ensures that other important forest functions (cf. MCPFE 2000) are fulfilled in addition, then the above yield relationships may become primary in controlling the decisions.

In comparison with annual systems, the lifespan of forests is longer by two orders of magnitude and the danger from risks consequently much greater. In addition, cyclic disturbances through silvicultural treatment take their toll. It is for this reason that the risk distribution in forests achieved in mixtures carries so much more weight than in short-lived ecosystems. Risk distribution through tree species diversity, however, need not necessarily imply a close mixture of tree species. The desired diversification could also be achieved by plot mosaics of pure stands of different species. Through this kind of species separation even likely unfavorable effects from species interactions and greater efforts required to facilitate inferior species in mixtures could be avoided. However, the above-mentioned beneficial interactions in mixtures, which may raise the productivity of commercial tree species in temperate and boreal zones up to 30% under "stable" conditions, as well as the higher resilience and superior productivity of mixed stands under disturbances, require a close spatial association of the mixed species.

Acknowledgments. The author wishes to thank the Deutsche Forschungsgemeinschaft for providing funds for mixed-stand research as part of Sonderforschungsbereich 607 "Growth and Parasite Defense" and the Bavarian State Ministry for Agriculture and Forestry for funding forest growth and yield research by the permanent project W 07. Thanks also go to the research scientists in charge of the long-term experimental areas at the Chair of Forest Yield Science, Martin Nickel and Gerhard Schütze for processing and evaluating the historical time series and Dr. Hermann Spellmann from the Forestry Research Institution in Lower Saxony for making data available to us from the former Prussian Forestry Research Station. Thanks are also due to H. Herling for preparation of graphs, and to Michael Scherer-Lorenzen, Christian Körner, and one anonymous reviewer, for constructive criticism.

References

Assmann E (1970) The principles of forest yield study. Pergamon Press, Oxford

Assmann E, Franz F (1963) Vorläufige Fichten-Ertragstafel für Bayern. Institut für Ertragskunde der Forstlichen Forschungsanstalt, Munich

Bartelink HH, Olsthoorn AFM (1999) Mixed forest in western Europe. In: Olsthoorn AFM, Bartelink HH, Gardiner JJ, Pretzsch H, Hekhuis HJ, Franc A (eds) Management of mixed-species forest: silviculture and economics. IBN Scientific Contribributions 15, Wageningen, pp 9–16

Bergel D (1985) Douglasien-Ertragstafel für Nordwestdeutschland. Niedersächsische Forstliche Versuchsanstalt, Göttingen

Bonnemann A (1939) Der gleichaltrige Mischbestand von Kiefer und Buche. Mitt Forstwirtsch Forstwiss X:1–45

Burger H (1941) Beitrag zur Frage der reinen oder gemischten Bestände. Mitt Schweiz Anst Forstl Versuchsw XXII:164–203

Cannell MGR, Malcom DC, Robertson PA (1992) The ecology of mixed-species stands of trees. Blackwell Scientific Publ, Oxford

Caspersen JP, Pacala SW (2001) Successional diversity and forest ecosystem function. Ecol Res 16:895–903

DeBell DS, Whitesell CD, Schubert TH (1989) Using N_2-fixing *Albizia* to increase growth of *Eucalyptus* plantations in Hawaii. For Sci 35:64–75

Dieterich V (1957) Langfristige Folgerungen des forstwirtschaftlichen Grundgebots der Nachhaltigkeit. Forstw Centralbl 76:193–207

Dittmar O, Knapp E, Zehler H (1986) Die langfristige Versuchsfläche Tornau im StFB Dübener Heide, ein Beispiel für den Weg vom Kiefernreinbestand zum Buchen-naturverjüngungsbetrieb. Soz Forstw 36:344–348

Ellenberg H, Mayer R, Schauermann J (1986) Ökoystemforschung – Ergebnisse des Sollingprojektes. Ulmer, Stuttgart

Frivold LH, Frank J (2002) Growth of mixed birch-coniferous stands in relation to pure coniferous stands at similar sites in South-eastern Norway. Scan J For Res 17:139–149
Frivold LH, Kolström T (1999) Yield and treatment of mixed stands of boreal tree species in Fennoscandia. In: Olsthoorn AFM, Bartelink HH, Gardiner JJ, Pretzsch H, Hekhuis HJ, Franc A (eds) Management of mixed-species forest: silviculture and economics. IBN Scientific Contribributions 15, Wageningen, pp 37–45
Gayer K (1886) Der gemischte Wald, seine Begründung und Pflege, insbesondere durch Horst- und Gruppenwirtschaft. Parey, Berlin
Hartig GL (1791) Anweisung zur Holzzucht für Förster. Neue Akademische Buchhandlung, Marburg
Hartig GL (1804) Anweisung zur Taxation und Beschreibung der Forste. Heyer, Gießen
Hausser K (1956) Tannen-Ertragstafel. In: Schober R (ed) Ertragstafeln wichtiger Baumarten. Sauerländer, Frankfurt/Main
Hector A, Schmid B, Beierkuhnlein C, Caldeira MC, Diemer M, Dimitrakopoulus PG, Finn JA, Freitas H, Giller PS, Good J, Harris R, Högberg P, Huss-Danell K, Joshi J, Jumpponen A, Körner C, Leadley PW, Loreau M, Minns A, Mulder CPH, O'Donovan G, Otway SJ, Pereira JS, Prinz A, Read DJ, Scherer-Lorenzen M, Schulze ED, Siamantziouras ASD, Spehn EM, Terry AC, Troumbis AY, Woodward FI, Yachi S, Lawton JH (1999) Plant diversity and productivity experiments in European grasslands. Science 286:1123–1127
Hector A, Bazeley-White E, Loreau M, Otway SJ, Schmid B (2002) Overyielding in grassland communities: testing the sampling effect hypothesis with replicated biodiversity experiments. Ecol Lett 5:502–511
Jensen AM (1983) Growth of Silver Fir (*Abies alba* Mill.) compared with the growth of Norway Spruce (*Picea abies* (L) Karst.) in pure and mixed stands on sandy soils in the Western parts of Denmark. Rep Dep For R Veterin Agric Univ 14:1–498
Jüttner O (1955) Eichen-Ertragstafel. In: Schober R (ed) Ertragstafeln wichtiger Baumarten. Sauerländer, Frankfurt/Main
Kelty MJ (1992) Comparative productivity of monocultures and mixed stands, In: Kelty MJ, Larson BC, Oliver CD (eds) The ecology and silviculture of mixed-species forests. Kluwer, Dordrecht, pp 125–141
Kennel R (1965) Untersuchungen über die Leistung von Fichte und Buche im Rein- und Mischbestand. Allg Forst Jagdz 136:149–161, 173–189
Knigge W, Schulz H (1966) Grundriss der Forstbenutzung. Parey, Hamburg
Lindner M, Cramer W (2002) German forest sector under global change. An interdisciplinary impact assessment. In: Pretzsch H, Lindner M, Suda M (eds) German forest sector under global change. Ger J For Sci 121(1):3–17
Langsaeter A (1941) Om tynning i enaldret gran- og furuskog Maddel. Det Norske Skogforoksvesen 8:131–216
Loreau M, Naeem S, Inchausti P, Bengtsson J, Grimr JP, Hector A, Hooper DU, Huston MA, Raffaelli D, Schmid B, Tilman D, Wardle DA (2001) Biodiversity and ecosystem functioning: current knowledge and future challenges. Science 294:804–808
MCPFE (2000) General declarations and resolutions adopted at the Ministerial Conferences on the protection of forests in Europe. Strasbourg 1990-Helsinki 1993-Lisbon 1998. Ministerial Conference on the protection of forests in Europe. Liaison Unit, Vienna
Mielikäinen K (1980) Mänty-koivusekametsiköiden rakenne ja kehitys. Summary: structure and development of mixed pine and birch stands. Commun Inst For Fenn 99:1–82
Mielikäinen K (1985) Koivusekoituksen Vaikutus Kuusikon Rakenteeseeen ja kehitykseen. Effect of an admixture of birch on the structure and development of Norway Spruce Stands. Commun Inst For Fenn 133:1–79

Mitscherlich G (1952) Der Tannen-Fichten-(Buchen)-Plenterwald. Schriften Bad Forstl Vers Freiburg im Breisgau 8:1–42

Möller A (1922) Der Dauerwaldgedanke. Sein Sinn und seine Bedeutung. Springer, Berlin

Olsthoorn AFM, Bartelink HH, Gardiner JJ, Pretzsch H, Hekhuis HJ, Franc A (eds) (1999) Management of mixed-species forest: silviculture and economics. IBN Scientific Contributions 15, Wageningen

Pfisterer AB, Schmid B (2002) Diversity-dependent production can decrease the stability of ecosystem functioning. Nature 416:84-86

Pretzsch H (1992) Konzeption und Konstruktion von Wuchsmodellen für Rein- und Mischbestände. Forstl Forschungsber Munich 115:358

Pretzsch H (2001) Modellierung des Waldwachstums. Blackwell Wissenschafts-Verlag, Berlin

Pretzsch H (2002) Grundlagen der Waldwachstumsforschung. Blackwell Wissenschafts-Verlag, Berlin

Pretzsch H (2003) The elasticity of growth in pure and mixed stands of Norway spruce (*Picea abies* [L.] Karst.) and common beech (*Fagus sylvatica* L.). J For Sci 49:491–501

Pretzsch H, Dursky J (2002) Growth response of Norway Spruce (*Picea abies* (L.) Karst.) upon expected climatic changes in Germany. In: Pretzsch H, Lindner M, Suda M (eds) German forest sector under global change. Ger J For Sci 121(1):145–154

Pretzsch H, Kahn M, Grote R (1998) Die Fichten-Buchen-Mischbestände des Sonderforschungsbereiches "Wachstum oder Parasitenabwehr?" im Kranzberger Forst. Forstw Centralbl 117:241–257

Pretzsch H, Biber P, Dursky J (2002) Single tree based stand simulator SILVA. Construction, application and evaluation. For Ecol Manage 162:3–21

Schober R (1950/1951) Zum jahreszeitlichen Ablauf des sekundären Dickenwachstums. Allg Forst Jagdz 122:81–96

Schober R (1967) Buchen-Ertragstafel für mäßige und starke Durchforstung. In: Schober R (ed) Die Rotbuche 1971. Sauerländer, Frankfurt/Main

Schober R (1975) Ertragstafeln wichtiger Baumarten. Sauerländer, Frankfurt/Main

Schwappach A (1912) Ertragstafeln der wichtigeren Holzarten in tabellarischer und graphischer Form. Neumann, Neudamm

Trendelenburg R, Mayer-Wegelin H (1955) Das Holz als Rohstoff. Hansa, Munich

Von Carlowitz HC (1713) Sylvicultura Oekonomica oder haußwirthliche Nachricht und Naturmäßige Anweisung zur Wilden Baum-Zucht. Braun, Leipzig

Von Cotta H (1828) Anweisung zum Waldbau. Arnoldische Buchhandlung, Dresden

Von Hagen O (1867) Die forstlichen Verhältnisse Preußens, Bd 1. Springer, Berlin

Von Lüpke B, Spellmann H (1999) Aspects of stability, growth and natural regeneration in mixed Norway spruce-beech stands as a basis of silvicultural decisions. In: Olsthoorn AFM, Bartelink HH, Gardiner JJ, Pretzsch H, Hekhuis HJ, Franc A (eds) Management of mixed-species forest: silviculture and economics. IBN Scientific Contribributions 15, Wageningen, pp 245–267

Weck J (1955) Forstliche Zuwachs- und Ertragskunde. Neumann, Radebeul Berlin

Wiedemann E (1942) Der gleichaltrige Fichten-Buchen-Mischbestand. Mitt Forstwirtsch Forstwiss 13:1–88

Wiedemann E (1943) Der Vergleich der Massenleistung des Mischbestandes mit dem Reinbestand. Allg Forst Jagdz 119:123–132

Wiedemann E (1948) Die Kiefer 1948. Schaper, Hannover

Wiedemann E (1949) Ertragstafeln der wichtigen Holzarten bei verschiedener Durchforstung. Schaper, Hannover

Wiedemann E (1951) Ertragskundliche und waldbauliche Grundlagen der Forstwirtschaft. Sauerländer, Frankfurt/Main

Wimmenauer K (1941) Zur Frage der Mischbestände. Allg Forst Jagdz 90:90–93
Yachi S, Loreau M (1999) Biodiversity and ecosystem productivity in a fluctuating environment: the insurance hypothesis. Proc Natl Acad Sci USA 96:1463–1488
Zöhrer F (1969) Bestandeszuwachs und Leistungsvergleich montan, subalpiner Lärchen-Fichten-Mischbestände. Forstw Centralbl 88:1–64

4 Confounding Factors in the Observational Productivity–Diversity Relationship in Forests

M. Vilà, P. Inchausti, J. Vayreda, O. Barrantes, C. Gracia, J. J. Ibáñez, and T. Mata

4.1 Introduction

Field experiments conducted in randomly assembled grassland communities have demonstrated that changes in plant species diversity affect ecosystem productivity over a range of environmental conditions (Hector et al. 1999; Tilman et al. 2001). However, there is still a controversy as to whether this causal relationship is also found in natural systems (Loreau et al. 2001; Schmid 2002), especially at regional scales (Bengtsson et al. 2002). For example, Troumbis and Memtsas (2000) found that Greek shrub lands were more productive in stands with high shrub diversity. However, the positive correlations between diversity and productivity may be confounded with other less conspicuous or unknown factors such as site quality or fertility that affect both variables and underlie the observed correlation (Huston 1997; Troumbis 2001; Wardle 2001). Furthermore, observational studies have not always found a positive relationship between diversity and productivity. The available evidence shows that multiple patterns exist and change with spatial scale. Exhaustive reviews on the observed relationship between vascular plant species richness and productivity have found that the hump-shaped (unimodal) relationship occurred more often than a monotonically increasing relationship, depending on the geographical scale and ecological organization (e.g., within or across community types; Waide et al. 1999; Mittelbach et al. 2001).

This controversy can be partially reconciled by superposing experimental results on observational patterns (Fig. 4.1). That is, observed diversity–productivity relationships compare diversity across sites of different productivities driven by environmental conditions. The observations (data points) fill the area below a humpbacked line. Instead, experiments compare productivities at different experimentally established levels of diversity at single sites and maintaining all environmental factors influencing productivity constant

Ecological Studies, Vol. 176
M. Scherer-Lorenzen, Ch. Körner, and E.-D. Schulze (Eds.)
Forest Diversity and Function: Temperate and Boreal Systems

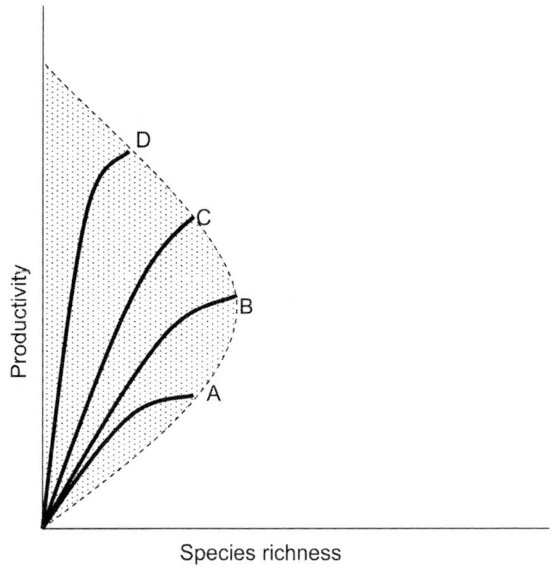

Fig. 4.1. Observed and experimental relationship between species richness and productivity and potential effects of disturbance and silvicultural practices in forests. The observations across sites fill the area below a humpbacked line. The experimental diversity–productivity relationship at specific sites can be represented as different ascending trajectories within the humpbacked area. From the initial *A* trajectory, *B*, *C*, and *D* trajectories indicate potential increasing slopes of the diversity–productivity relationship in a site at early stages of regeneration after disturbance. Sustainable forest management should avoid trajectories D and adopt trajectories B. (Adapted from Bengtsson et al. 2002)

(e.g., the same soil fertility, the same climate, the same topography). The experimental diversity–productivity relationship can be represented as different ascending trajectories within the humpbacked area (Loreau et al. 2001; Bengtsson et al. 2002; Schmid 2002). A rather unexplored issue concerns the mechanisms that underlie the different trajectories and the ecological processes that can shift one trajectory to another (but see Schmid 2002). The purpose of this chapter is to exemplify the difficulties of observational studies that rise while detecting diversity–productivity relationships in forests. Multivariate approaches are needed to separate the effects of covarying causal factors (Waide et al. 1999).

Monospecific forest stands and monocultures of the highly productive tree species have been extremely favored for pulp and timber production (Kelty 1992). At the same time, mixed forests in some regions have been maintained for landscape aesthetics, conservation of wildlife, recreational purposes, higher diversity of produces, and the belief that they are more resistant to disease and to disturbances such as wind or fire (Assman 1970; Kerr et al. 1992;

Dhôte, Chap. 14, this Vol.; Pautasso et al., Chap. 13, this Vol.; Wirth, Chap. 15, this Vol.). However, the concept of a potential increase in productivity in mixed tree stands has not generally been incorporated into forestry and conservation practice (but see Assman 1970). Recently, Caspersen and Pacala (2001) analyzed the Forest Inventory and Analysis (FIA) database in the United States (more than 20,000 plots) and have found a positive correlation between species richness and stand productivity. However, the lack of environmental description of stands hinders the interpretation of this association. Observational and experimental approaches in forests should examine the relationship between species diversity and productivity while controlling for the effect of other covariant factors that could underlie and confound the diversity–productivity relationship. In the boreal forest, comparisons between monospecific stands and adjacent mixed stands of similar age, tree density, soil characteristics, and management regime have revealed that whether mixed stands are more productive than monospecific stands depends on the identity of the species in the mixture. Mixtures of birch (*Betula* spp.) with spruce (*Picea abies*) are more productive than spruce stands, but mixtures of birch with *Pinus sylvestris* are not more productive than pure pine stands. Furthermore, the stage of stand development is influential: mixtures are more productive than monospecific stands in early stand development but not when trees are more than 17 m tall (Frivold and Frank 2002). Similarly, a regional survey conducted in Mediterranean forests dominated by Aleppo pine (*Pinus halepensis*) revealed that monospecific stands have lower wood production than mixed forests (two to five species). However, when climate, successional stage, bedrock type, and radiation were included in the analysis, stand tree-species richness was no longer a significant factor (Vilà et al. 2003).

In this chapter, we discuss some of the factors that can confound the tree diversity–productivity relationship in temperate and Mediterranean forests. We stress the role of seral stage, environmental factors, and management as such factors in the observational diversity–productivity relationship. As an example, we present a case study using the Ecological and Forest Inventory of Catalonia (IEFC), a large dataset that supports a positive association between tree species richness and stand stemwood production. Wood production in forests is somehow more relevant than total plant production for evaluating tree growth rates and competition interactions, since it is through the investment in the physical structure of wood (and also roots) that plants compete with one another (Huston 1994). However, we show that factors such as successional stage and climate overwhelm the positive relationship between tree species richness and stand stemwood production. Finally, we discuss the concept that in these forests past management practices are probably the major forces masking the natural patterns of forest diversity and productivity.

4.2 Covariant Factors Determining the Forest Diversity–Productivity Relationship

4.2.1 Successional Stage

There are considerable differences in structure and ecosystem functioning between young, mature, and old forests. In even-aged forests, for instance, growth declines after reaching a peak relatively early in a stand's life (which typically coincides with the maximum development of leaf area: Ryan et al. 1997). This decline is thus both size- and age related (Weiner and Thomas 2001). Tree height, diameter, and biomass increase through time while tree density decreases. As growth rates first increase and then decrease with tree size and age, the growth curves typically have a sigmoid shape. The physiological mechanisms causing this decline in the rate of carbon assimilation are related to reduced leaf area and reduced photosynthesis, mainly due to increasing hydraulic resistance in taller and older trees, lower nutrient availability, and maturation changes (Murty and McMurtrie 2000).

Patterns of age-related decline in forest productivity are well known in monospecific forests, but the timing, speed, and magnitude of decline vary between species and site quality (Ryan et al. 1997). However, the dynamics of mixed forests is much less understood and more difficult to predict than in single-species stands, because stand productivity is not only related to population dynamics of the dominant species but also to differences in resource use among species and to competitive interactions.

The age-related decline of tree growth implies that the successional stage also influences tree and forest productivity. Wood production is greater during early successional stages after natural or anthropogenic disturbances such as fire and clear cutting, than in late successional stages when canopy closure and competition typically prevent strong growth responses after disturbance (Whittaker and Woodwell 1969; Bormann and Likens 1979; Vilà et al. 2003).

Plant species diversity also changes during succession. Classically, diversity has been predicted to be higher with time after a disturbance because longer periods of time and spatial heterogeneity may be required in order to allow species to establish. For instance, overstory and understory plant species richness of Douglas fir forests of the Pacific Northwest builds up during succession after clear-cutting or burning in parallel with increasing vertical and horizontal spatial heterogeneity, despite the closure of the tree canopy (Halpern and Spies 1995; Franklin et al. 2002). However, the reverse pattern, more diversity at early and intermediate stages of regeneration than at late stages, can also be found. For example, a long-term analysis of permanent plots in the southern Appalachians (USA) found that, over a period of 14 years, tree diversity was the highest after clear-cutting and declined with canopy closure (Elliott and Swank 1994). It is also possible that species richness reaches a plateau and remains largely unchanged afterwards. This could be the situation of highly resilient Mediterranean forests where the same tree species are maintained after fire due to their regeneration strategy, such as the existence of serotinous fruits that open with heating, or of the ability to resprout when aboveground biomass is removed (Trabaud 1987).

Overall, after disturbance, the diversity–productivity relationships might result in different trajectories of ascending slopes (Fig. 4.1). Let us assume that before disturbance, the diversity–productivity relationship within a site is A. At early stages of regeneration after disturbance there can be an increase of the productivity from A to B, C, or D while species richness might increase (B), remain the same (C), or decrease (D). These changes in species richness would depend on the type and severity of the disturbance, the response of plant species to disturbances and colonization by opportunist species establishing through dispersal from nearby non-disturbed areas. Studies conducted in chronosequences of disturbed forests and permanent plots of forests before and after disturbances could elucidate how diversity–productivity relationships differ after disturbances from those before.

4.2.2 Environmental Correlates

In observational studies, the species diversity–productivity relationship cannot be easily separated from the effect of site conditions. If we take the unimodal pattern as the most common observed relationship across communities and focus on the ascending portion of the curve, species diversity is highest on sites conducive to high productivity (Huston 1994). In contrast, single species stands are often found in extreme environments. For example, among semi-natural forests in Europe, woods of *Pinus uncinata* at high altitudes, *Quercus petraea* in the Atlantic climate of west Britain, *Cupressus sempervirens* on the south-facing limestone-rocky slopes of Crete, etc. (Rackham 1992). Regional analyses across forests types have found a positive correlation between tree diversity and actual evapotranspiration (Currie and Paquin 1987). Increasing temperature, moisture availability, and soil fertility also favor tree growth. In California, rainfall explains 62 % of the variation of tree species richness (Richerson and Lum 1980). Similar patterns have been found in a whole-country analysis of tree diversity in New Zealand primary forests (Leathwick et al. 1998) wherein tree diversity increased with mean monthly temperature, solar radiation, and soil and atmospheric moisture, which in turn control forest productivity. However, on the descending portion of the curve (i.e., on the higher end of the productive gradient), wherein competition by dominant species reduces the availability of resources to other species, species diversity diminishes with productivity. In fact, within deciduous forest types in nearby humid sites, the relationship between normalized evaporation and tree richness is negative, possibly because dominant tree functional types or species may dominate the pools of available N and lower the amount for other tree species (Baldocchi, Chap. 7, this Vol.).

At the local scale, variation in site quality is thought to determine whether mixed forests are more productive than monospecific stands (Assman 1970; Kelty 1992; Pretzsch, Chap. 3, this Vol.). In the simplest case of two-species mixtures, higher productivity of mixed-species stands compared with monospecific stands would be expected whenever the two species are either more efficient or complementary in using limiting resources or when there is facilitation between species. For example, a mixture of an N-fixing tree species and a non-N-fixing tree species will be more productive than the monospecific stands of the component species in poor soils, but we might not expect a mixture advantage in stands with high N availability. Different light requirements of the species in a mixture may increase productivity in comparison with pure stands. A shade-tolerant understory tree species forming a dense canopy and a low-shade-tolerant overstory deciduous tree species can form stands of higher productivity when mixed than do monospecific stands in a sunny site, though not in a shady site (Kelty 1989; Pretzsch, Chap. 3, this Vol.).

We should also emphasize that most observational studies comparing productivity between monospecific and mixed-species stands have been carried out in climates where soil moisture is generally not critically limiting to stand productivity (e.g., Cannell et al. 1992; Kelty 1992; Caspersen and Pacala 2001). In regions where water and nutrients are the principal factors limiting stand productivity, we might expect a positive correlation between diversity and productivity, because both variables are limited by the same factors (Waide et al. 1999).

4.2.3 Management

Management can mask or alter the expected relationship between species diversity and stand productivity by directly or indirectly enhancing the presence and growth of some forest species and not others. Most monospecific stands in non-extreme environments are the result of human intervention favoring high-producing species. For example, *Betula papyrifera* stands in north Minnesota are replacing diverse pine forests. Similarly, stands dominated by *Pinus palustris* in Florida are artificially maintained by forestry practices (Rackham 1992). Monospecific stands can also be achieved by plantations or by deliberate elimination of unwanted species. While oaks were maintained for bark tanning, other accompanying species have been eliminated in Scottish forests after the eighteenth century (Rackham 1992). Conversely, in the same period, deciduous oaks were eliminated from mixed forests in Catalonia (Spain) for its high quality as firewood (Villaescusa 1993). It is well known that selection by grazing and browsing can also decrease tree species diversity of some forests. For example, fallow deer (*Dama dama*) preference in UK can change dominance from *Fraxinus* to *Populus*. Similarly, in Spain grazing by sheep and domestic goats can change patterns of woody species composition, diversity, and biomass (Cuartas and García-González 1992).

The species mixtures promoted by foresters (Pretzsch, Chap. 3, this Vol.) have been designed to find the best such mixtures. Therefore, it is not surprising to find that species tree composition is usually more important than tree richness in accounting for stand tree production. Not only might traditional silvicultural practices simplify forest diversity, but also they directly influence forest structure by maintaining even-aged stands, by keeping basal areas constant, or by removing wood. Stand wood production forecasts that do not take into account biomass export underestimate production.

All of these management practices aim mostly at increasing forest productivity over the short or long term. Management can have effects on diversity and productivity similar to those of disturbances. An extensive survey of boreal forests has shown that postlogging stands contain the same plant diversity and stand productivity as do plots burned by natural fires (Reich et al. 2001). This suggests that management practices and disturbances could influence the tree diversity–productivity relationship in the same way. Going back to Fig. 4.1, sustainable forest management should ensure that its structure maintains high diversity and productivity values (trajectory B) and avoids activities that increase sharp slopes of high productivity but low diversity (trajectory D). For example, even though the financial value of Norway spruce is overwhelmingly superior to that of beech, the recreational value, resistance to disturbance risks, and diversity of products are greater in mixed spruce/beech stands than in pure spruce stands (Assman 1970).

4.3 The Ecological and Forest Inventory of Catalonia (IEFC)

4.3.1 Characteristics of the IEFC

The IEFC (Gracia et al. 2000–2002) is an extensive forestry database comprising information from 10,644 sampling plots of 10 m radius randomly distributed throughout Catalonia, NE Spain. A subset of these plots (n=2,107) has more of the standard information gathered in a classic forestry survey than found elsewhere, and therefore was the one we used in the analysis. The IFEC includes the customary information of forest inventories and additional data related to functional aspects of forest ecosystems. In each of the 10,644 plots, and for each tree with a diameter at breast height (DBH) above 5 cm, species identities were noted and heights and DBHs measured. In all plots, core increments were taken for each representative live tree species (one or more) of a diameter class >5 cm to calculate age and annual tree growth over the last 5 years. The overall stemwood production of a plot per year was estimated as $P=(B_5-B_0)/5$, where B_0 is the tree plot stemwood biomass per area 5 years before the sampling and B_5 is the tree plot stemwood biomass per area during the sampling. Stemwood biomass was calculated by common silvicultural methods (see Gracia et al. 2003 for details). We did not include wood production of trees that died during the 5-year period because it was likely negligible. The IEFC does not include wood production of shrubs. Knowing that stemwood production values would underestimate total wood production in our study, an estimation of shrub cover per plot was included in the analysis to control for the effect of the shrub layer on tree stand stemwood production. Catalonia is the second most forested region of Spain (36 % of its area being covered by forests) and the one with the most acute climatic gradient (Gracia et al. 2000–2002). The region covered by the IFEC includes steep regional climatic, geological, and topographic gradients. Catalonia (ca. 31,900 km^2) is located in northeast of the Iberian Peninsula, bounded on the north by the Pyrenees and on the east by the Mediterranean Sea. Therefore, forests account for a large phytogeographic region, including Mediterranean, Sub-Mediterranean, Eurosiberian, and even Boreoalpine chorologies.

As a preliminary data exploration we tested the relationship between tree species richness (independent variable) and stand tree stemwood production (dependent variable) by ANOVA. We then further explored the effect of tree species richness and several forest structure and environmental parameters on stand tree stemwood production by a general linear model (GLM) analysis, following the JMP package (Anonymous 1992). The GLM analysis gives the significance of the full model (including all parameters) and for each parameter using *F* tests. A model that included all independent parameters was built for all stands and also for stands dominated by particular species, allowing for an approach across forests and within forests types, respectively.

By means of this approach, a previous analysis of pine forests using the IEFC database indicated that the positive relationship between tree species richness and stand stemwood production is confounded by the influence of macroenvironmental factors and successional stage (Vilà et al. 2003).

4.3.2 Productivity of Mixed Forests in Catalonia

Almost three-quarters (73.3 %) of Catalonian forests are mixed, with tree species richness ranging from two to five species, with a mode of two-species mixtures (29.9 % of plots) and five-species mixtures being the least common (6.2 %). Most dominant species do not form monospecific stands and all the species forming monospecific stands are also present in mixed stands. The IEFC includes observations on 95 tree species distributed into 43 genera.

Table 4.1. Stemwood production (in t ha^{-1} year^{-1}, mean±SE) of the dominant tree species when grown in pure stands and when mixed, and total stemwood production of mixed stands for Catalan forests according to the IEFC

Dominant species	Pure stands	Dominant sp mixed	Mixed stands
Castanea sativa	4.92±1.07	2.73±0.40	2.94±0.37
Abies alba	4.49±0.88	3.83±0.40	4.17±0.48
Fagus sylvatica	3.47±0.54	3.08±0.22	3.33±0.23
Pinus sylvestris	1.68±0.09	1.46±0.04	1.64±0.05
Pinus uncinata	1.48±0.15	1.50±0.09	1.95±0.15
Pinus nigra	1.47±0.15	1.31±0.06	1.53±0.06
Quercus ilex	1.27±0.12	1.36±0.06	1.68±0.08
Pinus halepensis	1.00 ±0.07	1.17±0.04	1.54±0.06
Quercus pubescens	0.80 ±0012	1.06±0.09	1.48±0.15
Quercus suber	0.70±0.11	0.75±0.06	0.63±0.06

Forty-two of these species are dominant (basal area >50 % of the total per plot): *Pinus halepensis* (present in 20 % of sampling plots), *Pinus sylvestris* (19 %), and *Quercus ilex* (16 %) being the three most dominant species.

On average, total stand stemwood production is low (mean±SE=1.65± 0.03 t ha^{-1} year^{-1}) compared to other temperate forests (http://www.efi.fi/databases/eefr; Tables 14.1 and 14.11 in Huston 1994). Of the ten most dominant species, the most productive species in monospecific stands are *Castanea sativa, Abies alba,* and *Fagus sylvatica. Quercus pubescens* and *Q. suber* are the least productive species (Table 4.1). Dominant species stemwood production was not significantly larger in monocultures than in mixed stands (ANOVA, $F_{4,1971}$=0.65, P<0.63). However, stemwood production of the whole stands was dependent on tree species richness (ANOVA, $F_{4,\,2076}$=8.85, P<0.001). While stands with five species were the most productive, monospecific stands where the least productive (Fig. 4.2A). Due to database constraints we could not analyze which mixture composition has the highest productivity. Therefore, we could not check whether mixed stands produce more than the best species in pure stands, what is known as overyielding (Vandermeer 1989).

4.3.3 Successional Stage and Other Biotic Correlates

As mentioned in Section 4.2.1, the effect of species diversity on tree production can be confounded with differences in seral stage that concomitantly influence tree size and vegetation structure. Unfortunately, the IFEC database

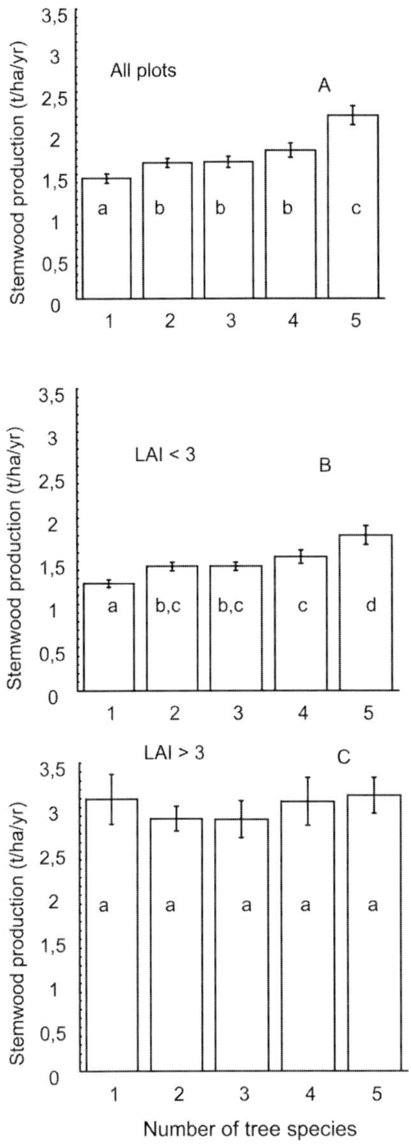

Fig. 4.2. Mean stand stemwood production (±SE) in relation to tree species richness in Catalonia for all IEFC plots (**A**), plots with LAI<3 (**B**), and plots with LAI>3 (**C**). Different *lowercase letters* indicate significant differences over tree species richness according to Fisher's test

does not contain quantitative or qualitative information on stand age or intervals between disturbances (each followed by regeneration). However, we used the age of the oldest tree sampled as a surrogate of successional stage, and found that this factor had a significant effect on stemwood productivity (Table 4.2). On average, Catalan forest are young, mean±SD being 61±34 years). Stands with one or two tree species are significantly older than mixed forests with higher species diversity ($F_{4, 1563}$=12.23, P<0.001; Fig. 4.3).

Table 4.2. Significance of the effects of several variables on stemwood production calculated from general linear models (GLM)

		All stands		Stands with LAI <3		Stands with LAI >3	
	df	*F*-ratio	*P*	*F*-ratio	*P*	*F*-ratio	*P*
Full model		r^2=0.52		r^2=0.54		r^2=0.53	
Parameter estimates							
Tree spp richness	4	1.87	0.11	3.00	0.02	1.83	0.13
Stand biomass	1	1171.07	<0.001	1200.56	<0.001	71.14	<0.001
Successional stage	1	241.88	<0.001	264.73	<0.001	24.74	<0.001
Shrub cover	1	0.46	0.50	0.02	0.90	0.18	0.67
Bedrock type	6	3.54	0.002	2.63	0.01	0.61	0.72
Radiation	1	0.59	0.44	0.36	0.55	0.29	0.59
Climate type	7	5.42	<0.001	3.34	0.002	8.36	<0.001

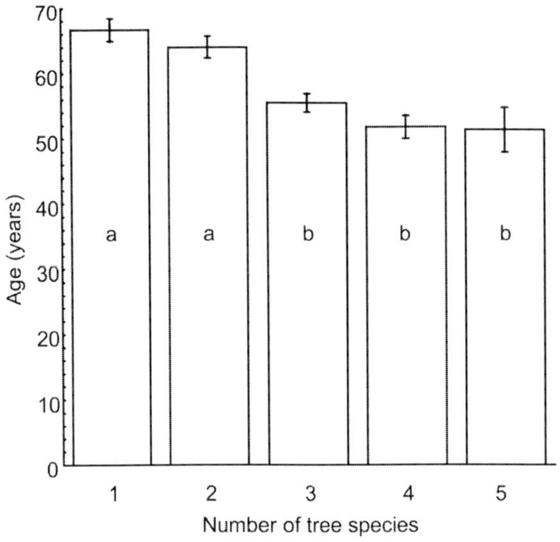

Fig. 4.3. Mean maximum age (±SE) of forests of different tree species diversity in Catalonia for all IEFC plots. Different *lowercase letters* indicate significant differences over tree species richness according to Fisher's test

The oldest forests are almost pure stands of *Abies alba* (mean±SD of 131±52 years) and *Pinus uncinata* (mean±SD. 105±47 years).

We included total stand stemwood biomass in the analysis as an estimation of wood size and found a significant positive correlation between stemwood biomass and production (r^2=0.42). This increase in stemwood production with biomass has also been found in other temperate regions. For example, an extensive survey across a 3,300-m altitudinal gradient in the central Himalayas disclosed patterns of forest productivity that matched those of stand biomass (Singh et al. 1994). The positive relationship between forest biomass and productivity indicates that on average these forests have quite open canopies. In fact, the increase in stand stemwood production was found only before canopy closure (LAI<3). In forests with closed canopies (LAI>3), stand stemwood production did not increase with tree species richness (Fig. 4.2B, C). The same trend was found if tree cover was used in the analysis instead of LAI. Low LAI values might result from young seral stages, environmental constraints (e.g., drought, poor or rocky soils), or silvicultural clearings.

One striking finding was that shrub cover did not modify stand tree stemwood production in any of the conducted analyses (Table 4.2). In Catalan forests variation in shrub layer across forests is large (mean±SD is 63.45±25.40 %). It was also surprising to find that shrub cover increased with increasing tree richness (ANOVA, $F_{4,2077}$=14.20, P<0.001, Fig. 4.4), suggesting that shrubs do not interfere with tree production and that, at the regional scale, factors leading to higher tree production and tree diversity might also be responsible for a better shrub development.

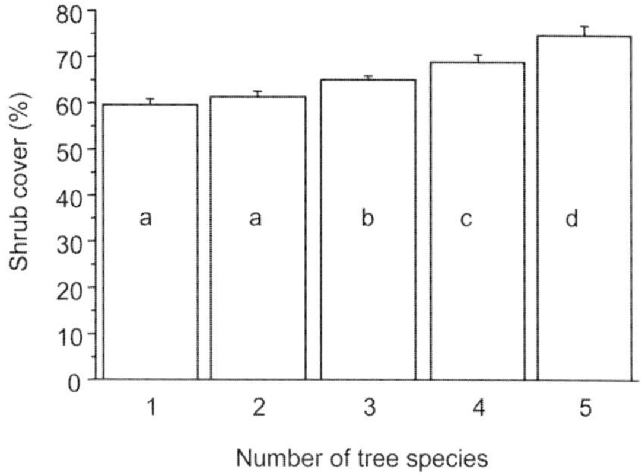

Fig. 4.4. Mean shrub cover (±SE) of forests of different tree species diversity in Catalonia for all IEFC plots. Different *lowercase letters* indicate significant differences over tree species richness according to Fisher's test

4.3.4 Climatic and Lithologic Correlates

To test for the effect of confounding environmental gradients, we selected three integrative parameters: climate type, bedrock type, and total spring solar radiation. Climate type and bedrock type were used as the main variables in regional environmental conditions. Total spring solar radiation was chosen as a measure of local environmental variation.

Each plot was assigned to one of the nine climate categories of the Thornthwaite index (Thornthwaite 1948). Climate had a significant effect on stemwood productivity (Table 4.2). The forests with the highest productivity were located in humid zones, while forests with the least production and the fewest tree species are located in semi-arid and arid zones (Fig. 4.5). Bedrock type had a significant effect on stemwood productivity of forests only before canopy closure (Table 4.2). The forests with the highest productivity were located in unconsolidated alluvium materials. Radiation had a non-significant effect on stemwood productivity. Therefore, our analysis reveals an emphasis on the influence of macro-environmental factors on forest productivity at the regional scale.

When we restricted the GLM analysis to stands located in humid climates or in warmer climates, tree species richness did not have a significant effect on stemwood production either ($F_{4,\ 593}$=0.57, P=0.69 for humid stands, $F_{4,969}$=1.52, P=0.19 for extreme stands). The model showed the same amount of variation (52–53 %), and the same variables were significant as they were when all stands were included in the analysis.

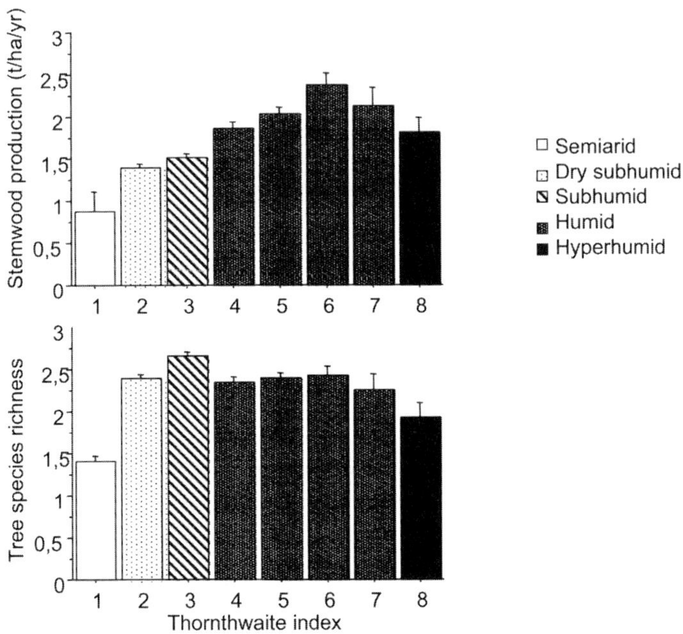

Fig. 4.5. Mean stand stemwood production (+SE) and tree species richness in relation to climate in Catalonia

4.3.5 Species Richness–Productivity Relationships Within Forest Types

That stand stemwood production increased with tree species richness was very striking when the dominant species was sclerophilous (ANOVA, $F_{4,420}$=7.05, P<0.001). When the dominant species was *Quercus ilex*, *Q. suber*, *Arbutus unedo* or the introduced *Eucalyptus globulus*, stemwood production in stands with four or five species was 19.6 and 45.8 % greater, respectively, than in monospecific stands or those with two or three species (Fig. 4.6A). Stemwood production was also significantly different within conifer forests in which monospecific stands had the lower productivity (ANOVA, $F_{4,1362}$=5.32, P<0.001; Fig. 4.6B). Stemwood production was not significantly different between forests dominated by deciduous species (ANOVA, $F_{4,291}$=0.34, P<0.85; Fig. 4.6C).

When we conducted a GLM for sclerophilous and conifer forests with the same forest structure and environmental variables as for all stands, tree species richness no longer had a significant effect on stemwood production ($F_{4,47}$=0.74, P=0.57 for sclerophilous; $F_{4,1240}$=1.95, P=0.10 for conifer forests). Stemwood production was dependent on tree biomass, stand age, and climate as for all plots. These results suggest that even within a forest type environ-

Fig. 4.6. Mean stand stemwood production (±SE) for forest dominated by sclerophilous (**A**), conifers (**B**), and deciduous (**C**) tree species in Catalonia. Different *lowercase letters* indicate significant differences over tree species richness according to Fisher's test

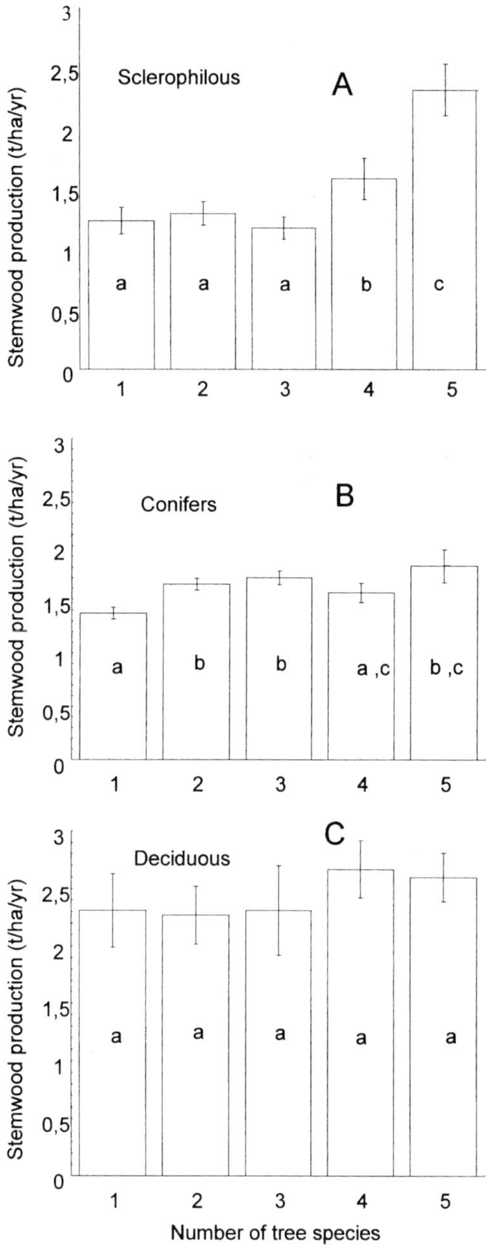

mental variables are more important in determining stemwood production than tree diversity. Only experiments conducted in the same site conditions and comparing plots with different tree species richness of the same functional type could determine if species richness has a causal effect on stemwood production.

4.3.6 Management Considerations

The IEFC included natural and semi-natural forests, plantations, and secondary woodlands, but unfortunately plots in the database were not classified into these categories. History and management have a great influence on tree diversity and productivity, and this is why climate sometimes does not predict tree diversity patterns (McGlone 1996). Our complete model had a weak predictive power (52–54 %) to explain differences in stemwood production, and this was probably due to the large effect of management practices on forest structure. For example, a different management of different productive sites can mask the diversity–productivity relationship. In addition, Catalan forests are young forests and the IEFC calculates tree growth for the last 5 years only. Hence, we do not have values for wood production over a longer period of time or within the whole rotation period, which would form the basis for management considerations from a forestry perspective.

Mediterranean forests have a long history of human intervention (e.g., tree planting, wood cutting, fire, pasture, charcoal production) that can mask natural spatial and temporal patterns of tree dominance and diversity (Rackham 1992; Villaescusa and Díaz 1998). In Catalonia, in general, monospecific forests have been deliberately favored over mixed forests. For example, some monospecific *Pinus halepensis* plots on poor soils might result from planting, while large areas of secondary forests dominated by *P. sylvestris* result probably from the harvesting of deciduous trees (mainly *Quercus humilis*), monospecific *Q. suber* stands have been favored for cork production (Aldomà 1988; Villaescusa and Díaz 1998), and so on. However, despite these efforts, in the last 20 years the extent of mixed forests has increased as a result of tree colonization of abandoned traditionally agricultural land, afforestation with fast growing species followed by reduced (low-intensity) silvicultural practices after planting, and a decrease in the commercial value of forest products (Raddi 1998; Villaescusa and Díaz 1998). We are also certain that in the last decades there has been no general management trend to maintain a certain tree basal-area value or to keep the shrub layer to a minimum in most forests, despite the fire risk that dense shrub cover can confer.

In general, Catalonian forests tend to have a low direct economic value, with only two thirds of the forests receiving any commercial value (Raddi 1998). Furthermore, the average timber quality is typically low: 90 % of conifer timber is accepted in only the packing industry, and 21 % of the overall timber production is used only as a source of energy (mass production; Raddi 1998). Mixed forests of hardwoods and conifers in Catalonia appear to be less exploited commercially (58 %) than pure conifer forests (67 %), but not significantly less than pure hardwood forests (58 %). As a result, we cannot say that mixed forests are less exploited than monospecific stands in general. Furthermore, silvicultural practices and type of use of forests depend on geography, forest ownership, and size of properties (Aldomà 1988). In Catalonia more than 90 % of the forested land is private and 85 % of the properties are smaller than 25 ha. Hence, the properties are smaller than the minimum size required to be subjected to forestry policy planning (Peix 1999). Overall, human driven activities have a great effect on tree species productivity, tree species diversity, and community composition, but human activities have not been directly quantified in the IEFC database. Therefore, their influence on the forest productivity–diversity relationship remains to be tested.

4.4 Discussion

The relationship between tree diversity and productivity that can be observed at the landscape scale across forests and within forest types can be confounded by environmental factors, seral stage, and management practices, directly and indirectly (Fig. 4.7). The analysis of the IEFC shows that the positive relationship between species tree richness and stemwood production occurs in sclerophilous and conifer forests before canopy closure. These results suggest that the relationship is only found in early successional forests or in stressed forests (e.g., water or soil-nutrient-limited, rocky soils) in which the effect of having more tree species in a stand is additive to that of the dominant species. Since most of the Catalan forests are functionally young resulting from a decrease of high-intensity management in the last decades (e.g., end of charcoal production and agricultural land abandonment) or because they are at an early regeneration stage after disturbances (e.g., fire), we encourage maintaining tree species richness as a security to sustain forest productivity and spread the risk between species after disturbances (Pretzsch, Chap. 3, this Vol.). Furthermore, our analyses suggest that on average, at the short term, afforestation with a mixture of tree species might increase stand tree productivity more than doing so with a single tree species.

The study of the relationship between plant diversity and productivity has been approached by direct investigation of plant communities assembled in natural systems or by conducting experiments of plant assemblages. These approaches should be viewed as complementary. As seen in this study, forest inventories have the central caveat of confounding factors underlying the diversity–productivity relationship. However, such factors provide the oppor-

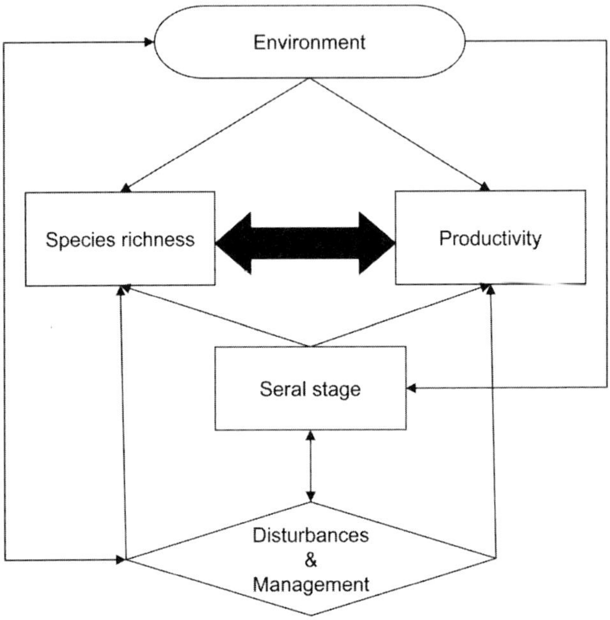

Fig. 4.7. Schematic representation of direct and indirect effects of environmental factors, seral stage, management, and disturbance on the diversity–productivity relationship in forests

tunity to compare natural systems within a matrix of biotic and environmental variation, from which specific hypotheses can be formulated that can be tested in more experimentally controlled conditions. In contrast, purposefully designed tree plantations provide causal information on the effect of tree diversity on tree production. However, due to the slow growth of tree species, several decades are needed to have a clear picture of the diversity–productivity relationship in tree plantations. There is also concern regarding how well experiments resemble the changes in species abundance and species losses that take place in real environmental and management conditions (Grime 2002; Schmid et al. 2002). From this perspective, most mixed tree plantations promoted by foresters contain tree varieties that have been genetically selected to have high production and to face environmental constraints (e.g., frost, drought). From the management point of view, well-designed experimental tree plantations are of interest to test which species mixtures are more productive. However, these plantations do not resemble natural conditions. Thus, to answer the central question of whether forest biodiversity increases forest productivity in the real world, experimental tree plantations should not select tree species at random from the local tree species pool, nor should the most productive species be selected, but selection should rather mimic the naturally occurring range of variation of tree species richness and abundance, taking into account that natural tree assemblages depend on a seral stage. These experiments should also consider the consequences of losing the tree species more at risk. Furthermore, as discussed by Schmid et al. (2002), experiments can investigate a broad spectrum of tree biodiversity components such as rarity, evenness, horizontal and vertical spatial patterns of species distribution within mixtures, age structure within species, etc.

4.5 Conclusions

The IEFC study case presented illustrates well some of the difficulties regional datasets raise in addressing the simple question of the effect of species mixtures on forest productivity. It is probably impossible to entirely divorce the association between species diversity and productivity from other biotic and environmental factors. Especially, in cultural landscapes such as large areas of Europe, disturbances and management have a great influence on forest composition and productivity (see Mund, Chap. 10, this Vol.). We advocate future analysis of forest inventories to include information on disturbance regimes (e.g., fire history, presence of domestic livestock) and silvicultural practices such as biomass removal either as wood export or understory clearings. In addition to inventory data collection to demonstrate causality, well-designed experiments with mixtures of naturally co-occurring tree species are needed. Up to now, such experiments in terrestrial ecosystems have mainly been limited to herbaceous assemblages.

Acknowledgments: We thank CREAF colleagues, Ch. Körner, M. Scherer-Lorenzen, E.-D. Schulze and an anonymous reviewer for stimulating comments and discussions. This paper is dedicated to those who conducted the IEFC fieldwork. The IEFC was financed by the "Departament d'Agricultura, Ramaderia i Pesca" and the "Departament de Medi Ambient de la Generalitat de Catalunya".

References

Aldomà I (1988) L'estructura de la superficie forestal a Catalunya. Quaderns d'Informació Técnica 2. Diputació de Barcelona, Barcelona
Anonymous (1992) JMP 2.0.5. SAS Institute, Belmont, CA
Assman E (1970) Principles of forest yield study. Pergamon Press, Oxford
Bengtsson J, Engelhart K, Giller P, Hobbie S, Lawrence D, Levine J, Vilà M, Weiner J, Wolters V (2002) Slippin' and slidin' between the scales: the scaling components of biodiversity-ecosystem functioning relations. In: Loreau M, Naem S, Inchausti P (eds) Biodiversity and ecosystem function. Oxford Univ Press, Oxford, pp 209–220

Bormann FH, Likens GE (1979) Pattern and process in a forested ecosystem. Springer, Berlin Heidelberg New York

Cannell MGR, Malcolm DC, Robertson PA (1992) The ecology of mixed-species stands of trees. Blackwell, Oxford

Caspersen JP, Pacala SW (2001) Successional diversity and forest ecosystem function. Ecol Res 16:895–903

Cuartas P, García-González P (1992) *Quercus ilex* browse utilization by Caprini in Sierra de Cazorla y Segura (Spain). Vegetatio 99/100:317–330

Currie DJ, Paquin V (1987) Large-scale biogeographical patterns of species richness of trees. Nature 329:326–327

Elliott KJ, Swank WT (1994) Changes in tree species diversity after successive clearcuts in the southern Appalachians. Vegetatio 115:11–18

Franklin JF, Spies TA, van Pelt R, Carey AB, Thornburgh DA, Berg DR, Lindenmayer DB, Harmon ME, Keeton WS, Shaw DC, Bible K, Chen J (2002) Disturbances and structural development of natural forests ecosystems with silvicultural implications, using Douglas-fir forests as an example. For Ecol Manage 155:399–423

Frivold LH, Frank J (2002) Growth of mixed birch-coniferous stands in relation to pure coniferous stands at similar sites in south-eastern Norway. Scand J For Res 17:139–149

Gracia C, Burriel JA, Ibáñez JJ, Mata T, Vayreda J (2003) Inventari ecològic i forestal de Catalunya. Mètodes. Ed. Centre de Recerca Ecológica i Aplicacions Forestals, Cerdanyola del Vallès

Gracia C, Burriel JA, Ibáñez JJ, Mata T, Vayreda J (2000–2002) Inventari ecològic i forestal de Catalunya. Ed. CREAF, Cerdanyola del Vallès (http://www.creaf.uab.es/iefc)

Grime JP (2002) Declining plant diversity: empty niches or functional shifts? J Veg Sci 13:457–460

Halpern CB, Spies TA (1995) Plant species diversity in natural and managed forests of the Pacific Northwest. Ecol Appl 5:913–934

Hector A, Schmid B, Beierkuhnlein C, Caldeira MC, Diemer M, Dimitrakopoulos PG, Finn J, Freitas H, Giller PS, Good J, Harris R, Högberg P, Huss-Danell K, Joshi J, Jumpponen A, Körner C, Leadley PW, Loreau M, Minns A, Mulder CPH, O'Donovan G, Otway SJ, Pereira JS, Prinz A, Read DJ, Scherer-Lorenzen M, Schulze E-D, Siamantziouras A-SD, Spehn E, Terry AC, Troumbis AY, Woodward FI, Yachi S, Lawton JH (1999) Plant diversity and productivity experiments in European grasslands. Science 286:1123–1127

Huston MA (1994) Biological diversity: the coexistence of species on changing landscapes. Cambridge Univ Press, Cambridge

Huston MA (1997) Hidden treatments in ecological experiments: re-evaluating the ecosystem function of biodiversity. Oecologia 110:449–460

Kelty MJ (1989) Productivity of New England hemlock/hardwood stands as affected by species composition and canopy structure. For Ecol Manage 28:237–257

Kelty MJ (1992) Comparative productivity of monocultures and mixed-species stands. In: Kelty MJ, Larson BC, Oliver CD (eds) The ecology and silviculture of mixed-species forests. Kluwer, Dordrecht, pp 153–170

Kerr G, Nixon CJ, Matthews RW (1992) Silviculture and production of mixed-species stands: the UK experience. In: Cannell MGR, Malcolm DC, Robertson PA (eds) The ecology of mixed-species stands of trees. Blackwell, London, pp 35–52

Leathwick JR, Burns BR, Clarkson BD (1998) Environmental correlates of tree alpha-diversity in New Zealand primary forests. Ecography 21:235–246

Loreau M, Naeem S, Inchausti P, Bergtsson J, Grime JP, Hector A, Hooper DU, Huston MA, Raffaelli D, Schmid B, Tilman D, Wardle DA (2001) Biodiversity and ecosystem functioning: current knowledge and future challenges. Science 294:804–808

McGlone MS (1996) When history matters: scale, time, climate and tree diversity. Glob Ecol Biogeogr Lett 5:309–314

Mittelbach GG, Steiner CF, Scheiner SM, Gross KL, Reynolds HL, Waide RB, Willig MR, Dodson SI, Gough L (2001) What is the observed relationship between species richness and productivity? Ecology 82:2381–2396

Murty D, McMurtrie RE (2000) The decline of forest productivity as stands age: a model based method for analysing causes for the decline. Ecol Mod 134:185–205

Peix J (ed) (1999) Foc verd II. Departament d'Agricultura, Ramaderia i Pesca. Generalitat de Catalunya, Barcelona

Rackham O (1992) Mixtures, mosaics and clones: the distribution of trees within European woods and forests. In: Cannell MGR, Malcolm DC, Robertson PA (eds) The ecology of mixed-species stands of trees. Blackwell, London, pp 1–20

Raddi A (1998) El mercat dels productes forestals a Catalunya. Quaderns d'Informació Técnica 1. Diputació de Barcelona, Barcelona

Reich PB, Bakken P, Carlson D, Frelich LE, Friedman SK, Grigal DF (2001) Influence of logging, fire, and forest type on biodiversity and productivity in southern boreal forests. Ecology 82:2731–2748

Richerson PJ, Lum KL (1980) Patterns of plant species diversity in California: relation to weather and topography. Am Nat 116:504–536

Ryan MG, Binkley D, Fownes JH (1997) Age-related decline in forest productivity: pattern and process. Adv Ecol Res 27:213–262

Schmid B (2002) The species richness-productivity controversy. Trends Ecol Evol 17:113–114

Schmid B, Hector A, Huston MA, Inchausti P, Nijs I, Leadley PW, Tilman D. (2002) The design and analysis of biodiversity experiments. In: Loreau M, Naem S, Inchausti P (eds) Biodiversity and ecosystem function. Oxford Univ Press, Oxford, pp 61–75

Singh SP, Adrikari BS, Zobel DB (1994) Biomass, productivity, leaf longevity, and forest structure in the central Himalaya. Ecol Monogr 64:401–421

Thornthwaite CW (1948) An approach toward a rational classification of climate. Geogr Rev 38:55–94

Tilman D, Reich PB, Knops J, Wedin D, Mielke T, Lehmann C (2001) Diversity and productivity in a long-term grassland experiment. Science 294:843–845

Trabaud L (1987) Fire and survival traits of plants. In: Trabaud L (ed) The role of fire in ecological systems. SPB Academic Publ, The Hague, pp 65–89

Troumbis AY (2001) No observational evidence for diversity enhancing productivity in Mediterranean shrublands? Oecologia 129:622–623

Troumbis AY, Memtsas D (2000) Observational evidence that diversity may increase productivity in Mediterranean shrublands. Oecologia 125:101–108

Vandermeer J (1989) The ecology of intercropping. Cambridge Univ Press, Cambridge

Vilà M, Vayreda J, Gracia C, Ibáñez JJ (2003) Does tree diversity increase wood production in pine forests. Oecologia 135:299–303

Villaescusa R (ed) (1993) Segundo Inventario Forestal Nacional (1986–1995). Cataluña. ICONA, Madrid

Villaescusa R, Díaz R (eds) (1998) Segundo Inventario Forestal Nacional (1986–1995). España. ICONA, Madrid

Waide R, Willig MR, Steiner CF, Mittelbach G, Gough I, Dodson SI, Juday GP, Parmenter R (1999) The relationship between productivity and species richness. Annu Rev Ecol Syst 30:257–300

Wardle DA (2001) No observational evidence for diversity enhancing productivity in Mediterranean shrublands. Oecologia 129:620–621

Weiner J, Thomas SC (2001) The nature of tree growth and the "age-related decline in forest productivity". Oikos 94:374–376

Whittaker RH, Woodwell GM (1969) Structure production and diversity of the oak-pine forest at Brookhaven, New York. J Ecol 57:157–176

5 Genetic Diversity Parameters Associated with Viability Selection, Reproductive Efficiency, and Growth in Forest Tree Species

G. Müller-Starck, M. Ziehe, and R. Schubert

5.1 Introduction

Biological diversity, "biodiversity" for short (e.g., Wilson 1988), encompasses all levels of the variability of life, i.e., the diversity within species, among species, among ecosystems, and among biomes. Our contribution addresses the genetic variation within species (intraspecific biodiversity, commonly designated as "genetic diversity") which is quantified as variation within populations (within and among individuals), among populations, and within metapopulations. The study of functions of biodiversity may follow a hierarchical structure, i.e., coding genes, individuals, ecotypes, species, and various other biological communities. Functionally relevant are also the dynamics of populations and species in time and space, as well as specific characteristics such as abundance, evenness, and reproducibility.

At the species level, genetic variation of populations – particularly their potential to create genetic variation ("genetic variability") – may play a major role in their ability to adapt to heterogeneous environmental conditions and unpredictable host–parasite interactions. Genetic variability is expected to determine the totality of adaptive abilities of populations (e.g., Gregorius 1991; Ziehe et al. 1999). In addition, genetic variation is expected to be correlated with fitness in various plant and animal species (Mitton and Grant 1984; Allendorf and Leary 1986). Biodiversity can also be considered as being redundant (e.g., Lawton and Brown 1993; Yachi and Loreau 1999). The question of whether or not genetic variation (variability) can be expected to be redundant refers to various components of the genetic system of a species including reproduction, gene flow, and the response to stress, but this question has not yet been studied in detail.

Concerning functions of biodiversity, forest ecosystems (i.e., natural forests and forest plantations) have a high indicative value because of their longevity compared to other plant species and the wide range of occupied

Ecological Studies, Vol. 176
M. Scherer-Lorenzen, Ch. Körner, and E.-D. Schulze (Eds.)
Forest Diversity and Function: Temperate and Boreal Systems

environmental niches from the lowlands to the sub-alpine regions all over the globe. The forest cover substantially affects atmospheric carbon exchange (Waring and Schlesinger 1985) and supplies various economic options such as utilization of timber and many other natural products for various industrial purposes and for energy production. In contrast to agricultural systems, the majority of forests still contain biodiversity in a non-domesticated status. Due to great environmental heterogeneity in space and in time, tree populations are exposed to a variety of biotic and abiotic stresses. Control of stress is not possible in the case of exposure to heat, frost, UV-B radiation, drought, and air pollution, and is strictly limited in the case of biotic stress following pathogen infection or insect attack. Prophylactic disease control such as in agriculture is not possible in forest systems. Adaptational and survival abilities are challenged predominantly by the discrepancy between long generation cycles of immobile trees in contrast to very short generation cycles of their mostly mobile parasites.

In addition, most forest tree species suffered from old genetic bottlenecks following post-glacial re-immigration and from severe exploitation, fragmentation, and devastation, particularly since the medieval period. Further challenges for adaptation and survival arise from air pollution and climate change (for survey see Karl et al. 1997; Geburek 2000) which particularly affect long-lived forest ecosystems and will result in uncertain future response of forest tree populations, particularly with respect to host–parasite interactions and corresponding changes in susceptibility.

In natural and long-lived tree populations, "stress" usually cannot be defined as a single component, but as a complex and dynamic system of highly variable abiotic and biotic factors that affects individuals, populations, and ecosystems alike in terms of dieback of individuals and corresponding reductions of density and size of populations. At the population level, individuals with different genotypes respond differently under stress conditions (e.g., Scholz et al. 1989; Müller-Starck 1993; Ziehe et al. 1999; Geburek 2000). Viability selection following stress causes genotype-dependent elimination of individuals and, consequently, induces changes in the frequency distribution of the corresponding population. Resulting modified frequency distributions are the necessary condition for populations to adapt to the given environmental conditions and to reproduce. The genetic response to stress is manifold such as by mutations, gene regulation, viability, and fertility selection, as well as by losses of genetic variation. Consequently, genetic markers can be employed as a tool for indication of stress ("bioindication") in various forest ecosystems (Müller-Starck and Schubert 2000).

In the marker development and corresponding assessments of genetic diversity in forest tree populations, problems arise from the fact that genome analysis in trees such as *Picea abies* (Norway spruce) is still rudimentary as compared to species such as *Arabidopsis thaliana*. One major reason for this gap in knowledge in *P. abies* is its outstandingly large genome size and pro-

Table 5.1. Survey of estimates of the number of gene loci (tentative for *Picea abies*) and the size of the genome (million bp) for a set of reference species

Species	Genome size (million bp)	Estimated number of gene loci	No. gene loci p. million bp
Escherichia coli	4.6[a]	4,300	935
Saccharomyces cervisiae	12[a]	5,800	483
Drosophila melanogaster	116[a]	13,600	117
Arabidopsis thaliana	125[b]	26,500	212
Oryza sativa	430[c]	30,000	70
Homo sapiens	2,690[a]	39,000	15
Picea abies	30,000[d, e]	30,000	1

[a] Bork and Copley (2001)
[b] Arabidopsis Genome Initiative (2000)
[c] Rice Genome Database, Beijing Genomics Institute: http://btn.genomics.org.cn/rice/
[d] Plant & Animal Gene VII Conference, 1999: http://www.intl-pag.org/7/tc.html
[e] Kirst et al. (2003)

portion of non-coding sequences as compared to *A. thaliana* (see Table 5.1). The *A. thaliana* genome is fully sequenced while the analysis of only a minor part of *P. abies* with low density linkage maps is in progress (e.g., Paglia et al. 1998). Furthermore, handling of individuals is easy in the small *A. thaliana* as compared to the extraordinarily large individuals in *P. abies*, with heights up to 35 and 40 m. Also, the reproduction in *A. thaliana*, with its regular annual flowering, is easy to observe in comparison with *P. abies*, with its non-regular reproduction starting at a late ontogenetic stage at the age of two to three decades. Natural selection is intensive and indicative of short- and long-term responses to complex environmental stress due to the natural longevity of *P. abies* populations. Generally, the research community is extremely small in the case of *P. abies* (not more than 20 molecular genetic groups worldwide), while nearly 1,000 groups study the model species *A. thaliana* – a situation reciprocal to the economic and also the ecological significance of these two species.

In Table 5.1, the genomic peculiarities of *Picea abies* are highlighted in contrast to a collection of reference species including humans. The specificity of this tree species can be seen by comparing the number of gene loci in relation to the size of the genome. The resulting quotient is, for instance, 935 in *Escherichia coli*, 212 in case of *A. thaliana*, and 1 in *Picea abies* (Table 5.1).

The objective of our contribution is to survey experimental studies in long-lived tree populations that illustrate differential aspects in the functional significance of genetic diversity. Our main focus is to demonstrate the indicative potential of diversity for vital functions such as reproductive efficiency, growth, and response to environmental stress.

5.2 Methodological Considerations

Genetic diversity can be measured, for example, in terms of variation between genes or between DNA or amino acid sequences, as well as by numbers of breeds, strains, and distinct populations. (For a recent compilation of methods and nomenclature, see e.g. Karp et al. 1998). In the case studies we present here, the following methods have been applied:

Geographical patterns of genetic variation were studied by using a set of DNA-markers from the chloroplast (cp) genome (cpSSR, cpPCR-RFLP; SSR = single sequence repeats, PCR = polymerase chain reaction, RFLP = restriction fragment length polymorphism) that was developed as a universal tool for forest tree species and shrubs (Demesure et al. 1995; Dumolin-Lapègue et al. 1997; Weising and Gardner 1999; Grivet et al. 2001).

The other empirical examples refer to three different types of co-dominant nuclear markers: Firstly, isoenzyme coding gene loci that are widely used for genotyping with respect to expressed genes (for survey see Paule 1990; Fineschi et al. 1991; Baradat et al. 1995; Müller-Starck 1998); secondly, nuclear microsatellites (nSSRs; La Scala 2000); and finally, a set of EST markers (expressed sequence tags) that were newly developed from a cDNA library of Norway spruce and tested under various greenhouse and field conditions (Schubert et al. 2001; Schubert and Müller-Starck 2002). Most of the EST markers correspond to coding nuclear genes with partially or fully known function and were proven to follow a Mendelian mode of inheritance: PA0002 (A-like cyclin), PA0005 (cyclophilin), PA0034, (non-identified gene), PA0038 (halotolerance protein HAL3), PA0043 (high-molecular heat-shock protein), PA0055 (ATP synthase beta chain), PA0066 (60S ribosomal protein L13–2), and CAD (cinnamyl alcohol dehydrogenase).

Measures of genetic variation within populations include the number of alleles per gene locus (A_L), the gene diversity (v; Gregorius 1978, 1987), or the intra-populational genetic differentiation (δ_T; Gregorius 1987). Heterozygosity is measured by the observed proportion of heterozygotes.

For cohorts of the same generation, comparisons between genetic frequencies at different ontogenetic stages allow study of selection regimes (see Müller-Starck 1993; Gregorius and Ziehe 1995) by estimating survival parameters or at least selection coefficients.

In order to determine whether phenotypically relevant expressions are also genetically controlled, the "pair method" was applied. Pairs of individual neighbors representing contrasting phenotypes were chosen and genetic structures between phenotypic groups were compared (Gregorius 1989; Müller-Starck 1993).

Fertility selection was studied by comparing zygotic structures of the offspring generation with those of its parental population. Rare alleles in particular help to determine seed and/or pollen parents and can be used to estimate contributions of different genotypes to the subsequent generation (see Müller-Starck and Ziehe 1984).

Relations between genetic characters and growth traits of individual trees were studied by using a linear regression and a two-factorial analysis of variance, respectively. The genetic parameters were transformed by accounting for heterozygosity (homo-, heterozygosity, transformed 0/1), allelic frequency (rare, not rare, transformed 0/1), as well as for the representation of the two most frequent alleles (n.a., allele 1 homozygous, heterozygous, allele 2 homozygous, transformed 0/1/2/3).

5.3 Case Studies of Genetic Diversity in Forest Trees

Case studies are compiled that demonstrate the high genetic diversity within tree species and the functional significance of this diversity. The example given in below in Section 5.3.1 describes existing patterns of diversity in tree populations and illustrates the value of gene markers in detecting these patterns and inferring phylogenetic relationships. The subsequent sections address the potential of genetic diversity as an indicator of functions related to adaptation and survival under specified environmental conditions (Sect. 5.3.2); reproduction, fertility selection, and genetic loads (Sect. 5.3.3); and finally, with respect to growth in various tree populations (Sect. 5.3.4).

5.3.1 Geographic Patterns of Genetic Diversity

Geographic patterns of genetic variation are exemplified by means of chloroplast DNA markers (haplotypes): In a recent study of 21 natural populations of sycamore (*Acer pseudoplatanus*), 17 different haplotypes were identified by means of cpPCR-RFLPs, and 5 by cpSSRs (Bittkau 2002). As can be seen from Fig. 5.1A, the distinct haplotypes detected show a geographic differentiation within western, central, southern, and eastern regions of Europe. The study of phylogeographic structures based on a statistical parsimony criteria resulted in relations between haplotypes included in Fig. 5.1B. In most cases, haplotypes differ from related haplotypes by only one mutation step, except in the group from Corsica and the French Alps. Within the main group, clusters were observed consisting of haplotypes from southern Italy, western, central, and southeastern Europe. Three lineages, i.e., those from southern Italy, southeastern Europe, and the Alps, confirm fossil records (Huntley and Birks 1983; Brewer et al. 2001). For further explanation see Bittkau (2002).

On the basis of these data, three major refugia of *Acer pseudoplatanus* are suggested, i.e. southern Italy, southeastern Europe, and regions close to the

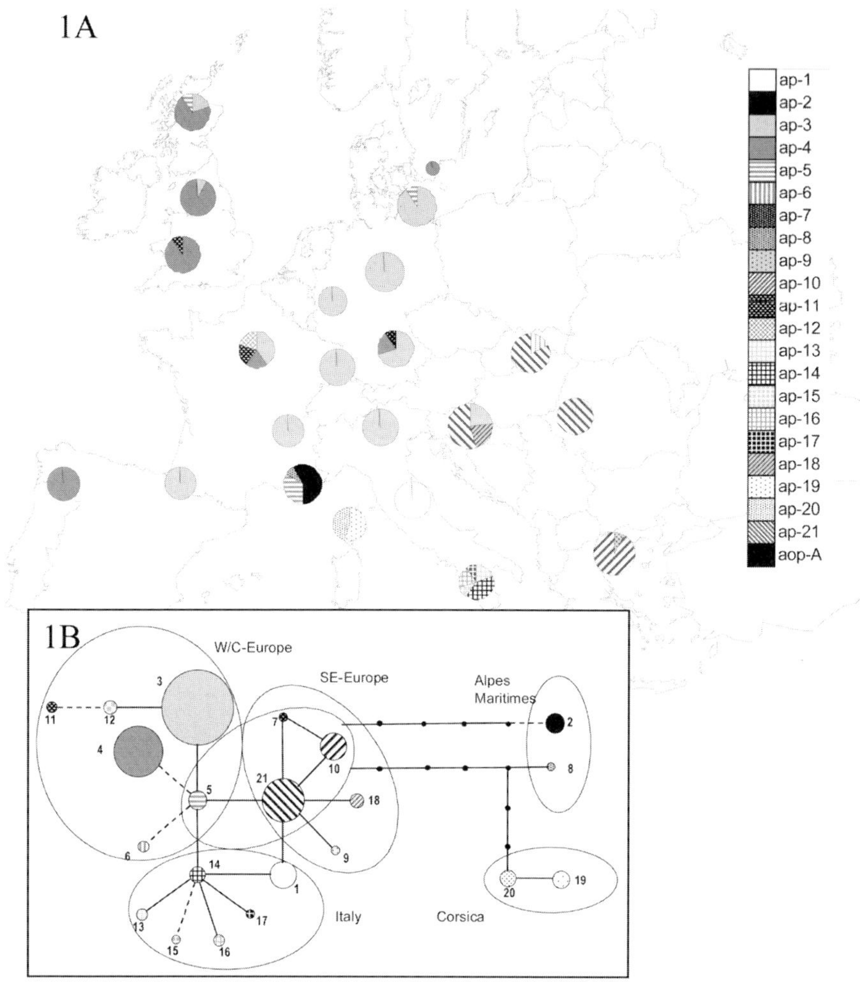

Fig. 5.1. **A** Distribution of haplotypes (cpSSRs in combination with cpPCR-RFLPs) of *Acer pseudoplatanus* (*ap*) in Europe (Bittkau 2002, modified). **B** In the corresponding phylogeographic structure, *numbers* refer to mutation steps and *diameters of the circles* to the haplotype frequency

Alps (Huntley and Birks 1983). The Italian lineage could not cross the Alps due to the alpine lineage which already had occupied this region. Furthermore, the alpine haplotype could not colonize entire central and western Europe. Finally, results indicate that *Acer pseudoplatanus* has been present in the island of Corsica for a long time, which corresponds to pollen data of Reille et al. (1997), but does not fit to the re-colonization pattern postulated by Hewitt (2000, 2001). In contrast to molecular data, pollen records are generally biased in the genus *Acer* because a species-specific classification is not possible (Huntley and Birks 1983) and pollen dispersal is limited in this insect-pollinated tree species.

This example confirms that genetic diversity is geographically differentiated among tree populations in Europe and that genetic markers allow inferences on phylogenetic relations and post-glacial re-immigration. Further evidence is supplied by a recent multi-species study on genetic diversity in relation to European glacial refugia of forest tree species (Petit et al. 2003).The majority out of a total of 22 widespread European tree and shrub species had genetically divergent populations in the Mediterranean refugial regions, but the genetically most diverse populations were located further north. This is interpreted by the authors as a consequence of the mixture of divergent lineages colonizing the continent from separate refugia particularly located on the Iberian peninsula and in Italy and the Balkans. However, such analyses may explain the regional structure of genetic variation but not the actual patterns in response to local environmental stresses or peculiarities of the genetic system of any given species, particularly its reproduction.

5.3.2 Viability Selection and Other Responses to Stress

An efficient way to demonstrate the significance of genetic variation with respect to viability selection is to compare two ontogenetic stages. For instance, germinating seeds and survivors that were growing under various environmental field conditions, implying heterogeneous components of stress, can be compared in terms of genotypes frequencies. In the study we present here as an example, main emphasis was put on acidification within the upper soil layer (for details see Müller-Starck 1993). At a total of eight locations all over Germany, germinating seeds of European beech (*Fagus sylvatica*) were transferred to experimental plots (6×6 m) in adult beech stands at only lightly shaded locations in spring time, and covered by leaf litter from the ground in order to imitate conditions of natural regeneration. The plots were protected from mice by close meshed fences and from birds by nets.

Figure 5.2 shows some results of a genotyping at the gene locus LAP-A (LAP = leucine aminopeptidase, E.C. 3.4.11.1) in germinating seeds in contrast to corresponding survivors in the open field (standard garden soil) and forest stand conditions, respectively. The material refers to provenance samples representing the open-pollinated offspring of a couple of forest stands located close to each other. Substantial changes in genotype frequencies were found. In particular, frequencies of those genotypes which carry the allele LAP-A_4 drastically increased among survivors. Since LAP serves to make amino acids available, the special catalytic efficiency of this allelic variant A_4 may play a particular metabolic role during this early ontogenetic phase of beech trees, and may be involved in the coding of stress tolerance (mainly against soil acidity).

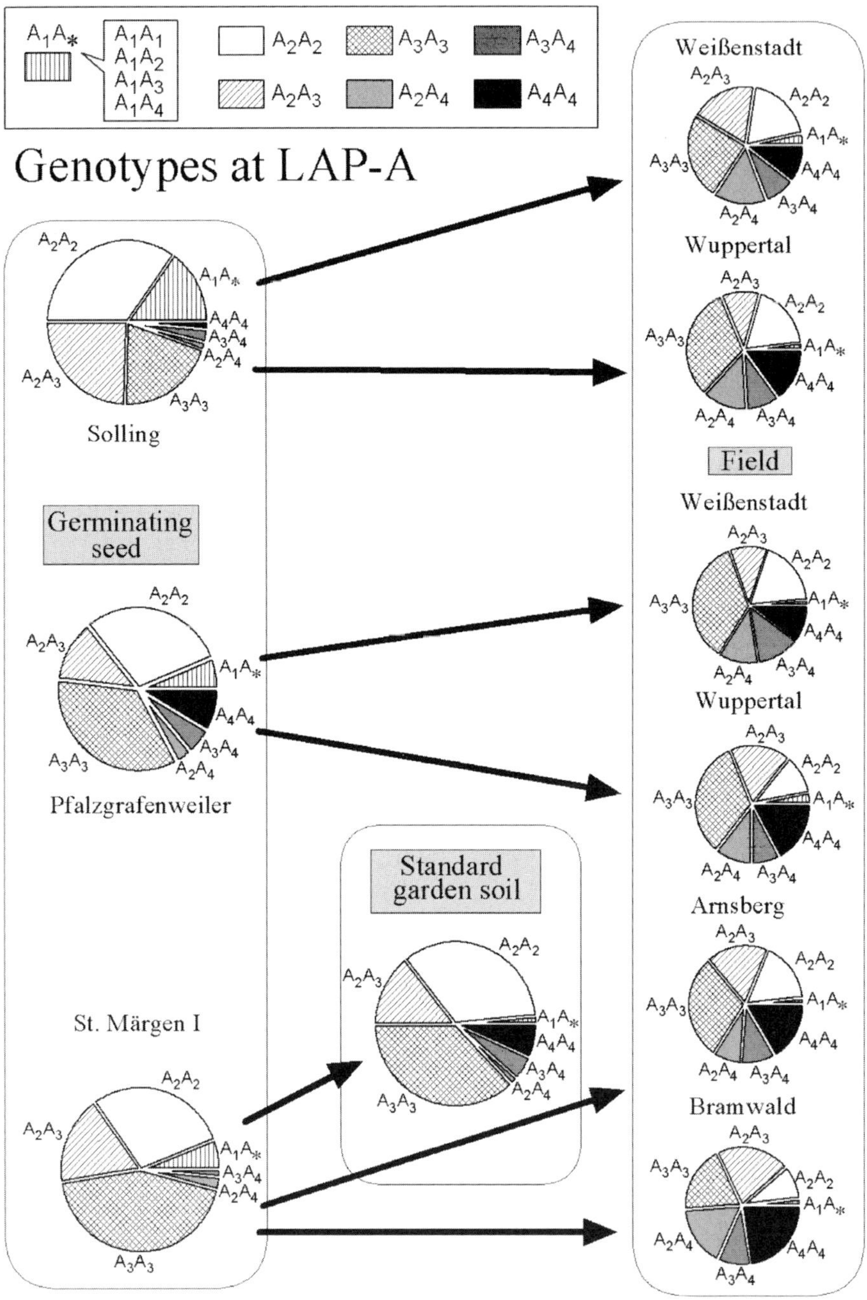

Fig. 5.2. Changes in genotype frequencies at LAP-A in three provenances of germinating seeds of European beech and corresponding survivors under open field conditions in garden soil and in six forest stands. (Data taken from Müller-Starck 1993)

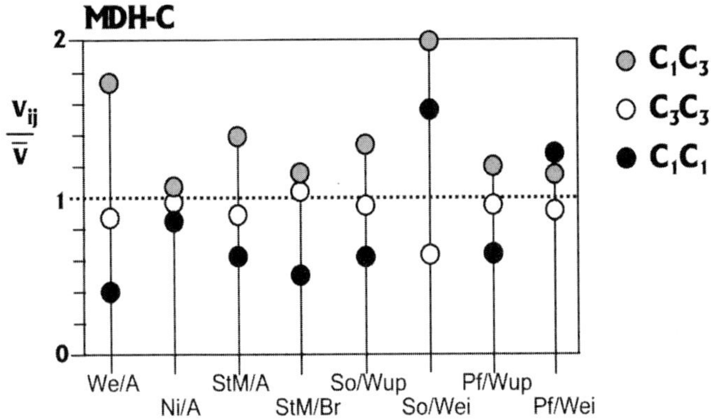

Fig. 5.3. Viability coefficients of genotypes at the gene locus MDH-C among different samples of germinating seedlings of European beech (*We* Wesel, *Ni* Nidda, *StM* St.Märgen, *So* Solling, *Pf* Pfalzgrafenweiler) for a period of 2-years' exposure to field stress dominated by soil acidification (locations: *A* Arnsberg, *Br* Bramwald, *Wup* Wuppertal, *Wei* Weißenstadt). (Modified after Müller-Starck 1993)

As can be seen from Fig. 5.3, single locus effects are also obvious with respect to the gene locus MDH-C (MDH = malatedehydrogenases, E.C.1.1.1.37). Comparing genetic frequencies of two different stages, i.e., germinating seeds and survivors after 2-year exposure under forest stand conditions, viability coefficients (presented as ratios of relative frequencies) have been derived for different genotypes. Superior viabilities are obvious for the heterozygote C_1C_3 as compared to the homozygotes C_3C_3 and C_1C_1, clearly indicating a case of overdominance. In contrast, viabilities of heterozygotes at the gene locus PER-B (PER = peroxidases, E.C. 1.11.1.7) are located between both homozygotes, representing the classical case of an intermediate trait expression of the heterozygote (Müller-Starck and Schubert 2000).

Based on the same set of experiments, viabilities of two-locus combinations were studied (see Fig. 5.4). Results demonstrate that the viability superiority is greatest for the double heterozygote MDH-C_1C_2/IDH-A_2A_3, which could not be expected from single locus observations (IDH = isocitrate dehydrogenase, E.C. 1.1.1.42). In addition to impacts of certain alleles on survival abilities as indicated in Fig. 5.3, the results given in Fig. 5.4 illustrate coincidence of heterozygosity, which is a measure of individual genetic variation, with the viability of the studied individuals.

The above observations suggest that in the case of juvenile European beech populations, genetic variation strongly affects the viability response of individuals represented by specified genotypes to complex stress, especially to soil acidification.

In adult trees, the study of differential survival between genotypes would require long-term experiments. However, an alternative is the pair method,

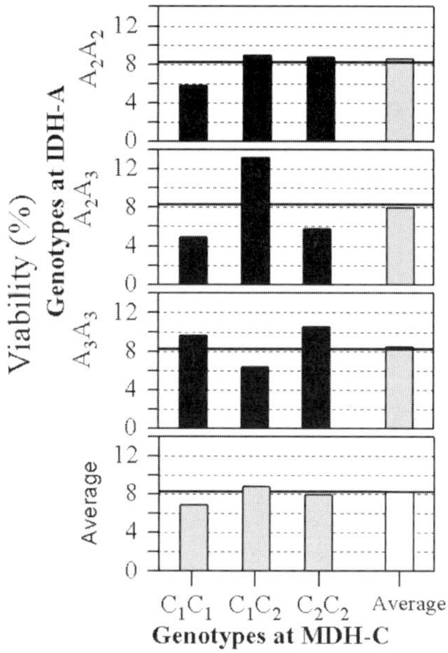

Fig. 5.4. Viability coefficients of two-locus genotypes at the gene loci *MDH-C* and *IDH-A* in samples of germinating seedlings of European beech and corresponding survivors (material Nidda and location Arnsberg). The *solid lines* indicate the mean viability of the complete material

where neighboring trees of contrasting phenotypes such as sensitive (damaged) and tolerant (healthy) are compared genetically. Significant differences between genetic structures of the different phenotypic groups indicate that genetic information at such a gene locus or other loci may be involved in controlling the phenotypic response. Results of homogeneity tests (Fig. 5.5) show some significant genotypic differences between tolerant and sensitive beech trees. However, with larger numbers of different genotypes, as is the case for LAP-A, it becomes difficult to establish statistical significance although genotypic differences are relatively large. In case of PGM-A (PGM = phosphoglucomutase, E.C. 2.7.5.1), substantial genetic differences between sensitive and tolerant beech trees have been found: particular genotypes as the heterozygote A_2A_3 at PGM-A were always observed with remarkably greater frequencies among tolerant trees than among sensitive trees. This trend is evident for each of the studied populations and the pooled subsets of the tolerant trees of all populations in contrast to the corresponding pooled sensitives. The superior indicative value of PGM isoenzyme gene markers for tolerance and sensitivity, respectively, with respect to environmental stress is confirmed in a recent study on the resistance of agricultural crops to nematodes, which clearly points out that genotypes at PGM-coding gene loci are useful markers for resistance (Yu et al. 2001).

In a second set of examples, we want to illustrate the genetic response of Norway spruce (*Picea abies*) to two types of environmental stress, revealing genetic differences between tolerant and sensitive tree subsets under field

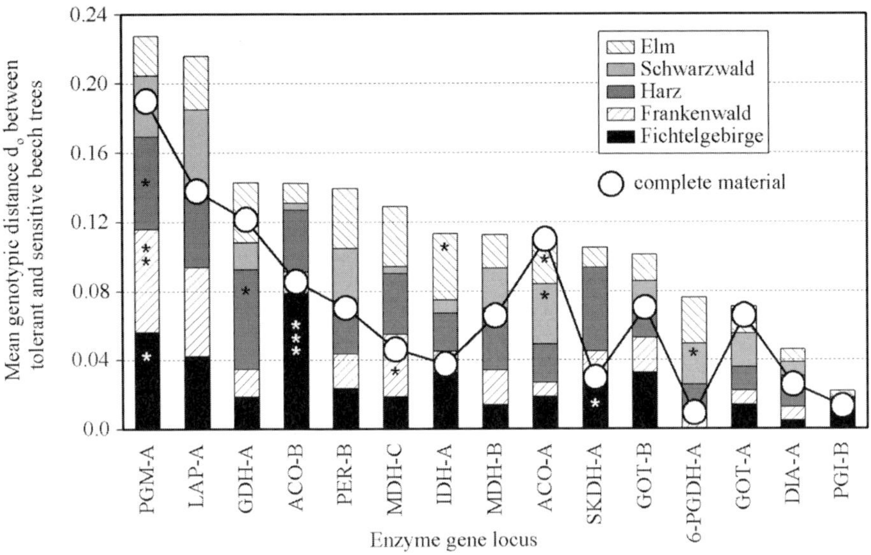

Fig. 5.5. Mean single-locus genotypic distances d_0 between tolerant and sensitive beech trees averaged over five different sites. The total height of the columns reflects the mean values, the different patterns indicate contributions of the observation under single site conditions. The *line* represents the genotypic distances d_0 between all tolerant and all sensitive beech trees (total sample size is 486, ranging between 86 and 106 trees per site). Statistically significant differences between genotypic structures among tolerant and sensitive beech trees are indicated

conditions. Different categories of DNA markers were utilized, namely, nuclear co-dominant EST markers and chloroplast microsatellites (haplotypes).

In the first study, environmental conditions were dominated by excessive soil salinity. In the forest district of Gunzenhausen, Bavaria, Germany, a 10-year-old and a 30-year-old population were sampled, both suffering from the use of salt (NaCl) used to remove ice from an adjacent state road (Schömig 1988). Thirty-five and 36 pairs of trees, respectively, were sampled (making a total of 142 trees) with each pair consisting of a sensitive tree with obvious needle decoloration and dieback, and its nearest unaffected neighbor with green needles. The frequency distributions of two-locus genotypes with respect to the EST markers PA 0005 and PA 0066 deviate with statistical significance between the two subsets (Bozhko et al. 2003: see Fig. 5.6).

The second study refers to clonal progenies affected by high-peak SO_2 exposure at the location of Altenberg and Bärenfels (Ore Mountains, Saxony, Germany; Wolf 2001). The material had originally been selected for exhibiting long-term SO_2 resistance under field conditions (Tzschacksch 1981). For each location, the 24 most sensitive clones as indicated by visible needle decol-

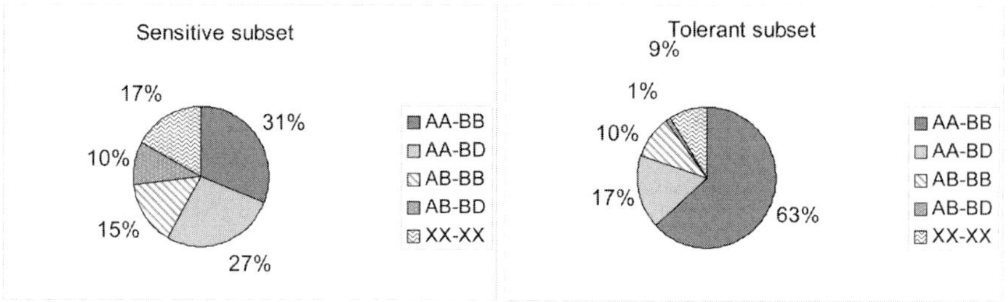

Fig. 5.6 Frequencies of two-locus genotypes (EST-markers PA 0005, PA 0066;) observed in two Norway spruce (*Picea abies*) stands under field stress (soil borne salinity) in the forest district of Gunzenhausen, Germany (n=142); genotypes below 5 % frequency were pooled (*XX–XX*). (Modified after Bozhko et al. 2003; data taken from Riegel 2001).

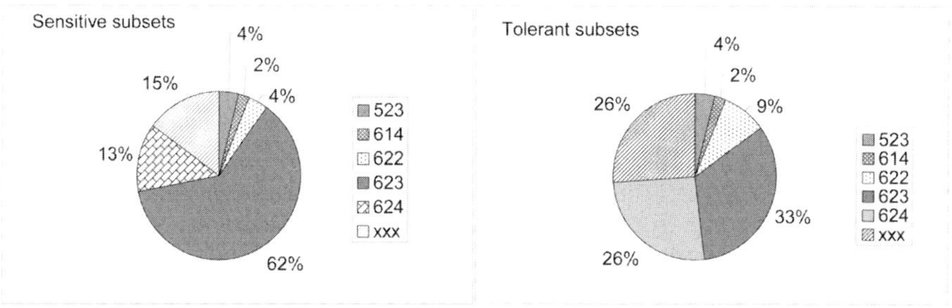

Fig. 5.7. Frequencies of three-locus haplotypes (Pt26081, Pt63718, Pt71936) observed under field stress (SO_2) within sensitive and tolerant subsets (n=48) in two Norway spruce stands in the Ore Mountains, Germany. Haplotypes below 5 % frequency were pooled (*XXX*). (Data taken from Riegel 2001; Riegel et al. 2004)

oration, and 24 healthy-looking clones with green needles were sampled. Based on the chloroplast microsatellite markers Pt26081, Pt63718, and Pt71936 (designed by Vendramin et al. 1996), the frequency distribution of the 3-locus genotypes significantly differed between the subsets (Riegel et al. 2004; see Fig. 5.7). In addition, haplotypic diversity and differentiation were found to be higher within tolerant in comparison to sensitive population subsets (Riegel 2001).

In both examples, frequency differences between tolerant and sensitive subsets of each experimental plot were statistically significant. Results indicate that genetic variation, monitored by both marker types, i.e. nuclear and chloroplast markers, allows differential characterization of tolerant and sensitive subsets of Norway spruce trees. Furthermore, these findings emphasize the significance of genetic variation with respect to adaptation and to survival in complex environmental conditions. The field conditions studied involved some of the most prominent cases of both soil- and airborne pollution.

5.3.3 Reproduction, Fertility, and Inbreeding Depression

By newly recombining the parental genetic information, sexual reproduction is a major driving force in the creation of novel genetic types. The potential of creating genetic variation can easily be demonstrated in a population in which on average three genes (alleles) are observed per gene locus. Considering diploid organisms, a locus with three alleles can reveal a maximum of six genotypes, i.e., three homozygotes and three heterozygotes. In combination with a second locus, the resulting number of two-locus genotypes is 36. With ten gene loci, i.e., a very small portion of many thousand polymorphic loci (see Table 5.1), more than 6×10^7 different ten-locus genotypes can be created via recombination following sexual reproduction.

Survival of individuals is the precondition to reach the reproduction phase and eventually to contribute their genetic information to the next generation. Often, when long-term consequences of viability selection across generations are discussed, regular reproduction systems such as panmixia are assumed (i.e., random mating or at least random fusion of gametes). Hence, selective effects occurring during reproduction are often ignored, although differential success of genotypes during reproduction appears to be a widespread phenomenon.

Even if the genotype of reproducing individual trees is known, it is generally impossible to trace gamete production (fertilities), pollen release and its spatial distribution (pollen gene flow), or mechanisms controlling fusion of pollen grains with ovules (mating system). However, the observation of genetic structures among the offspring generation may provide the identification of those allelic types that are particularly successful during reproduction.

For seed material of conifers, separate investigations of macrogametophytes and embryos allow one to distinguish genetic contributions via ovules from those via pollen (Müller-Starck 1976). This has been applied to determine female and male gametic contributions of single clones among bulked seeds in two Scots pine (*Pinus sylvestris*) seed orchards established with the same set of clones at two locations in the forest district of Ebrach (I, II), Bavaria. From these seed orchards, open-pollinated samples originating from two different years, each with intense female and male flowering and a superior seed crop, were analyzed. A total number of 500 seeds per year per location, hence 2,000 seeds, were genotyped separately for endosperm and corresponding embryo. If a gamete type can be produced by only a single clone, the female and male contributions of trees representing this clone can be estimated. Figure 5.8 illustrates substantial deviations among the gametic contributions of ten identifiable clones in two reproduction periods at the location Ebrach I (for results concerning Ebrach II, see Müller-Starck 1985). Evidently, equal contribution and sexual symmetry in the reproductive success of single clones are the exceptions rather than the rule.

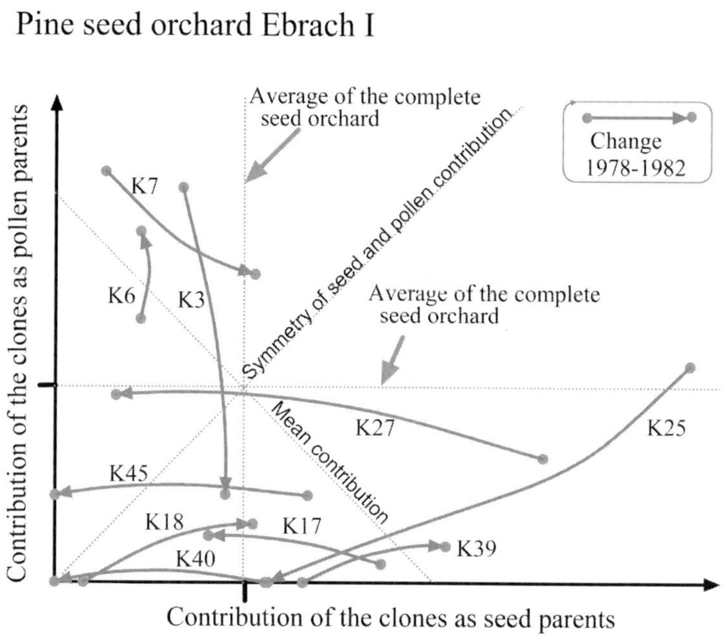

Fig. 5.8. Mean female and male contributions of pine clones (*Pinus sylvestris*) to the orchard seed as compared to average gametic contributions in the orchard. *K3–K45* denote ten different clones, the *arrows* indicate the change from reproduction period 1978 to 1982, both selected because of abundant flowering. (Modified after Müller-Starck 1985)

In self-fertile but predominantly outcrossing tree species, detrimental effects of inbreeding on individual viabilities at different life stages and the formation of empty seeds have been well known for decades (e.g., Koski 1973). Figure 5.9 presents an example of severe growth depression among progenies of Norway spruce (*Picea abies*) after self pollination and open pollination.

Inbreeding effects such as partial selfing promote homozygosity. As compared to expectations under random combination of alleles (random mating leading to Hardy-Weinberg proportions), low levels of heterozygosity provide evidence for inbreeding, in particular if this is consistently observed for various gene loci. As an example, in Fig. 5.10 results of a study of La Scala (2000) are compiled that refer to genotyping of open-pollinated seed samples originating from stands of sessile oak (*Quercus petraea*) in Germany, by means of five co-dominant nuclear microsatellites. Substantial deviations between the observed frequencies of homozygous genetic types ("homozygotes") and the corresponding Hardy-Weinberg proportions indicate an excess of homozygotes that is not more evident when the corresponding inbreeding structure is

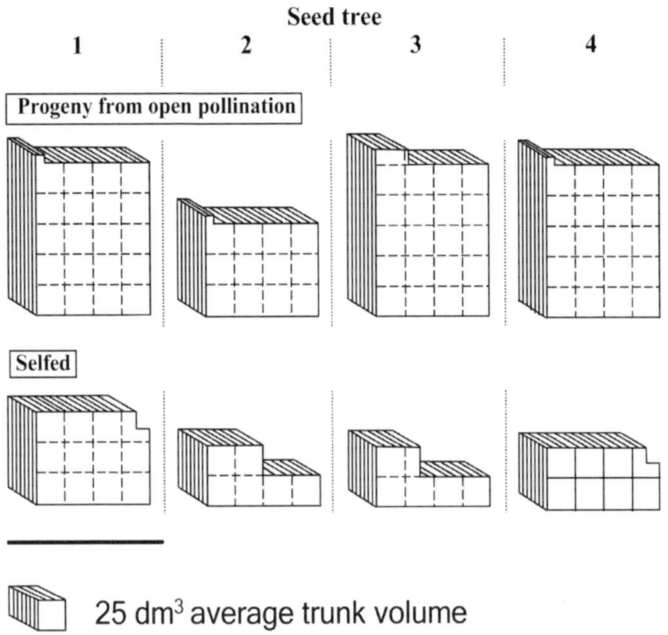

Fig. 5.9. Average trunk volume at the age of 61 years from open pollination and from self-pollination ("selfed") progeny of four seed trees from an experimental plantation of *Picea abies*. (Modified after Eriksson et al. 1973)

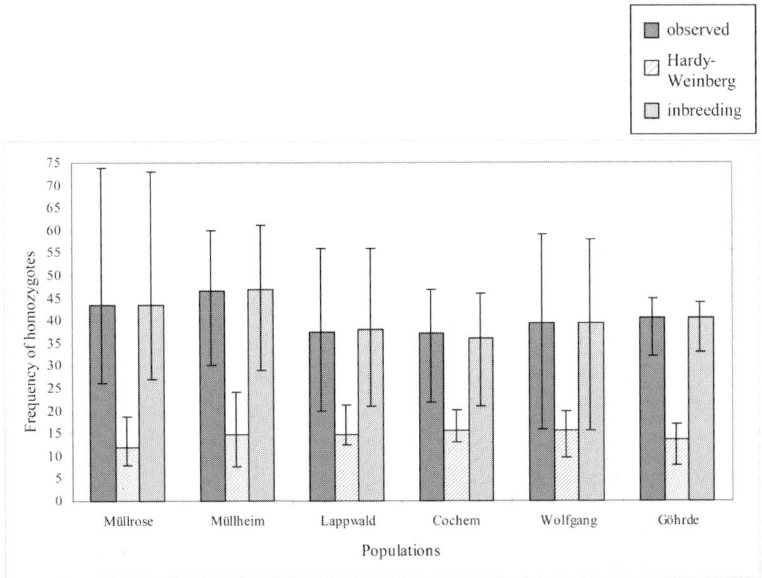

Fig. 5.10. Observed homozygote frequency, Hardy-Weinberg structure and inbreeding structure, calculated for six open-pollinated seed samples of sessile oak stands as a mean value of five nSSR loci. (Modified after La Scala 2000)

taken into account. This reference structure exclusively reveals effects of inbreeding, i.e., it implies an equally distributed excess of all homozygotes and a corresponding deficiency of all heterozygotes (Hattemer et al. 1993). Consequently, it allows verification of inbreeding in contrast to selective effects with respect to specific genotypes (e.g., viability or fertility selection), and helps explain complex overlays.

In the present case, the outstandingly good fit of the inbreeding structure to the observed structure clearly indicates that inbreeding is the predominant component of the mating system under the given field conditions.

5.3.4 Growth

Environmental impacts on growth of trees can be expected to be substantial under heterogeneous and long-lasting forest field conditions. Consequently, heritability estimates concerning growth, i.e., the estimation of the proportion of genetic variance of the phenotypic variance of tree growth ranging between 0 and 1, are in most cases low in tree populations. For example, in the analysis of growth curves in *Pinus taeda*, Gwaze et al. (2001) estimated heritabilities within a range of 0.06 to 0.26 with high correlations among growth-curve parameters. The authors conclude that these values are only slightly lower than those for age-specific tree heights.

In another recent study, correlations between genetic characters and growth parameters (diameter growth) were investigated in a 47-year-old provenance trial of Scots pine (*Pinus silvestris*) analyzing four different provenance samples. Two samples with superior growth were contrasted with other two others selected because of contrastingly weak growth (Blumenröther et al. 2001). Each provenance sample was represented by 100 trees. Genotyping comprised 16 isoenzyme-coding gene loci. In the study of correlations between genotypes and growth, the authors applied two-factorial analyses of variance with respect to isoenzyme-coding gene loci. Two examples are represented in Fig. 5.11.

Based on the pooled provenance data (n=400), the crossing lines in Fig. 5.11A indicate interactions between genotypes at the gene loci MDH-C and 6PGDH-B (6PGDH = 6-phosphogluconate dehydrogenase, E.C. 1.1.1.44). Factor loads of these two gene loci can be combined in four ways, two corresponding with lower values of diameter at breast height (dbh), i.e., both gene loci are either homozygous or heterozygous, and two with higher values of dbh, i.e., combinations of homozygous and heterozygous gene loci.

In Fig. 5.11B, the gene locus AAT-A (AAT = aspartate amino-transferase, E.C. 2.6.1.1) is particularly indicative in verifying correlations between genetic and metric traits: heterozygosity at this gene locus (factor load 2, representing A_2A_4) corresponds to a remarkable reduction of diameter growth dimension.

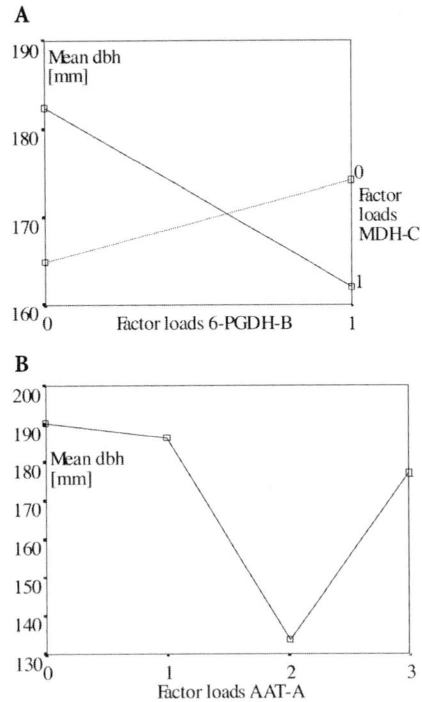

Fig. 5.11. Two-factorial analysis of variance with respect to the gene loci 6PGDH-B and MDH-C (graph **A**) and AAT-A (graph **B**). Mean diameters at breast height (mean dbh) for different combinations of factor loads (**A,** *load 0* "homozygosity"; *load 1* "heterozygosity"; **B,** *load 0* rare alleles pooled; *load 1* first allele homozygous, *load 2* "heterozygosity"; *load 3* second allele homozygous); data refer to the Scots pine provenance trial "Bodenwöhr". (Blumenröther et al. 2001)

Generally, these results reveal evidence of relations between genetic parameters and diameter growth. Moreover, these examples demonstrate that even when traits are strongly influenced by environmental conditions, isoenzyme-coding gene loci can help to predict growth parameters.

5.4 Conclusions

Functioning of forest ecosystems has been widely studied in the past by analyzing primary production, nutrient cycling and retention, decomposition, herbivory impacts and pathogen interactions, or by estimating biodiversity (for a survey see Körner et al., Chap. 2, this Vol.). It appears, however, that the role of genetic diversity, described in terms of genetic variation within and among populations of a single species, has so far not sufficiently been the focus of forest ecosystem research.

The present chapter contributes to the characterization of the functional role of genetic diversity and its dynamics in forest ecosystems. Besides a presentation of a species-specific geographic distribution of genetic markers, a set of case studies reflects the strong influence of distinct abiotic stress on the genetic diversity observed. This stress includes long- and short-term exposure of populations to extreme environmental conditions. The results presented here demonstrate the high indicative potential of both conventional isoenzyme markers and newly developed DNA marker technologies in analyzing genetic diversity within stress-affected field populations. The following main conclusions can be drawn:

Firstly, genetic diversity can be applied to determine geographic patterns of variation within distinct tree species and to explain how they have evolved after post-glacial re-immigration.

Secondly, adaptation and survival under conditions of site-specific abiotic stress factors directly affect intrapopulational genetic diversity patterns since viabilities (defined as survival probabilities) are proven to depend strongly on genotypic information (at single or multiple loci) of their carriers. Using appropriate markers for stress-responding genome regions, selection favoring tolerant genotypes at single gene loci either becomes directly obvious or can be traced with a number of loci linked at the same chromosome or scattered over the whole genome.

Thirdly, our results clearly demonstrate that genetic markers are highly informative when analyzing reproduction within seed orchards or native tree populations. Since reproduction controlled by particular environmental components affects the genetic structure of the next generations, the detection of possible genetic loads (for example inbreeding) or distortions among certain types of gametes are relevant for adaptation and survival.

Finally, current knowledge on gene identification and detailed genome organization is still fragmentary for the majority of tree species (eucalyptus, poplar, and loblolly pine may count as exceptions), so that chances to verify correlations between metric traits such as growth and any genetic character are strictly limited at present. Quantitative trait loci (QTL) detected in various crop plants have also been identified for distinct traits in some tree species (for a recent survey see Schubert and Müller-Starck 2002), but large parts of their genomes have still not been characterized by genetic markers. Information gaps are also obvious with respect to regulatory genes, controlling the switch between primary and secondary metabolism and therefore triggering among growth and defense reactions.

Evidently, the indicative potential of genetic markers strongly depends on the marker type applied. An expanding variety of markers is available now, tagging the nuclear genome as well as different parts of organelle genomes (chloroplast and mitochondrion). They are dominant/co-dominant inherited or simply show uniparental inheritance and are known to exhibit extremely different evolutionary rates. Whereas non-coding genome regions in most cases are considered to be neutral for any adaptation, coding parts of stress-defense genes are a likely target for adaptation and survival under specified environmental conditions, as indicated here for spruce EST markers PA 0005 and PA 0066. "Which marker for which purpose?" (Gillet 1999) is therefore a key question, clearly pointing out that the choice of an inappropriate marker may result in ineffective research.

In general, genetic diversity is crucial for adaptation and survival, as exemplified above by viability and fertility selection, inbreeding depression, and growth. Viability selection, for instance, induces genetically selective removal of individuals and thus directional changes of genetic structures of populations. The remaining part of the population transmits its genetic information to the next generations where more favorable genes may become enriched and eventually guarantee persistence.

Well-adapted key species of an ecosystem such as forest tree species represent a significant precondition for the persistence of the whole ecosystem. Consequently, genetic variability, i.e., the potential to create and maintain genetic variation, is crucial for adaptation, particularly under heterogeneous and changing environmental conditions. Moreover, genetic markers that are capable of indicating viability and fertility selection or differentially expressed dynamics of adaptively relevant characters like growth, competition abilities, and defense strategies are fundamental for verifying the vital functions of forest ecosystems. In this sense, genetic parameters generally appear appropriate as indicators for both the ability of an ecosystem to persist and its response to disturbances induced by environmental stress, genetic erosion, or specific genetic loads that result from inbreeding. Consequently, genetic variation as a substantial component of biodiversity and its dynamics appears to be a significant prerequisite for survival and persistence of forest ecosystems. It can therefore be expected to supply one of several criteria for the management of sustainable forest ecosystems.

Acknowledgments: We greatly appreciate very useful comments of H.H. Hattemer, Göttingen, M. Scherer-Lorenzen and two anonymous referees, as well as the help of E. Röschter and B. Kain, Freising, and S. Fehrmann, Göttingen, in the compilation of results and the preparation of the manuscript.

The experimental studies surveyed in this manuscript were financially supported by the Deutsche Forschungsgemeinschaft, Bad Godesberg (SFB-607/A9, Mu 774/2-2), the European Union (96-0706 "Tree Biodiversity," 97-3795 "Cytofor") and the Umweltbundesamt, Berlin (106 07 046/23).

References

Allendorf W, Leary RF (1986) Heterozygosity and fitness in natural populations of animals. In: Soulé ME (ed) Conservation biology: the Science of Scarcity and Diversity, Sinauer Associates, Sunderland, MA, pp 57–76

Arabidopsis Genome Initiative (2000) Analysis of the genome sequence of the flowering plant *Arabidopsis thaliana*. Nature 408:796–815

Baradat Ph, Adams WT, Müller-Starck G (eds) (1995) Population genetics and genetic conservation of forest trees. PSB Academic Publ, The Hague, 480 pp

Bittkau C (2002) Charakterisierung der genetischen Variation europäischer Populationen von *Acer* spp. und *Populus tremula* auf der Basis der Chloroplasten-DNA: Rückschlüsse auf die postglaziale Ausbreitung und Differenzierung forstlicher Provenienzen. Doctoral Diss, Technische Universität München, Freising, 154 pp

Blumenröther M, Bachmann M, Müller-Starck G (2001) Genetic characters and diameter growth of provenances of Scots pine (*Pinus sylvestris* L.). Silv Gen 50:212–222

Bork P, Copley R (2001) Comparative genome analysis: Annotation and context. Nature 409:818–820

Bozhko M, Riegel R, Schubert R, Müller-Starck G (2003) A cyclophilin gene marker confirming geographical differentiation of Norway spruce and indicating viability response on excess soil born salinity. Mol Ecol 12:3147–3155

Brewer S, Chedddadi R, de Beaulieu JL, Reille M and data contributors (2001) The migration of deciduous *Quercus* throughout Europe since the last glacial period. For Ecol Manage 156(1–3):27–48

Demesure B, Sodzi N, Petit RJ (1995) A set of universal primers for amplification of polymorphic non-coding regions of mitochondrial and chloroplast DNA in plants. Mol Ecol 4:129–131

Dumolin-Lapègue S, Pemonge MH, Petit RJ (1997) An enlarged set of consensus primers for the study of organelle DNA in plants. Mol Ecol 6:393–397

Eriksson G, Schelander B, Åkebrand V (1973) Inbreeding depression in an old experimental plantation of *Picea abies*. Hereditas 73:185–193

Fineschi S, Malvolti ME, Cannata F, Hattemer HH (eds) (1991) Biochemical markers in the population genetics of forest trees. SPB Academic Publ, The Hague, 251 pp

Geburek T (2000) Effects of environmental pollution on the genetics of forest trees. In: Young A, Boshier D, Boyle T (eds) Forest conservation genetics. Principles and practice. CSIRO, Collingwood, Australia, pp 135–157

Gillet EM (ed) (1999) "Which marker for which purpose?" Final compendium of the research project CT96–0706 "Development, optimisation and validation of molecular tools for assessment of biodiversity in forest trees", EU-DGXII Biotech. FW IV Res Progr "Molecular tools for biodiversity". http://webdoc.sub.gwdg.de/ebook/y/1999/whichmarker/index.htm

Gregorius HR (1978) The concept of genetic diversity and its formal relationship to heterozygosity and genetic distance. Math Biosci 41:253–271

Gregorius HR (1987) The relationship between the concepts of genetic diversity and differentiation. Theor Appl Genet 74:397–401

Gregorius HR (1989) The attribution of phenotypic variation to genetic or environmental variation. In: Scholz F, Gregorius HR, Rudin D (eds) Genetic effects of air pollutants in forest tree populations. Springer, Berlin Heidelberg New York, pp 2–16

Gregorius HR (1991) Gene conservation and the preservation of adaptability. In: Seitz A, Loeschke V (eds) Species conservation: a population-biological approach. Birkhäuser, Basel, pp 31–47

Gregorius HR, Ziehe M (1995) Detecting independence in viability selection on two traits. Heredity 74:70–79

Grivet D, Heinze B, Vendramin GG, Petit RJ (2001) Genome walking with consensus primers: application to the large single copy region of chloroplast DNA. Mol Ecol Notes 1:345–349

Gwaze DP, Bridgewater FE, Williams CG (2001) Genetic analysis of growth curves for a woody perennial species, *Pinus taeda* L. Theor Appl Gen 105:526–531

Hattemer HH, Bergmann F, Ziehe M (1993) Einführung in die Genetik für Studierende der Forstwissenschaft. Sauerländer, Frankfurt/Main, 492 pp

Hewitt GM (2000) The genetic legacy of the Quaternary ice ages. Nature 405:907–913
Hewitt GM (2001) Speciation, hybrid zones and phylogeography – or seeing genes in space and time. Mol Ecol 10:537–549
Huntley B, Birks HJB (1983) An atlas of past and present pollen maps for Europe. Cambridge Univ Press, Cambridge
Karl TR, Nicholls N, Gregory J (1997) The coming climate. Sci Am 276:78–83
Karp A, Isaac PG, Ingram DS (eds) (1998) Molecular tools for screening biodiversity: plants and animals. Chapman and Hall, London, p 498
Kirst M, Johnson AF, Baucom C, Ulrich E, Hubbard K, Staggs R, Paule C, Retzel E, Whetten R, Sederoff R (2003) Apparent homology of expressed genes from wood-forming tissues of loblolly pine (*Pinus taeda*, L.) with *Arabidopsis thaliana*. Proc Natl Acad Sci USA 100:7383–7388
Koski V (1973) On self-pollination, genetic load and subsequent inbreeding in some conifers. Comm Inst For Fenn 78(10):1–42
La Scala S (2000) Microsatellite markers as a tool for the characterisation of genetic variation in populations of sessile oak (*Quercus petraea* (Matt.) Liebl.). Dissertation, Technische Universität München, Freising, p 73
Lawton JH, Brown VK (1993) Redundancy in ecosystems. In: Schulze ED, Mooney HA (eds) Biodiversity and ecosystem function. Springer, Berlin Heidelberg New York, pp 265–270
Mitton JB, Grant MC (1984) Associations among protein heterozygosity, growth rate and developmental homeostasis. Annual revue of ecology and systematics, vol 15. Academic Press, New York, pp 479–499
Müller(-Starck) G (1976) A simple method of estimating rates of self fertilization by analysing isozymes in tree seeds. Silv Gen 25:15–17
Müller-Starck G (1985) Reproductive success of genotypes of *Pinus sylvestris* L. in different environments. In: Gregorius HR (ed) Population genetics in forestry. Lecture notes in biomathematics 60. Springer, Berlin Heidelberg New York, pp 118–133
Müller-Starck G (1993) Auswirkungen von Umweltbelastungen auf genetische Strukturen von Waldbeständen am Beispiel der Buche (*Fagus sylvatica* L.). Sauerländer, Frankfurt/Main, 163 pp
Müller-Starck G (1998) Isozymes. In: Karp A, Isaac PG, Ingram DS (eds) Molecular tools for screening biodiversity: plants and animals. Chapman and Hall, London, pp 75–81
Müller-Starck G, Schubert R (2000) Genetic markers as a tool for bioindication in forest ecosystems. In: Young A, Boshier D, Boyle T (eds) Forest conservation genetics: principles and practice. CSIRO Publ, Collingwood, Australia an CABI Publ, Wallingford, pp 227–237
Müller-Starck G, Schubert R (eds) (2001) Genetic response of forest systems to changing environmental conditions, vol 70. Forestry science. Kluwer, Dordrecht, 363 pp
Müller-Starck G, Ziehe M (1984) Reproductive systems in conifer seed orchards. 3. Female and male fitnesses of individual clones realized in seeds of *Pinus sylvestris* L. Theor Appl Gen 6:173–177
Paglia GP, Olivieri AM, Morgante M (1998) Towards second-generation STS (sequence-tagged sites) linkage maps in conifers: a genetic map of Norway spruce (*Picea abies* K.). Mol Gen 258:466–478
Paule L (1990) Bibliography: isozymes and forest trees. Swedish University of Agricultural Sciences, Umeå, 82 pp
Petit R, Aguinagalde I, Beaulieu JL, Bittkau C, Brewer S, Cheddadi R, Ennos R, Fineschi S, Grivet D, Lascoux M, Mohanty A, Müller-Starck G, Demesure-Musch B, Palmé A, Martin J, Rendell S, Vendramin G (2003) Glacial refugia: hotspots but not melting pots of genetic diversity. Science 300:1563–1565

Reille M, Gamisans L, Beaulieu JL, Andrieu V (1997) The late-glacial at Lac de Creno (Corsica, France): a key site in the western Mediterranean basin. New Phytol 135:547–559

Riegel R (2001) Entwicklung molekulargenetischer Marker bei der Fichte (*Picea abies* (L.) Karst) und deren Anwendung für genetische Erhebungen in umweltbelasteten Populationen. Doctoral Diss, Technische Universität München, Freising, 105 pp

Riegel R, Schubert R, Metzger HG, Müller-Starck G (2004) Chloroplast microsatellite markers monitoring genetic differentiation of *Picea abies* (L.) Kart. In response to air and soil pollution. Ecologia (Bratislava) (in press)

Schömig M (1988) Das Wachstum von Fichten-Kiefern-Beständen unter dem Einfluss von im Grundwasser angereicherten Straßenauftausalzen im Forstamt Gunzenhausen. Diploma Thesis, LMU, Munich

Scholz F, Gregorius HR, Rudin D (eds) (1989) Genetic effects of air pollutants in forest tree populations. Springer, Berlin Heidelberg New York, 201 pp

Schubert R, Müller-Starck G (2002) Development and application of molecular markers in conifers. In: Jain SM, Brar DS, Ahloowalia BS (eds) Molecular techniques in crop improvement. Kluwer, Dordrecht, pp 139–160

Schubert R, Müller-Starck G, Riegel R (2001) Development of EST-PCR markers and monitoring their intrapopulational genetic variation in *Picea abies* (L.) Karst. Theor Appl Gen 103:1223–1231

Tzschacksch O (1981) Stand und Perspektiven der forstlichen Rauchresistenzzüchtung in der DDR. Beitr Forstwirtsch 15:134–137

Vendramin GG, Lelli L, Rossi P, Morgante M (1996) A set of primers for the amplification of 20 chloroplast microsatellites in *Pinaceae*. Mol Ecol 5:595–598

Waring RH, Schlesinger WH (1985) Forest ecosystems: concepts and management. Academic Press, Orlando, 338 pp

Weising K, Gardner RC (1999) A set of conserved PCR primers for the analysis of simple sequence repeat polymorphisms in chloroplast genomes of dicotyledonous angiosperms. Genome 42:9–19

Wilson EO (ed) (1988) Biodiversity. National Academy Press, New York, 495 pp

Wolf H (2001) Effects of extreme SO_2 air pollution in winter 1995/1996 on vitality and growth of SO_2-tolerant Norway spruce (*Picea abies* [L.] Karst.) clones in the Ore Mountains. In: Müller-Starck G, Schubert R (eds) Genetic response of forest systems to changing environmental conditions. Kluwer, Dordrecht, pp 35–49

Yachi S, Loreau M (1999) Biodiversity and ecosystem productivity in a fluctuating environment: the insurance hypothesis. Proc Natl Acad Sci USA 96:1463–1468

Yu MH, Pakish LM, Zhou H (2001) An isozyme marker for resistance to root-knot nematode in sugarbeet. Crop Sci 41:1051–1053

Ziehe M, Hattemer HH, Müller-Starck R, Müller-Starck G (1999) Genetic structures as indicators for adaptation and adaptational potentials. In: Mátyás C (ed) Forest genetics and sustainability. Kluwer, Dordrecht, pp 75–89

Part C Biogeochemical Cycles

6 Functioning of Mixed-Species Stands: Evidence from a Long-Term Forest Experiment

H.E. Jones, N. McNamara, and W.L. Mason

6.1 Introduction

The attitude of grant-giving bodies to long-term research programs over the past two decades has been ambivalent and, on occasions, such research has been equated to "stamp collecting". Yet, when challenges to the environment have occurred such as acid rain, climate change, N deposition, land use change, it is to the "stamp collectors" that modelers have turned for reliable, temporal data on complex ecosystems. Nowhere is such complexity better illustrated than in old-growth native forests, with their mixture of different species and ages in the understory, intermediate, and canopy layers. In contrast, plantation forestry has traditionally consisted of even-aged monoculture stands, although evidence has accumulated that mixtures of tree species may have advantages over monocultures. The benefits include the potential for increased yield (e.g., Bartsch et al. 1996; von Lupke and Spellmann 1997; Luis and Monteiro 1998; Pretzsch, Chap. 3, this Vol.; Vilà et al., Chap. 4, this Vol.), resistance to pests and diseases (e.g., Su et al. 1996; Müller and Hallaksela 1998; Jactel et al., Chap 12, this Vol.; Pautasso et al., Chap. 13, this Vol.), reduced risk of wind-throw (e.g., Burkhart and Tham 1992; Dhôte, Chap. 14, this Vol.) and increased biological diversity (Lahde et al. 1999; Hartley 2002). However, while the area of mixed forest stands in western Europe has increased considerably in the last few decades (Gardiner 1999), few long-term experiments have attempted to provide clear answers about the potential advantages of mixtures over monocultures in well-defined and replicated conditions.

Forest mixture experiments are often set up with the aim to understand mechanisms that maximize yield, and a number have studied the potential advantages of growing nitrogen-fixing and non-nitrogen-fixing tree species together (e.g., Tarrant 1961; Tarrant and Trappe 1971; Brozek 1990; Binkley 1992; Bi and Turvey 1994; DeBell et al. 1997). Kelty (1992) described this mechanism of interaction of species as facilitation (one species improving the

Ecological Studies, Vol. 176
M. Scherer-Lorenzen, Ch. Körner, and E.-D. Schulze (Eds.)
Forest Diversity and Function: Temperate and Boreal Systems

growth conditions for the other), as opposed to complementarity, in which the two species exhibit differences in characteristics which cause interspecific competition to be significantly less than intraspecific. These would include, for example, differences in canopy structure, phenology, and root distribution. There have been some studies that have examined the potential benefits of mixing deciduous and coniferous species (e.g., Lockow 1998). Mixture effects have also been examined by comparing a variety of established native forest stands of different compositions of species (e.g.. Schuler and Smith 1988; Freist 1991; Nüsslein 1993; Longpré et al. 1994; Oyen and Tveite 1998). Vilà et al. (Chap. 4, this Vol.) have examined the impact of confounding factors when comparing such established stands.

This chapter reviews the existing information in relation to one such experiment conceived in 1955, when the former Nature Conservancy (now Centre for Ecology and Hydrology) and the UK Forestry Commission (now Forest Research) set up a tree-planting program at Gisburn in the North Pennine area of NW England (Holmes and Lines 1956). The experiment was planned with two aims:

- to compare tree growth in several two-species mixtures with their component monoculture stands, and
- to measure and identify changes in a range of soil factors under the different species.

As much of the information about Gisburn relates to that published by Brown (1992) on tree performance and soil changes in the first rotation, we have not repeated the detailed description of the methods and statistical analyses. However, Gisburn is perhaps unique in that the design, replication, and length of time over which the data have been collected have not been undertaken elsewhere.

6.2 Background of the Gisburn Experiment

6.2.1 Topography and Climate

The site (54° 02′N, 2° 22′W) ranges from 260–290 m in altitude. It has a 3° slope to the SW, and is very exposed to prevailing winds. Mean annual rainfall for a 30-year period from 1961–1990 (which covers most of the first rotation) was 1,510 mm. At this elevation and rainfall, conditions are predisposed to peat formation, and on a European scale, the climate can be described as oceanic.

6.2.2 Experimental Design

Sessile oak, *Quercus petraea*; alder, *Alnus glutinosa*; Scots pine, *Pinus sylvestris*; and Norway spruce, *Picea abies*, were planted in 1955 into a species-poor *Festuca–Agrostis* pasture, containing some *Nardus stricta*, *Juncus* spp., and *Deschampsia caespitosa*. The four tree species were laid out in three blocks, each with single-species plots and all combinations of two-species mixtures, plus an unplanted control; 11 treatments in all. The mixed plots were planted in a checkerboard pattern of alternating groups of 18 of each species (Fig. 6.1). This was a deliberate decision to permit the portion of the

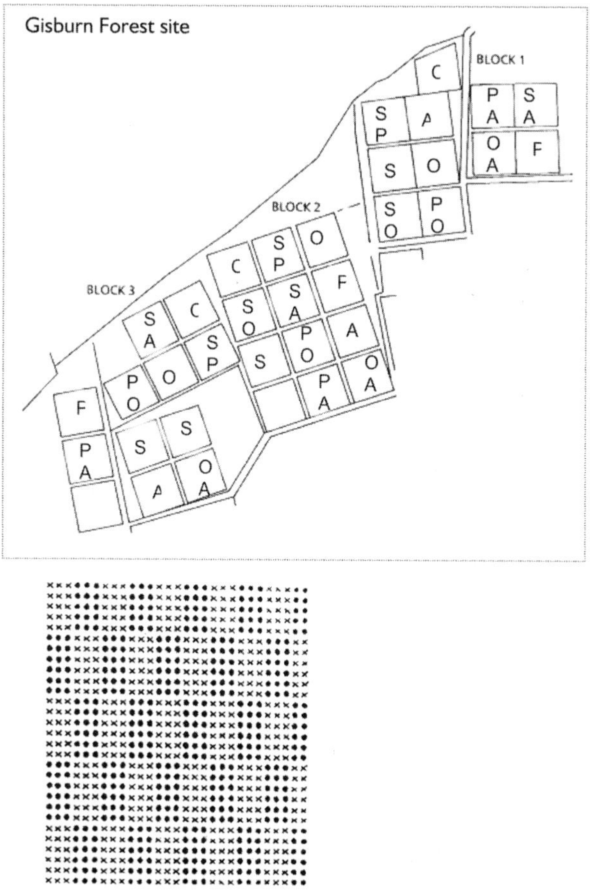

Layout of a mixed plot

Fig. 6.1. Layout of the Gisburn forest mixture experiment, and the checkerboard design of the mixed plots. *S* Norway spruce; *P* Scots pine; *A* alder; *O* oak; *C* unplanted control. *Blank spaces* are discontinued sheep-grazed plots

stand beneath the canopy of each component species to be sampled separately and compared with the pure stand. Although a more intimate mixture would have been more valuable for any early "nursing" effects, the checkerboard design increased the possibility of maintaining the mixture if one species were to outperform the other in the longer term. Each plot was 0.2 ha, containing 45 groups of each species and trees at 1.5-m intervals, although oak was planted at double density at 0.75-m intervals. Signs of wind throw in the late 1980s led to the site being clear felled after 34 years (between January and March 1989) and replanted in the same design early in 1991. Studies have therefore continued for a period spanning nearly 50 years. Franklin (1989), in Likens' book on long-term ecological experiments, referred to the experiment as a beautifully simple and accommodating design, which fulfilled the requirement that experimental treatments should be straightforward and unambiguously repeatable. However, some difficulties became apparent early in the first rotation because of the heterogeneous nature of the site. The soil in block I had higher peat content, reflected by some differences in the vegetation, with more *Holcus lanatus* and *Rumex acetosa*, and it was also more heterogeneous. As a result, block 1 was omitted from many of the subsequent studies, which were confined to sampling blocks 2 and 3 only, or to a single block, usually 2. Thus, many of the potential advantages of the original block design were not fully realized.

6.2.3 Geology and Soils

The regional geology is Ordovician mudstone and shales with extensive glacial deposits, and the underlying rocks are grits, sandstones, and shales of the Millstone Grit series (Carboniferous), overlain in most places by a clay-rich till (Robertson et al. 2000). The soils are generally thought to be transitional between stagnohumic gleys and cambic stagnogleys (Avery 1980). No fertilizers were applied either at planting or later, so any soil changes recorded could be directly attributed to the tree species and mixture effects. After 32 years in the first rotation, Moffat and Boswell (1990) were able to show some small but significant differences in soil properties relating to soil organic matter accumulation and incorporation. The soil under the conifers and alder was slightly more acid than under the oak, and the conifers had thicker F and H but thinner A horizons.

6.3 Tree Performance

6.3.1 First Rotation

Growth in height was assessed in all three blocks using the two tallest trees in each group of 18, and a consistent pattern emerged between the three replicates. However, this method of assessment was restricted to comparisons among the most vigorous specimens. Heights of these tallest trees, measured at 4, 7, 10, 20, and 26 years, showed significant treatment differences ($P<0.01$) after 7 years. Mixtures that included pine grew better than the pure stands of the component species, those with oak generally performed poorly. The effects persisted over the rotation, as Fig. 6.2 demonstrates, for heights at 26 years. In terms of timber production, a more valid estimate may be that of timber volume. This was assessed after 27 years' growth in 1982, from the product of top height and basal area, and the beneficial effect of pine in the mixtures was also here indicated. Analysis of the growth data identified three types of mixture effects (Brown 1992):

- Enhancement: seen in all the mixtures with pine where the yield of both components was greater relative to that in pure stands, so the result was an overall benefit to both species.
- Compensation: in mixtures with alder (other than those with pine) which showed better growth of the admixed species, at the expense of the alder.
- A negative interaction: seen in the spruce/oak mixture.

In contrast to the method used to analyze the mixture effects described above, Yanai and Malcolm (1992) tested the hypothesis that pine was outcompeting rather than aiding spruce at Gisburn by measuring tree diameters in the respective monoculture and mixed stands. Under this hypothesis, tree height would be an indication of competition for light rather than enhanced performance. They measured and mapped all the trees in a plot rather than the two of greatest height and showed that there was reduced survival and smaller basal area of the spruce in mixture which suggested that the interaction was not entirely beneficial. However, while competition for light may have induced better spruce height in the mixture, it would not account for the greater timber volume which was also seen in the mixed plots, so this explanation would need to be examined further by studying the performance of the admixed species in the other pine mixtures. One method used to measure the outcome of performance in mixtures of species compared with monocultures is to compute the relative yield total (RYT; see de Wit and van den Bergh 1965), where:

RYT = (yield of species A in mixture with B)/(yield species A pure) + (yield of species B in mixture with A)/(yield species B pure)

The expected value is the weighted average of the monoculture yields of the component species (e.g., Hector 1998; Loreau and Hector 2001), and

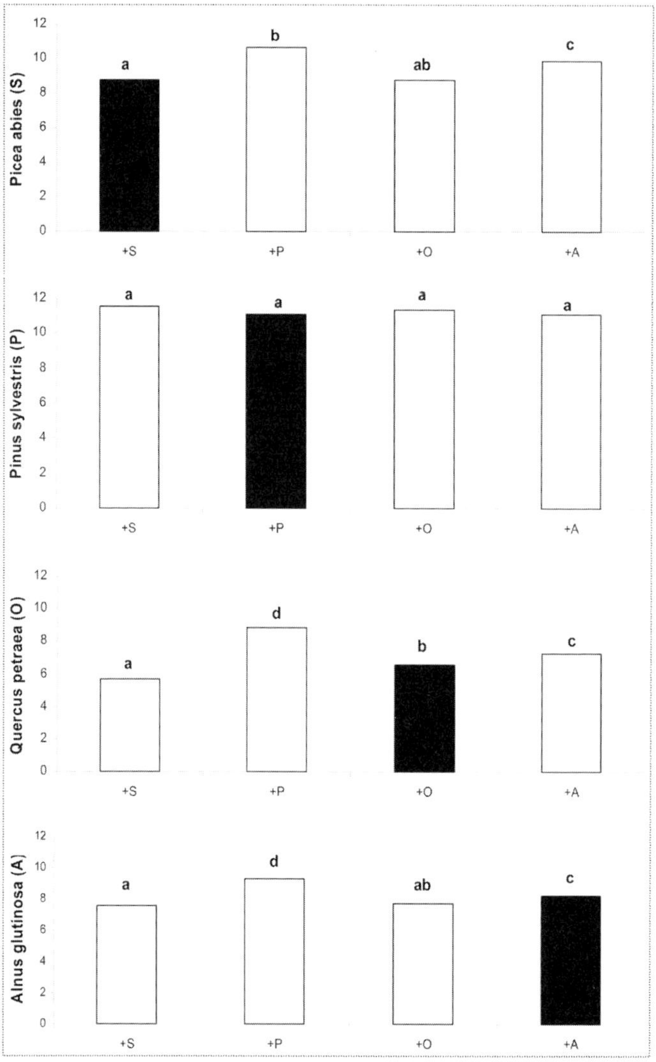

Fig. 6.2. Mean top heights (m) of the four tree species at 26 years (1981). *Black bars* represent monocultures, *white bars* represent mixtures. Different *lowercase letters* indicate significant differences within a given species. *+S* with Norway spruce, *+P* with Scots pine, *+A* with alder, *+O* with oak

RYT>1 suggests some complementary resource use. Brown (1992) used this concept to compare the relative yields of the various mixed stands: the pine mixtures all exceeded one, the other alder mixtures were about one, and the spruce/oak was less than one. It may need care to interpret RYT values in long-lived species, as the ratios may change over time and an observed response at one stage may disappear by another. For example, results published by Morgan et al. (1992) in studies on the dynamics of spruce/pine mixtures would show an RYT>1.0 at an early stage when the pine was "nursing" the spruce, but RYT<1.0 later when the spruce suppresses the pine.

6.3.2 Second Rotation

The mixture effect on tree performance was also studied in the second rotation. The method used to assess growth has been different from that used in the first, due to concerns (I.M.S White, Forest Research, pers. comm.) that only recording height of the two tallest trees might obscure long-term stand dynamics. The different recording hampers direct comparisons between rotations, however. A central plot of 12 of the 45 groups of 18 trees was selected, with an equivalent area in the monocultures. Within this area all the trees are being measured. As the alder that was felled in the first rotation had coppiced vigorously from the stumps, 2 years' growth had occurred by the time of replanting of the other species; a further difference between first and second rotations. Thus, mean height of alder 6 years after replanting was 3.1 m (Table 6.1), compared with

Table 6.1. Mean height of trees in pure and mixed stands at Gisburn after 6 years in the second rotation. Measurements were made on a central plot of 12 of the 45 groups of trees in the checkerboard layout in the mixed plots, and an equivalent area in the pure stands

Stand type[a]	Height (cm)	Significance	SED
S pure	113		
S/P	137	n.s.	15
S/A	127		
S/O	107		
P pure	170		
P/S	187	n.s.	16
P/A	160		
P/O	190		
A pure	307		
A/S	227	n.s.	39
A/P	213		
A/O	293		
O pure	150		
O/S	117	*	15
O/P	123		
O/A	183		

* $P<0.05$

[a] *S* Norway spruce, *P* Scots pine, *A* alder, *O* oak

1.7 m after 7 years in the first rotation. The main mixture effect identified at this stage was height growth in oak, which was significantly better with alder than with the two conifer species. The beneficial effect of the Scots pine in mixtures, noted in the first rotation, was not apparent at this relatively early stage of growth. Forest Research is continuing to measure tree performance at the site, so it will be possible to compare longer-term mixture effects with those of the first rotation.

6.3.3 Comparisons with Other Experiments Involving Mixed Species

Data in other experiments show some trends that are similar to those described above, although few cover a trial period as long as that at Gisburn. Malcolm and Mason (1999), however, reported on long-term trials of Scots pine/birch (*Betula pendula* and *B. pubescens*) mixtures and pure stands at two sites in the UK to determine whether the inclusion of the birch in pine stands would affect the pine yield, measured as top height and diameter, and whether there would be changes in the soil that might differ between the mixtures and the pure-species stands. The experiments included different ratios of 3:1 and 1:1 Scots pine/birch as well as the pure stands. After 32 years, the volume of pine timber in the 3:1 ratio did not differ significantly from that in the pure stand, and pine diameter was greater, while the height growth of birch was greater in mixture. The mean height of the pine in pure stands at this stage was 13.1 m at both sites, compared with mean height of 11.1 m at Gisburn after 26 years, which suggests that the growth rate was similar despite different climatic and soil conditions. Norokorpi (1994) also recorded a higher total yield when birch was planted with Norway spruce in northern Finland. More commonly and conversely, compensation effects, similar to those with the Gisburn alder mixtures, have been observed, though some of these effects relate to two-storied stands, where the nature of mixing may be different from single-storied stands such as those at Gisburn, so that Valkonen and Valsta (2001) recorded a yield loss in Norway spruce in Finland, compensated by increased production in inter-planted birch. Similarly, Mård (1996) showed that, while young birch trees present in a stand of Norway spruce significantly reduced the yield of the spruce, it was more than compensated for by the yield of birch wood. This is also the principle behind so-called nursing mixtures where a higher-yielding species is inter-planted with a lower-yielding one that is less sensitive to established conditions; the latter is removed at an earlier stage in the rotation once canopy closure is achieved. Cameron and Watson (2000) describe Japanese larch (*Larix leptolepis*)/Sitka spruce (*Picea sitchensis)* mixtures where the spruce has equivalent growth to that in pure spruce stands with N fertilizer applied. Numerous studies have shown benefits for the growth of the main crop, particularly Sitka spruce (on N-deficient soils in the UK) interspersed with an alternative species such as lodgepole pine (*Pinus contorta)* or Japanese larch (Morgan et al. 1992).

The results of other studies in European plantations have also demonstrated the importance of the role of canopy structure in mixed stands. Kelty (1992) reviewed the results from a series of experimental plantations set up in the late 19th century on various sites in Germany and Switzerland, and described by Assmann (1970). As at Gisburn, the design of each was limited to two-species mixtures and 50:50 ratios, but was a long-term comparison of stands growing on adjacent plots, with detailed measures of biomass yields. Mixtures included Scots pine/Norway spruce, Scots pine/beech, sessile oak/beech, European larch (*Larix decidua*)/beech (*Fagus sylvatica)* and Norway spruce/silver fir (*Abies alba*). In each case, the canopies developed a stratified structure with the less shade-tolerant species forming the upper layer. These mixtures consistently showed greater yields than the monocultures of the less-tolerant upper-canopy species and, where it was recorded, greater than the more shade-tolerant monoculture too (the complementarity mechanism already discussed, as identified by Kelty 1992). See also Pretzsch (Chap. 3, this Vol.) for further examples and a discussion on the importance of site conditions on the outcome of the mixture effects.

6.4 Underlying Mechanisms Linked with Mixture Effects

6.4.1 Plant Nutrients – Foliar Concentrations

One possible reason for the beneficial effects of planting mixed stands of trees might be an impact on the nutritional status of the component species, which is likely to be reflected in foliar concentrations, particularly where tree performance is improved in mixture because the site has marginal mineral nutrition for the more "sensitive" species. At Gisburn, foliar nutrients were only analyzed in Norway spruce needles. A sample of needles was taken from a range of trees of different heights, from both pure and mixed stands in November 1982 (Brown 1992). Based on standard methods described by Everard (1973), needles were analyzed for N, P, and K. Tree heights and foliar K were not correlated, but there was a significant positive relationship between height and both N ($P<0.001$) and P ($P<0.05$) concentrations in the foliage. Thus, the smallest trees from pure stands or mixtures with oak had near-deficient concentrations of about 1.2 % N, but those with the greatest height with pine had the highest, 1.6 %, near-optimal concentrations as defined by Binns et al. (1980). While this is an interesting finding, it is possible that the higher nutrient concentrations are less an effect of the admixed species and more an effect of canopy position. This explanation could only be clarified by analyzing the nutrients in needles from trees of a range of heights in one mixture type or in a pure stand.

Foliar nutrient concentrations have been used as a diagnostic aid in determining the nutrient status of other forest mixtures (see review by Rothe and Binkley 2001). For example, Thelin et al. (2002) showed, in paired comparisons of mixed and pure stands of Norway spruce and deciduous species at 30 sites across southern Sweden and Denmark, that the mixed stands had higher spruce needle P and, particularly, K concentrations, than the pure stands. They surmised that the mixture effect was due to the more nutrient-rich litter of the deciduous trees, but noted that the improved nutritional status was not reflected in a significant improvement in the tree growth. In mixtures of paper birch (*Betula papyrifera*) and subalpine fir (*Abies lasiocarpa*), the birch foliage had higher N, P, K, Ca, and Mg concentrations than the fir, although the latter had higher concentrations of P and Ca in the branches and stembark (Wang et al. 2000). While there was no information on comparable concentrations in pure stands, the different resource quality of the foliage and woody components would influence the nutrient cycling in the mixed species forest, particularly if different species contained different relative amounts of nutrients (Swift et al. 1979). Direct evidence to link better tree growth in mixed stands with enhanced nutrient status is not shown directly in any of these studies, however. It would require more exhaustive and directed sampling to establish any relationship.

6.4.2 Soil Nutrients – Lysimeter Studies

While not the only form of nutrient available for plant uptake, soluble ions are readily accessible and measures of their concentrations will be an index of soil fertility. These soluble ions will be derived from direct inputs in throughfall, stemflow, and precipitation and, indirectly, by mineralization of organic matter. At Gisburn, lysimeters were used in several studies of the variation in soil nutrients, both seasonally and under the different species (Chapman 1986; Brown and Iles 1991). Chapman (1986) showed significantly enhanced release of N and P in the spruce/pine mixture relative to the pure stands, thus providing further evidence of a positive interaction involving the pine. This might be attributed to increased nitrification, seasonal release of P, enhanced decomposer activity, or a mixture of all three factors. Differences in ion exchange capacity in the soil from the pure and mixed stands could also be a cause. Whatever the balance of these processes, the impact on tree growth would depend on the timing of the mineralization. Possibly mycorrhizal interactions are involved, although evidence supporting this comes from pot experiments with single species (Downes et al. 1992; Cairney and Alexander 1992a,b).

In contrast, there was reduced mineralization in the spruce/alder stands, so the spruce appeared to "switch off" the activity associated with N and P release in the alder. Chapman (1986) thought that reduced nitrification rates might be responsible for the low level of available inorganic N in this mixture. However, he considered this to be surprising, as alder is traditionally regarded as a "nurse" species (Tarrant and Trappe 1971) and, therefore, concluded that further research was required. On the other hand, there is separate evidence of inhibition of alder-related nitrification by coniferous litter (Hendrickson and Chatarpaul 1984). This is possibly caused by an inhibitory effect on the N-fixing *Frankia* spp. in the root nodules, although this is purely speculative. Chapman (1986) also recorded reduced mobilization of inorganic N in both portions of the spruce/oak mixture, i.e., a negative interaction. Brown and Iles (1991) noted that the marked species effects on the throughfall chemistry was not reflected in changes in the water chemistry lower in the soil profile. Nutrient losses in drainage waters may come from any or all the soil horizons but, based on the concentrations recorded, oak was associated with the smallest losses, and pine the greatest.

The impact of forest mixtures on soil solution properties has been studied by Brantberg et al. (2000) and Rothe et al. (2002), among others. The former determined how soil chemistry was affected by an admixture of birch (*Betula pendula* and *B. pubescens*) with Norway spruce. Where birch was present, Ca^{2+} and Mg^{2+} were significantly higher in the LF layers, and K^{+} showed a similar increase in the H-horizon. Rothe et al. (2002) showed that leachates sampled below the rooting horizons were significantly higher in NO_3^- and SO_4^- in Norway spruce stands than beech, suggesting losses of these ions to groundwater in the conifer stands. In mixtures, there was a nonlinear effect of the spruce, so that losses only declined where the spruce component was less than 20 %. The general conclusion that can be drawn from these studies confirms the perceived wisdom that the nutrient status of soils where deciduous trees are present is likely to be higher than under pure conifer stands, with fewer losses to the system.

6.4.3 Soil Nutrients – Extractable Ions

If mineralization is the main source of soluble N and P in the forest soils, then the release of the two elements is likely to occur in parallel in soil solution. Chapman (1986) demonstrated a strong positive correlation (r=0.78; P<0.01) between concentrations of inorganic N and PO_4-P in soils. He also measured extractable ions in the top 5 cm of the mineral soil in all three blocks at Gisburn from the pure and mixed spruce stands, but obtained conflicting evidence. KCl-extractable NH_4-N and NO_3-N measured on one occasion in September 1986 showed the same patterns as in the soluble ions in the forest-floor leachates. However, 2.5 % acetic acid-extractable ions showed fluctuating patterns and no relationship with the spruce mixtures when measured over a longer period in all three blocks. Brown and Harrison (1983) also looked at the N status in the soils of the spruce mixtures, but in relation to tree performance. Extractable NO_3^- in the 0–5 cm mineral layer provided the best relationship with spruce height (r=0.838; P<0.01), but the relationship with NH_4^+ was not significant. This is somewhat surprising as it might be expected that the latter would be the better indicator under these acid soil conditions. The evidence available to date, therefore, suggests that spruce growth varies more in relation to NO_3^- than to NH_4^+ concentrations, and the NO_3^- levels are enhanced by admixed pine and alder, but not by oak. Analysis of the LFH horizon in the Scottish experiments set up by Malcolm and Mason (1999) did not show significant differences in extractable ions between different birch/pine mixtures. Although % N content showed a increasing but non-significant trend with more pine, they did not measure extractable NO_3^-. When Turner et al. (1993) studied mixed old-growth forests in Oregon they found a strong influence of different tree mixtures on availability, and uptake of inorganic N nitrification, NO_3-N concentrations, and N uptake rates were all higher under red cedar (*Thuja plicata*) than hemlock (*Tsuga heterophylla*) or Douglas fir (*Pseudotsuga menziesii*). There is also an interesting link between N-fixing alders and soil P. In the Pacific northwest, Giardina et al. (1995) noted that mixed stands of Douglas fir/red alder (*Alnus rubra*) contained higher P levels in the litter, and soil phosphatase activity was nearly three times greater than in adjacent pure fir stands. However, although the alder increased the availability of P, the authors considered it might not be sufficient to prevent P limitation of either the alder or the fir.

6.4.4 Soil Nutrients- Litter Decomposition and Mineralization

A possible benefit of the use of tree mixtures is that mixed litters can result in interactions leading to higher decomposition, enhanced mineralization and hence greater availability of soil nutrients (Hättenschwiler, Chap. 8, this Vol.). McTiernan et al. (1997) used litters from Gisburn in microcosm experiments and found that mixtures of spruce/pine, spruce/oak, pine/alder, pine/oak, and oak/alder all had significantly higher respiration than observed in single species samples, although there was often a lag phase in release of NH_4-N in the spruce mixtures. This would further suggest enhanced mineralization in mixed litters. However, evidence from other studies on litter decomposition in tree mixtures and pure stands is conflicting. Generally, data show that mixtures of deciduous litters with conifers increase decomposition in Norway spruce and spruce/birch stands (Saetre et al. 1999), in red pine (*Pinus resinosa*), beech (*Fagus grandifolia*), hemlock (*Tsuga canadensis*), and mixed broadleaves (*Acer saccharum* and *Quercus alba*; Elliott et al. 1993). In this latter paper, the authors showed that the relative importance of bacteria and fungi to total metabolism varied among forest types and was related to the pH of the forest floor. The importance of bacteria to decomposition declined successively from the mixed broadleaves to beech to red pine to hemlock. Carlyle and Malcolm (1986) demonstrated the beneficial effect of hybrid larch (*Larix eurolepis*) litter compared with Sitka spruce on the growth of Sitka seedlings. Higher levels of P in the larch appeared to be the key, as N and K levels were similar. Prescott et al. (2000) studied rates of decomposition in pure and mixed litters of three common forest types of British Columbia comprising white spruce (*Picea glauca*)/trembling aspen (*Populus tremuloides*), Douglas fir/red alder and Douglas fir/paper birch/lodgepole pine. Although initial decomposition in the deciduous trees was greater than in the spruce or fir, it declined after the first year, so there was no significant difference after 3 years, nor was there evidence that decomposition was enhanced when the litters were mixed. However, general conclusions are rather difficult to draw from these examples because of the wide range of different species, site characteristics (as to both climate and soil), and forest structures.

Cotton strip decomposition, measured by loss in tensile strength, can provide a useful comparative index of cellulolytic decomposer activity in the field, and Brown (1988) used the method at Gisburn to compare spruce mixtures and the associated pure stands. Loss in tensile strength in strips buried in the spruce/pine mixture was greater than that in the pure spruce; and in the alder mixture the strips in the spruce component showed greater rotting than those in the alder. That pattern supports the evidence of the different responses in the tree growth.

6.4.5 Role of Decomposer Organisms

Soil animals aid decomposition through comminution. Chapman (1986) studied the earthworm population at Gisburn and found that the numbers of the two acid-tolerant species present (*Bimastos eiseni* and *Lumbricus rubellus)* were significantly higher in the spruce/pine mixtures than expected from the numbers in the pure stands. Numbers in the spruce/alder mixture were not significantly different from those in the pure stands, and none were extracted from soil in the spruce/oak mixture. A parallel pattern was noted in enchytraeid populations, small annelid worms that tend to replace earthworms in upland acid soils and are known to make a significant contribution to the breakdown of organic matter (Briones et al. 1998). Other studies generally confirm the greater numbers and activity of soil animals in mixed than pure stands, particularly those with a deciduous component. Kautz and Topp (1998) studied forest plots in Germany with pure Scots pine, pine/birch, pine/oak, and pure beech. Litter decomposition was significantly higher in the pine/oak than the pure pine stands, and the abundance of soil fauna (Lumbricidae, Enchytraeidae, Collembola, and Oribatidae) was greater in the mixed forests than the coniferous ones. There was also a significant correlation between high biotic activity, increased pH and lower Al^{3+} concentrations in the deciduous forest plots. Ponge and Prat (1982) studied collembolan populations at eleven sites under coniferous, deciduous, or mixed forest. Deciduous tree litter, particularly that of hornbeam (*Carpinus* spp.), stimulated the degradation of pine litter leading to a type of humus that differed from that under the pure conifer stand. The differences appeared to be related to the composition of the collembolan fauna; acidophilic species predominated in the coniferous forest. Poursin and Ponge (1982) looked at the species composition of collembola and oribatids under oak, pine, and mixed oak/pine stands. The specific composition was analyzed by the analysis-of-correspondents method and found to be very similar under the oak and mixed stands, but markedly different under the pine. The difference was associated with pH and the type of humus. Saetre et al. (1999) extracted higher numbers of Coleoptera, Diptera, and Collembola from mixed birch/Norway spruce stands than from pure spruce.

Chapman (1986) recorded a significant effect of tree species on the soil respiration rate per unit weight from the LF layers. Highest rates were in the spruce and alder, followed by oak and pine. However, total respiration depends on the quantity of organic matter in the litter layer, and when that was taken into account no significant differences were detected. More interestingly, the corrected rate in the spruce/pine mixture was about 40 % higher than would be expected if it were the mean of the rates measured in the two pure stands ($P<0.001$), whereas respiration in spruce/alder and spruce/oak was significantly lower than expected ($P<0.001$ for both). A thorough discussion of the role of soil fauna in decomposition is given by Scheu (Chap. 11, this Vol.).

6.4.6 Influence of Different Rooting Patterns

One potential advantage of tree mixtures over monocultures is the ability to exploit soil resources in a complementary way, by means of different rooting patterns. Sykes and Robertson carried out a detailed study of the pure stands of spruce and pine species at the Gisburn site (Brown 1992). They analyzed cores sampled in 1983 for the distribution of live, fine roots <1 mm diameter down the profile. These would be the active feeder roots. While 44 % of the spruce roots occurred in the LFH layer, they found fewer than 10 % of the pine roots here. Only 20 % of the spruce roots occurred below the top 5 cm of the mineral soil, compared with 43 % of the pine. Chapman (1986) also looked at root distribution in the LFH horizon, in the spruce/pine mixture as well as the monocultures, and found that there were more than twice as many spruce roots in the mixture compared with the monoculture. However, the biomass of pine roots was the same in mixed and pure stands. It appeared that the spruce had proliferated to take advantage of the forest floor, whereas most of the pine's feeding roots were in the mineral soil. This provides a further mechanism to explain the mixture effect of these two species at Gisburn, and also indicates how the spruce in the mixture can benefit from the extra release of N and P without detriment to the pine. Other research has found different patterns of rooting competition or synergy, depending on the species mix and root fraction extracted. In contrast to the Gisburn results, in an upland heath in Scotland, McKay and Malcolm (1988) recorded almost twice the standing crop of Sitka spruce roots <2 mm diameter in pure stands than in mixtures with Scots pine, although there was a similar distinction in the rooting distribution, with more pine roots at depth. Their category of fine root was <2 mm diameter, which might explain the contradiction with the Gisburn analysis of samples below 1 mm. Kalinin and Zakharchuk (1983) noted that two pine species, *Pinus banksiana* and *P. sylvestris* in mixed stands, did not compete because of the development of deeper tap roots in the latter, and better development of vertical ramifications from the horizontal roots. Leuschner et al. (2001) introduced experimental root chambers to investigate the interaction between beech and oak roots, in mixed stands where stem densities were very similar. The fine beech root biomass grew more rapidly, so that the root:shoot ratio was greater than in the oak, although the opposite was true of the coarse root fraction. They concluded that the difference in fine root production gave beech a competitive advantage due to the nutritional benefits. The impact of competitive below-ground interactions was also demonstrated by Schmid and Kazda (2001), who showed enhanced beech root growth in mixtures with Norway spruce, so that the spruce root system was even shallower in the mixed than the pure stand. This would suggest that the effect of tree mixtures on Norway spruce observed at Gisburn is likely to occur with other species, too.

6.4.7 Possible Inhibitory Effects in Mixtures

Studies on the role of inhibition as a possible mechanism for mixture effects are sparse. Chapman (1986) used microcosms at Gisburn to study the effect of freshly collected throughfall waters from spruce, oak, and alder on the mineralization of NO_3-N, NH_4-N, PO_4-P, and on soil respiration from spruce litter and compared the results with those obtained with distilled water. There were few significant effects except from the spruce throughfall, which significantly reduced the levels of the three nutrients by 97, 40, and 55 %, respectively, relative to the distilled water treatment. Although this was a study of the effect of spruce on spruce, it might reasonably be assumed that a similar impact would be seen on the litter of the other species and, hence, could be a reason for the reduced N mineralization in the spruce/alder mixtures.

6.4.8 Role of Invertebrate Herbivores

There also appear to be few studies on the impact of herbivorous insects that might affect the relative performance of the tree species in mixed and pure stands. Watt (1992) reviewed the literature in this field and felt the evidence was inconclusive on how tree species composition might be effectively manipulated to reduce the incidence of pest outbreaks in forests (see also Jactel et al., Chap. 12, this Vol.). Moore et al. (1991) carried out some investigations of the feeding damage by a range of oak herbivores in mixed and pure plots at Gisburn. They found that the densities of the insects were different, although the diversity of insect pests was the same whether the tree was growing in a pure or a mixed stand. The number of all herbivore groups which attacked the oak was highest in the oak/alder mixture and lowest in that of oak/spruce. The damage caused by different organisms also differed depending on the composition of the stand, so that leafhoppers caused little damage in pure oak or oak/spruce but caused almost as much damage as caterpillars on oak in oak/alder mixtures. Su et al. (1996) showed that increasing the broadleaf content to >40 % in mixed stands with balsam fir (*Abies balsamea*) reduced the susceptibility and vulnerability of the fir to spruce budworm, but suggested the need for further research.

6.5 Conclusions

- Practically all the well-planned experiments designed to study tree mixture effects have been limited to comparisons between two-species mixtures, more often than not as 50:50 ratios, and pure stands, so cannot themselves represent the complexities of an old-growth forest. There is a requirement, therefore, for further work on well-designed multi-species mixture experiments, which should be set up to examine long-term effects on relative yield, soil changes, etc. (see Scherer-Lorenzen et al., Chap. 16, this Vol.).
- A long-term experiment to measure tree growth requires a simple, robust design and faith on the part of those who establish the experiment that their successors will have the resources and commitment to continue the work. The Gisburn experiment shows that this can be achieved.
- Different mixtures have their own specific characteristics in terms of yield response and impact depending on species traits and stand composition and mediated by site characteristics, so it can be difficult to draw general conclusions. However, coming from one of the few long-term forest mixture experiments, data from Gisburn suggest that tree mixtures can outperform monocultures where one species modifies conditions in a way

that benefits the other. At this site, soil nutrient availability (N and P) appears to be a key factor, related to organic matter turnover brought about by microbial and faunal activity.

- Some areas which would benefit from more detailed study include the differential impacts of insect herbivores on trees in mixed and pure stands, and belowground interactions, e.g., root competition versus complementarity, the impact of root interactions on the long-term stability of mixed stands, and the effect of root exudates of different species.

References

Assmann E (1970) The principles of forest yield study. Pergamon Press, Oxford, 506 pp

Avery BW (1980) Soil classification for England and Wales (higher categories). Tech Monogr no 14. Soil Survey of England and Wales, Harpenden

Bartsch N, Petercord R, von Lupke B (1996) Growth of sessile oak in mixture with Scots pine. Forst Holz 51:195–200

Bi H, Turvey ND (1994) Inter-specific competition between seedlings of *Pinus radiata*, *Eucalyptus regnans* and *Acacia melanoxylon*. Aust J Bot 42:61–70

Binkley D (1992) Mixtures nitrogen-fixing and non-nitrogen-fixing tree species. In: Cannell MGR, Malcolm DC, Robertson PA (eds) The ecology of mixed-species stands of trees. Blackwell, Oxford, pp 99–124

Binns WO, Mayhead GJ, MacKenzie JM (1980) Nutrient deficiencies of conifers in British forests. Forestry Commission Leaflet no 76. HMSO, London

Brandtberg P-O, Lundkvist H, Bengtsson J (2000) Changes in forest-floor chemistry caused by a birch admixture in Norway spruce stands. For Ecol Manage 130:253–264

Briones MJI, Ineson P, Poskitt JM (1998) Climate change and *Cognettia sphagnetorum*: effect on carbon dynamics in organic soils. Funct Ecol 12:528–535

Brown AHF (1988) Discrimination between the effects on soils of 4 tree species in pure and mixed stands using cotton strip assay. In: Harrison AF, Latter PM, Walton DWH (eds) Cotton strip assay: an index of decomposition in soils. ITE symposium no 24 Institute of Terrestrial Ecology, Grange-over-Sands, pp 80–83

Brown AHF (1992) Functioning of mixed-species stands at Gisburn, N W England. In: Cannell MGR, Malcolm DC, Robertson PA (eds) The ecology of mixed-species stands of trees. Blackwell, Oxford, pp 125–150

Brown AHF, Harrison AF (1983) Effects of tree mixtures on earthworm populations, and nitrogen and phosphorus status in Norway spruce (*Picea abies*) stands. In: Lebrun P, Andre HM, de Medts H, Gregoire-Wibo C, Wauthy G (eds) New trends in soil biology. Ottignies-Louvain-la Neuve, Dieu-Brichart, pp 101–108

Brown AHF, Iles MA (1991) Water chemistry profiles under four tree species at Gisburn, NW England. Forestry 64:169–187

Brozek S (1990) Effect of soil changes caused by red alder (*Alnus rubra*) on biomass and nutrient status of Douglas-fir (*Pseudotsuga menziesii*) seedlings. Can J For Res 20:1320–1325

Burkhart HE, Tham A (1992) Predictions from growth and yield models of the performance of mixed-species stands. In: Cannell MGR, Malcolm DC, Robertson PA (eds) The ecology of mixed-species stands of trees. Blackwell, Oxford, pp 21–34

Cairney JWG, Alexander IJ (1992a) A study of the ageing of spruce (*Picea sitchensis* (Bong) Carr.) ectomycorrhizas II Carbohydrate allocation in ageing *Picea sitchensis/Tylospora fibrillosa* (Burt.) Donk ectomycorrhizas. New Phytol 122:153–158

Cairney JWG, Alexander IJ (1992b) A study of the ageing of spruce (*Picea sitchensis* (Bong) Carr.) ectomycorrhizas III Phosphate absorption and transfer in ageing *Picea sitchensis/Tylospora fibrillosa* (Burt.) Donk ectomycorrhizas. New Phytol 122:159–164

Cameron AD, Watson BA (2000) Growth and wood properties of Sitka spruce (*Picea sitchensis* (Bong.) Carr.) in nursing mixtures established on nitrogen-deficient mineral soils. Scand J For Res 15:237–246

Carlyle JC, Malcolm DC (1986) Larch litter and nitrogen availability in mixed larch-spruce stands II. A comparison of larch and spruce litters as a nitrogen source for Sitka spruce seedlings. Can J For Res 16:327–329

Chapman K (1986) Interaction between tree species: decomposition and nutrient release from litters. PhD Thesis, University of Lancaster

DeBell DS, Cole TG, Whitesell CD (1997) Growth, development and yield of pure and mixed stands of *Eucalyptus* and *Albizia*. For Sci 43:286–298

De Wit CT, van den Bergh JP (1965) Competition between herbage plants. Neth J Agric Sci 13:212–221

Downes GM, Alexander IJ, Cairney JWG (1992) A study of the ageing of spruce (*Picea sitchensis* (Bong.) Carr.) ectomycorrhizas I Morphological and cellular changes in mycorrhizas formed by *Tylospora fibrillosa* (Burt.) Donk and *Paxillus involutus* (Batsch. ex Fr.) Fr. New Phytol 122:141–152

Elliott WM, Elliott NB, Wyman RL (1993) Relative effects of litter and forest type on rate of decomposition. Am Midl Nat 129:87–95

Everard J (1973) Foliar analysis: sampling methods, interpretation and application of the results. Q J For 67:51–66

Franklin JF (1989) Importance and justification of long-term studies in ecology. In: Likens GE (ed) Long-term studies in ecology. Springer, Berlin Heidelberg New York, pp 3–19

Freist H (1991) Is it sensible to mix four tree species (beech, oak, spruce, larch)? Forst Holz 46:501–502

Gardiner J J (1999) Changing forests, management and growing conditions. In: Olsthoorn AFM, Bartelink HH, Gardiner JJ, Pretzsch H, Hekhuis HJ, Franc A (eds) Management of mixed-species forest: silviculture and economics. IBN scientific contributions, vol 15. DLO Institute for Forestry and Nature Research, Wageningen, pp 17–19

Giardina CP, Huffman S, Binkley D, Caldwell BA (1995) Alders increase soil phosphorus availability in a Douglas-fir plantation. Can J For Res 25:1652–1657

Hartley MJ (2002) Rationale and methods for conserving biodiversity in plantation forests. For Ecol Manage 155:81–95

Hector A (1998) The effect of diversity on productivity: detecting the role of species complementarity. Oikos 82:597–599

Hendrickson O, Chatarpaul (1984) Nitrification potential in an alder plantation. Can J For Res 14:543–546

Holmes GD, Lines R (1956) Mixture experiments. Forestry Commission Rep Forest Res, London

Kalinin MI, Zakharchuk IA (1983) Differences in the structure of the root systems of *Pinus banksiana* and *P. sylvestris*. Lesnoi-Zhurnal 1:28–31

Kautz G, Topp W (1998) Sustainable forest management for improving soil quality. Forst Centralbl 117:23–43

Kelty MJ (1992) Comparative productivity of monocultures and mixed species stands. In: Kelty MJ, Larson CB, Oliver CD (eds) The ecology and silviculture of mixed-species forests, vol 40. Forestry sciences. Kluwer, Dordrecht, pp 131–141

Lahde E, Laiho O, Norokorpi Y (1999) Diversity oriented silviculture in the Boreal Zone of Europe. For Ecol Manage 118:223–243

Leuschner C, Hertel D, Coners H, Buttner V (2001) Root competition between beech and oak: a hypothesis. Oecologia 126:276–284

Lockow KW (1998) Long-term experimental plots at Eberswalde pine open stand experiment with beech underplanting at Schonholz 16. Beitr Forstwirts Landschaftsökol 32:145–154

Longpré M-H, Bergeron Y, Paré D, Béland M (1994) Effect of companion species on the growth of Jack pine (*Pinus banksiana*). Can J For Res 24:1846–1853

Loreau M, Hector A (2001) Partitioning selection and complementarity in biodiversity experiments. Nature 412:72–76

Luis JFS, Monteiro M do L (1998) Dynamics of a broadleaved (*Castanea sativa*) conifer (*Pseudotsuga menziesii*) mixed stand in northern Portugal. For Ecol Manage 107:183–190

Malcolm DC, Mason WL (1999) Experimental mixtures of Scots pine and birch: 30 year effects on production, vegetation and soils. In: Olsthoorn AFM, Bartelink HH, Gardiner JJ, Pretzsch H, Hekhuis HJ, Franc A (eds) Management of mixed-species forest: silviculture and economics. IBN scientific contributions, vol 15. DLO Institute for Forestry and Nature Research, Wageningen, pp 79–87

Mård H (1996) The influence of a birch shelter (*Betula* spp) on the growth of young stands of *Picea abies*. Scand J For Res 11:343–350

McKay HM, Malcolm DC (1988) A comparison of the fine root component of a pure and a mixed coniferous stand. Can J For Res 18:1416–1426

McTiernan KB, Ineson P, Coward PA (1997) Respiration and nutrient release from tree leaf litter mixtures. Oikos 78:527–538

Moffat AJ, Boswell RC (1990) Effect of tree species and species mixtures on soil properties at Gisburn Forest, Yorkshire. Soil Use Manage 6:46–52

Moore R, Warrington S, Whittaker JB (1991) Herbivory by insects on oak trees in pure stands compared with paired mixtures. J Appl Ecol 28:290–304

Morgan JL, Campbell JM, Malcolm DC (1992) Nitrogen relations of mixed-species stands on oligotrophic soils. In: Cannell MGR, Malcolm DC, Robertson PA (eds) The ecology of mixed-species stands of trees. Blackwell, Oxford, pp 65–85

Muller MM, Hallaksela AM (1998) Diversity of Norway spruce needle endophytes in various mixed and pure Norway spruce stands. Mycol Res 102:1183–1189

Norokorpi Y (1994) Admixture of birch in planted Norway spruce stands enhances total yield. In: Pinto da Costa ME, Preuhsler T (eds) Mixed stands: research plots, measurements and results. Proceedings of the symposium of the IUFRO working group S 4.01. Lousa/Coimbra April 1994 Instituto Superior de Agronomia, Lisboa, pp 97–103

Nüsslein S (1993) Comparison of performance of pure spruce stands and mixed spruce/beech stands. Allg Forst Z 48:682–684

Oyen B H, Tveite B (1998) A comparison of site index class and potential stem volume yield between different tree species growing on equivalent sites in western Norway. Rapport fra Skogforskningen no 15, 32 pp

Ponge JF, Prat B (1982) Collembola, indicators of the humification process in coniferous, deciduous and mixed plantations: results form the forest of Orleans, Franc. Rev Ecol Biol Sol 19:237–250

Poursin JM, Ponge JF (1982) Comparison of three forest plantations (broadleaves, conifers and mixed) by means of their soil fauna composition (Collembola and Oribatids). CR Hebdom Seances Acad Sci 22:1021–1024

Prescott CE, Zabek LM, Staley CL, Kabzems R (2000) Decomposition of broadleaf and needle litter in forests of British Columbia: influences of litter type, forest type and litter mixtures. Can J For Res 30:1742–1750

Robertson SMC, Hornung M, Kennedy VH (2000) Water chemistry of throughfall and soil water under four tree species at Gisburn, northwest England, before and after felling. For Ecol Manage 129:101–117

Rothe A, Binkley D (2001) Nutritional interactions in mixed species forests: a synthesis. Can J For Res 31:1855–1870

Rothe A, Kreutzer K, Küchenhoff H (2002) Influence of tree species composition on soil and soil solution properties in two mixed spruce-beech stands with contrasting history in southern Germany. Plant Soil 240:47–56

Saetre P, Brandtberg P -O, Lundkvist H, Bengtsson J (1999) Soil organisms and carbon, nitrogen and phosphorus mineralisation in Norway spruce and mixed Norway spruce-Birch stands. Biol Fertil Soils 28:382–388

Schmid I, Kazda M (2001) Vertical distribution and radial growth of coarse roots in pure and mixed stands of *Fagus sylvatica* and *Picea abies*. Can J For Res 31:539–548

Schuler TM, Smith FW (1988) Effect of species mix on site/density and leaf area relations in southwest pinyon/juniper woodlands. For Ecol Manage 25:211–220

Su Q, Maclean DA, Needham TD (1996) The influence of hardwood content on balsam fir defoliation by spruce budworm. Can J For Res 26:1620–1628

Swift MJ, Heal OW, Anderson JM (1979) Decomposition in terrestrial ecosystems. Studies in ecology, vol 5. Blackwell, Oxford

Tarrant RF (1961) Stand development and soil fertility in a Douglas-fir/red alder plantation. For Sci 7:238–246

Tarrant RF, Trappe JM (1971) The role of *Alnus* in improving the forest environment. Plant Soil Spec Vol 335–348

Thelin G, Rosengren U, Callesen I, Ingerslev M (2002) The nutrient status of Norway spruce in pure and in mixed-species stands. For Ecol Manage 160:115–125

Turner DP, Sollins P, Leuknig M, Rudd N (1993) Availability and uptake of inorganic nitrogen in a mixed old-growth coniferous forest. Plant Soil 148:163–174

Valkonen S, Valsta L (2001) Productivity and economics of mixed two-storied spruce and birch stands in Southern Finland simulated with empirical models. For Ecol Manage 140:133–149

Von Lupke B, Spellmann H (1997) Aspects of stability and growth of mixed spruce-beech stands as a basis of silvicultural decisions. Forstarchiv 68:167–179

Wang JR, Letchford T, Comeau P, Kimmins JP (2000) Above- and below-ground biomass and nutrient distribution of a paper birch and subalpine fir mixed-species stand in the Sub-Boreal zone of British Columbia. For Ecol Manage 130:17–26

Watt AD (1992) Insect pest population dynamics: effects of tree species diversity. In: Cannell MGR, Malcolm DC, Robertson PA (eds) The ecology of mixed-species stands of trees. Blackwell, Oxford, pp 267–275

Yanai RD, Malcolm DC (1992) Competitive interactions between Norway spruce and Scots pine at Gisburn Forest, NW England. Forestry 65:435–451

7 The Role of Biodiversity on the Evaporation of Forests

D.D. BALDOCCHI

7.1 Introduction

On visiting an arboretum or walking through a mixed-species forest, an impression one receives is that there is much diversity in the form and function of trees. The leaves of angiosperm trees can be thin or fleshy, shiny, dull, or hairy. They can be large or small. Their shape can be simple or compound, linear or lobed, cordate, deltate, ovate, or elliptical, among many examples of variation. On a stem, the leaves can be grouped in clumps, arranged in whorls or extend from individual isolated petioles. As for the regulation and transport of water, angiosperms may possess stomata on one or both sides of a leaf and have either ring porous (e.g., *Quercus* or *Ulmus* spp.) or diffuse porous (e.g., *Betula* or *Acer* spp.) xylem. Gymnosperms, by contrast, have either needles (e.g., *Pinus* spp.) or scales (*Junipers* spp.). Their phytoelements can be arranged in shoots, as with spruce (*Picea* spp.), be comprised of groups of needles on fascicles, as with pine (*Pinus* spp.). From these simple observations, one may surmise that biodiversity could affect rates of transpiration of trees and their annual water budget. But does it?

The answer to this simple question is complicated due to interactions and competition among species for light energy, water and nutrients (e.g., Allen et al. 2002), evolution (Beerling et al. 2001), and to the space and time scale at which it is asked (Waide et al. 1999). Genetic diversity, combined with biogeochemical forcings, produce plant species that differ in physiological and morphological features (Mooney 2001). A specific set of plant features contributes to a ranking of transpiration among tree species because such elements affect the energy balance of leaves and plants. For example, biodiversity, achieved through competition and evolution, is responsible for morphological differences in leaf size, thickness, shape, and reflective properties. These features are important because they affect the aerodynamic resistance and radiative balance of leaves. Species-dependent differences in physiological factors affect transpiration by altering the demand for and supply of water. For exam-

Ecological Studies, Vol. 176
M. Scherer-Lorenzen, Ch. Körner, and E.-D. Schulze (Eds.)
Forest Diversity and Function: Temperate and Boreal Systems

ple, differences in surface resistance arise through species differences in photosynthetic capacity, lifespan (Reich et al. 1997), and maximum stomatal conductance (Schulze et al. 1994; Pataki et al. 2000). Species-dependent differences in the supply of water are mediated by differences rooting depth (Lewis and Burgy 1964; Ehleringer and Dawson 1992; Jackson et al. 1996), timing of physiological activity (Xu and Baldocchi 2003), and drought tolerance or avoidance (Stephenson 1998).

The second question to ask is: do all species in a mixed-species landscape make an independent and proportional contribution to canopy evaporation? The answer to this question can be debated, as there is evidence for and against. On the pro side, an analysis by Currie and Paquin (1987) shows that species richness of trees in North America increases with annual evaporation. A second line of evidence comes from grassland field studies, indicating that aboveground biomass and net primary productivity increase with species richness (Tilman et al. 1997; Hector et al. 1999; Waide et al. 1999; Roy 2001). By inference one could conclude that increasing biodiversity increases evaporation rates and amounts because evaporation of forests scales with net primary productivity (Law et al. 2002).

A contrary view can be derived on the basis of biometeorology and ecological theory. As groups of trees come together and form a closed canopy, the coupling of individual plants with the atmosphere changes (Jarvis and McNaughton 1986). Theoretical and experimental studies show that rates of evaporation, normalized by available energy and temperature, increase with leaf area up to a threshold (a leaf area index of about 3; Jarvis and McNaughton 1986; Saugier and Katerji 1991; Kelliher et al. 1995; Baldocchi and Meyers 1998; Eamus et al. 2001). On ecological grounds, the number and combination of functional factors are limited due to convergent evolution. So the total number of species across a landscape may not be as important as the different number of functional types (Hooper and Vitousek 1997; Tilman et al. 1997; Mooney 2001; Roy 2001). An intermediate view can also be drawn, based on a recent analysis of net primary productivity by Waring et al. (2002). They found that greatest species diversity, along a transect of forests in Oregon, occurs at sites of intermediate productivity.

As a micrometeorologist, I am presupposed to favor the view that the impact of biodiversity on evaporation is realized by how it alters the structural and functional properties of a plant stand, such as its aerodynamic roughness, the amount of transpiring plant material, its physiological capacity to transpire, and its ability to intercept solar radiation. However, I leave the answer to this question to be drawn at the end of this essay. To arrive at a final conclusion, I explore the question of how biodiversity may or may not affect water use of plants by examining theory and experimental data across the scales of leaf, tree, and canopy.

7.2 Leaf Scale

A combination of two physical laws can be used to evaluate evaporation from leaves (Jarvis and McNaughton 1986; Campbell and Norman 1998). One is an analogy to Ohm's Law (Eq. 7.1). It states that evaporation rates are proportional to the product of the leaf conductance and the potential difference in humidity between the leaf surface and the atmosphere:

$$E \sim \frac{g_s g_b}{g_s + g_b}\left(e_s(T_l) - e_a\right) \qquad (7.1)$$

The other important law is the leaf energy balance relationship (Eq. 7.2). It defines how available solar and terrestrial energy is partitioned into sensible heat (H) and latent heat exchange (λE):

$$Q = (1 - \rho)R + \varepsilon L = \varepsilon\sigma T_l^4 + H + \lambda E \qquad (7.2)$$

In these equations, g_s and g_b are the stomatal and boundary layer conductances, respectively, T_l is leaf temperature, e_s is the saturation vapor pressure, e_a is the atmospheric vapor pressure, ρ is leaf reflectance, R and L are flux densities of incoming short- and long-wave energy, respectively, ε is emissivity, σ is the Stefan-Boltzmann coefficient, and λ is the latent heat of evaporation; note that Eq. (7.2) only applies to one side of a leaf.

From Eqs. (7.1) and (7.2), we can identify the plant-specific and climate factors that control leaf evaporation. The plant-specific factors that vary with plant biodiversity and alter leaf evaporation are the stomatal and boundary layer conductances and leaf reflectance. The other variables in Eqs. (7.1) and (7.2) are influenced by weather and the leaf microclimate (Campbell and Norman 1998). Since this paper is focusing on biodiversity, we will discuss the only plant-dependent variables.

On the basis of engineering theory, it can be shown that leaf boundary layer conductance (Eq. 7.3) varies with leaf size, d, is a function of wind speed, u, and is inversely related to the resistance, r_b (Campbell and Norman 1998):

$$g_b = \frac{1}{r_b} \sim const\left(\frac{u}{d}\right)^{1/2} \qquad (7.3)$$

By inspection of Eq. (7.3) we deduce that g_b will increase with a decrease in leaf size, thereby facilitating the transfer of vapor from the leaf to atmosphere.

To discuss and quantify the impact of biodiversity on stomatal conductance, we employ the empirical stomatal conductance model of Collatz et al. (Eq. 7.4; 1991):

$$g_s \sim m\frac{A\cdot rh}{C_s}+b_0 \quad (7.4)$$

where g_s is a function of leaf photosynthesis, A, and the relative humidity, rh, and CO_2 concentration, C_s, at the leaf's surface; m and b_0 are constants. On the basis of this equation and a survey of the ecophysiological literature one can conclude that maximum stomatal conductance for a leaf scales with leaf nitrogen, a correlate with photosynthetic capacity (Fig. 7.1; Schulze et al. 1994; Kelliher et al. 1995; Reich et al. 1997). Hence, leaves with greater amounts of nitrogen, N, attain greater rates of A and g_s. Consequently, they have the potential to transpire at greater rates than nutrient-poor leaves.

The positive relationship between g_s and N is not only a function of species, but is dependent upon climate, soil fertility, and leaf acclimation. For example, leaf thickness, which is positively correlated with photosynthesis, stomatal conductance and leaf nitrogen, increases with light exposure and climatic dryness (Niinemets 2001). At the canopy scale, leaves at the top of a plant stand are thicker and possess more N than do those near the bottom of the canopy (Reich et al. 1997).

To examine and quantify how variations in leaf boundary layer and stomatal resistances interact to alter leaf evaporation, a set of computations with a leaf energy balance model (Paw 1987; Campbell and Norman 1998) were performed. For the case of wide-open stomata (r_s equals 32 s m^{-1}), computations plotted in Fig. 7.2 show that increasing the boundary layer resistance for vapor transfer, r_v, from 2 to 1,024 s m^{-1} forces latent heat exchange, λE, to

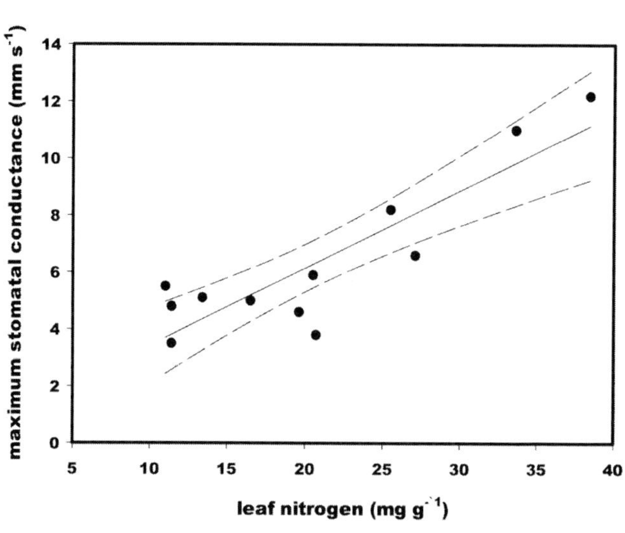

Fig. 7.1. Relationship between maximum stomatal conductance and leaf nitrogen. (Adapted from Schulze et al. 1994)

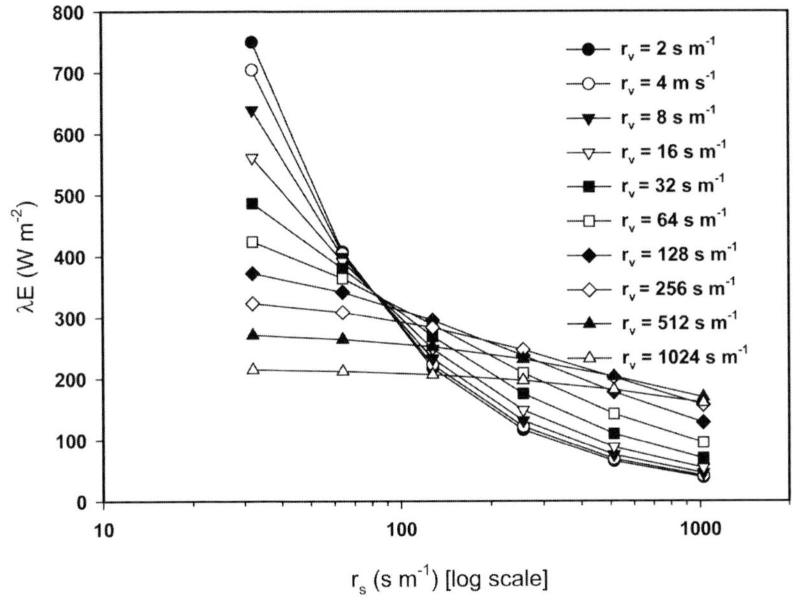

Fig. 7.2. Calculations of leaf latent heat exchange for a range of leaf size, as quantified in terms of the leaf boundary layer resistance for vapor, r_v, and stomatal resistance, r_s. These computations are for a fully sunlit leaf (incoming radiation, Q, equals 1,500 W m^{-2}, air temperature, T_a, is 25 °C and humidity, q, is 15 mg g^{-1})

decrease by 75 %. The sensitivity of λE to changes in leaf size is different when stomata are nearly closed (r_s equal 1,024 s m^{-1}). Under this situation evaporation rates – which are relatively low – increase by a factor of four as r_v increases from 2 to 1,024 s m^{-1}. Evaporation rates increase with high boundary layer resistances, when the stomata are relatively closed, because elevated leaf temperatures strengthen the vapor pressure gradient between the leaf and atmosphere and promote evaporation.

With a theoretical framework in hand, we can make additional assessments on how changes in biodiversity can influence evaporation. In Table 7.1, we catalogue the links between various species-dependent plant characteristics and the energy-balance variables they affect.

Table 7.1. A catalogue of species-dependent variables and the energy-balance variable they influence

Characteristic	Species-dependent attribute	Energy-balance variable
Photosynthetic pathway	C_3,C_4, or CAM photosynthetic pathway, maximum stomatal conductance	C_s, g_s
Leaf size/shape	Needle/planar/shoot	g_b
Stomatal distribution	Amphistomatous/hypostomatous	g_s, C_s
Leaf thickness	Photosynthetic capacity, CO_2 diffusion to chloroplast, light interception	C_s, g_s, ϱ
Leaf surface property	Waxes, pubescence	ϱ, ε, g_b

7.3 Tree Scale

At the tree scale many investigators, using sap flow measurement methods, report that trees of different species transpire at different rates (Granier et al. 1996; Pataki et al. 1998; Wullschleger et al. 1998, 2001; Catovsky et al. 2002). In Fig. 7.3, we see that transpiration rates from pine exceed those of oak (Pataki et al. 1998). In another study, Catovsky et al. (2002) report that red oak and red maple had two- and fourfold, respectively, greater annual water fluxes than eastern hemlock. In a third study, Granier et al. (1996) report a difference by a factor of 4 among transpiration rates of eight tropical tree species in French Guiana. However, is species the independent variable or are other factors such as size, microclimate, and age the cause of the differences in transpiration? In a recent review on tree transpiration, Wullschleger et al. (1998) concluded:

"The relative contribution of each species to stand transpiration was driven largely by sapwood area per unit ground area and to a lesser extent by species-specific differences in daily water use."

Their conclusion is based on the data compiled in Fig. 7.4. Across two orders of magnitude, sapwood area accounts for over 53 % of the variation in tree water use.

What other sources of variation exist with regard to the data in Fig. 7.4? In another paper, Wullschleger et al. (2001) report that there are significant differences in the sapwood area of ring-porous and diffuse-porous trees. For similarly sized individuals, species with diffuse-porous xylem had greater sapwood area, by factors of 3 to 4. So species differences still play a role on limiting transpiration.

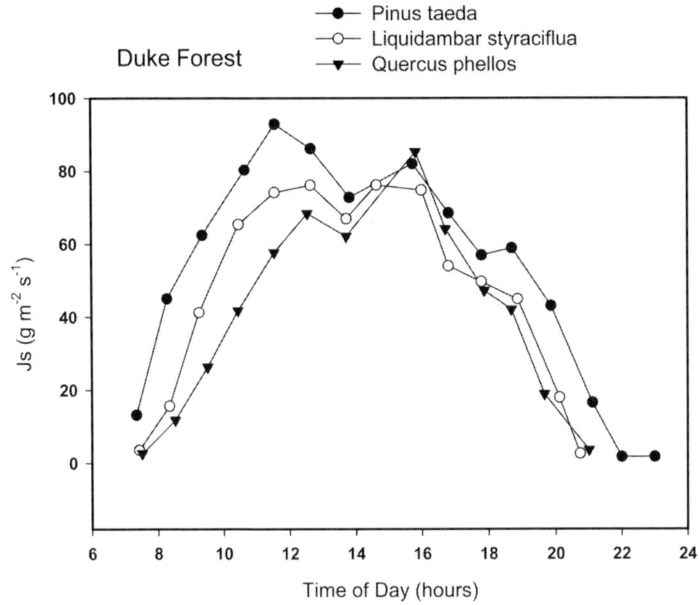

Fig. 7.3. Diurnal pattern of stem sap flow, J_s, of three species growing at Duke forest. (Adapted from Pataki et al. 1998)

On the other hand, one has to be careful about drawing broad conclusions because other groups of investigators have reported that transpiration varies with tree age and height (Vertessy et al. 1995; Ryan et al. 2000; Zimmerman et al. 2000). Acclimation can also be a source of variation, as the leaf to sapwood area ratio area within a species varies with climate factors such as maximum temperature (Mencuccini and Bonosi 2001).

The tree-based synthesis of Wullschleger et al. (1998) is supported by the scaling theory of Enquist et al. (1998). They conclude that xylem sapflow velocity (Q_{xylem}) has a squared power law dependence on diameter (D) and a 3/4 power law dependence upon mass (M), based on the following equations (Eq. 7.5a–c):

$$M_{leaf} \propto M_{stem}^{3/4}$$

$$D \propto M^{3/8}$$

$$Q_{xylem} \propto D^2 \propto M^{3/4} \quad (7.5a, b, c)$$

An appeal of the theory of Enquist et al. (1998) revolves around its ability to successfully predict xylem sapflow rates across four orders of magnitude of tree size, five orders of magnitude in sapflow, and for 37 species (Fig. 7.5). On the other hand, one needs to carefully consider data relating transpiration

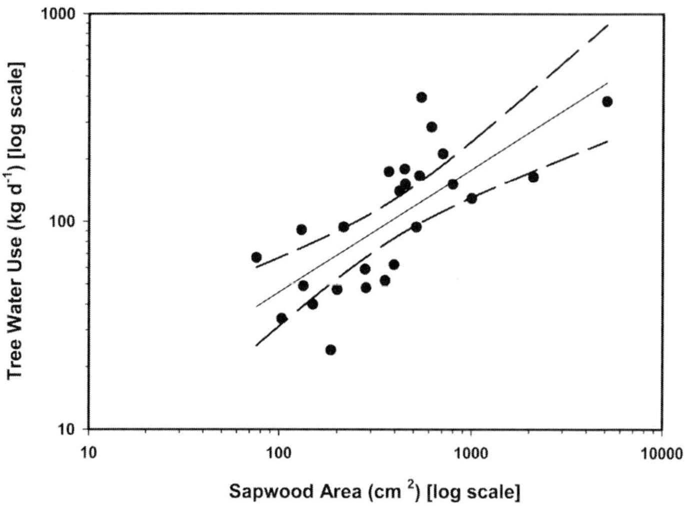

Fig. 7.4. Relation between tree water use and sapwood area (adapted from Wullschleger et al. 1998). Coefficients for the linear regression (*solid line*) are: 0.481 for the intercept, 0.589 for the slope, and 0.529 for the coefficient of determination

sums and stem size or cross section area carefully for they may suffer from autocorrelation (E. Falge, pers. comm.).

7.4 Canopy Scale

While sap flow on a tree basis scales with size and sapwood cross section, which is linked to species differences, what happens when you add up all the evaporation from trees on a landscape?

As noted earlier, Currie and Paquin (1987) reported that annual evaporation of eastern US forests increases with species richness. However, their finding was drawn by taking data from a north-south gradient of forests, which also differed in available energy, temperature, and rainfall. Furthermore, their estimates of evaporation were inferred from the temperature-dependent Thornthwaite equation, as noted in the cited *Climate Atlas of North America*. If one could normalize these data for climate differences, and if one used measured rates of evaporation, would one arrive at the same answer?

New theory on scaling tree information to the landscape scale suggests that species may have little or no effect on transpiration. Enquist (2002) shows that xylem flow on the land surface basis, Q_{land}, is scale invariant. The result occurs because self-thinning and competition for limited resources causes the maximum number of trees, N_{max}, to scale with the –3/4 power of plant mass, M, and

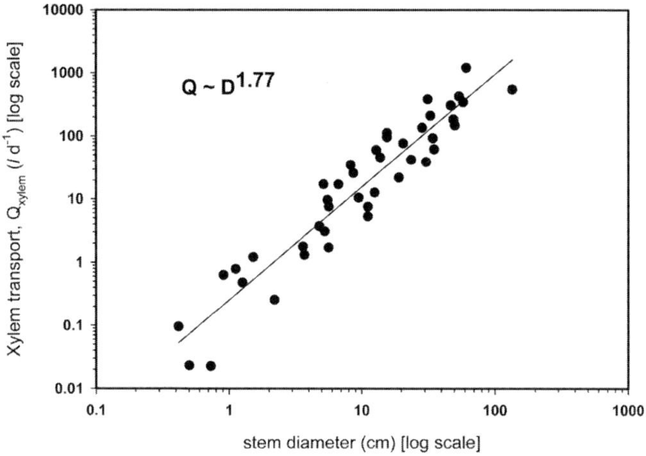

Fig. 7.5. Relationship between the rate of xylem transport of water, Q_{xylem}, and stem diameter, D. Data are from 37 species. (Adapted from Enquist et al. 1998)

sap-flow rates of individual trees scale with the 3/4 power of plant mass (Eq. 7.6):

$$N_{max} \propto M^{-3/4}$$

$$Q_{xylem} \propto M^{3/4} \therefore$$

$$Q_{land} \propto N_{max} \cdot Q_{xylem} \propto M^{0} \tag{7.6}$$

Enquist and coworkers (Enquist et al. 1998; Enquist 2002) have validated this theory using data across 12 orders of magnitude of plant size (Fig. 7.6). However, close inspection of Fig. 7.6 shows that for a given size class the range of data on xylem flux ranges across two orders of magnitude. So the explanation for these sources of variation remains unsolved by scaling theory. There is also a need to normalize transpiration data for temperature and available energy.

To address the source of variation in transpiration within the class of a landscape, we draw on canopy evaporation data from micrometeorological field studies and models. Working with colleagues in the boreal forest of Canada during the BOREAS experiment, we reported that dissimilar rates of evaporation occurred from aspen/hazel, black spruce, and jack pine forest stands when they were exposed to the same weather (the forest stands were located within 50 km of each other; Baldocchi et al. 2000). The greatest evaporation rates emanated from the broadleaved aspen/hazel stand, due to its greater leaf area index. Next in rank was evaporation from a black spruce stand growing on a water-logged habitat, followed by evaporation from a jack

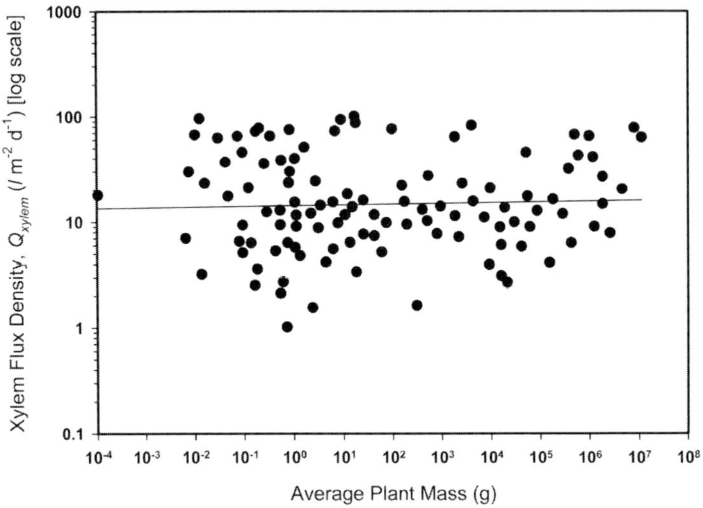

Fig. 7.6. The relationship between landscape scale xylem flux and plant size. (Adapted from Enquist et al. 1998)

pine forest, which grew on a dry, sandy upland habitat. Drawing on evaporation data from the Siberian boreal forest produces a similar conclusion. Kelliher et al. (2001) reported that differences in tree life form and nitrogen contributed to differences in evaporation. They report that greatest evaporation rates were associated with deciduous broadleaved trees, followed in sequence by deciduous needle-leaved and evergreen needle-leaved trees.

At the canopy scale, the Penman-Monteith Equation can be used to identify and quantify the factors governing rates of evaporation to the atmosphere (Jarvis and McNaughton 1986). Consequently, it (Eq. 7.7) can be used as a tool to inquire how biodiversity may or may not affect stand evaporation rates.

$$\lambda E = \frac{s(R_n - S) + \rho \cdot C_p \cdot G_H \cdot D}{s + \gamma + \gamma \dfrac{G_H}{G_s}} \tag{7.7}$$

In this equation, R_n is net-radiation flux density, S is the soil-heat flux density, G_h and G_s are the canopy-scale conductances for boundary layer and surface, D is vapor pressure deficit, s is the slope of the saturation vapor pressure–temperature curve, and γ is the psychrometric constant. The Penman-Monteith equation can also be inverted to compute the canopy surface conductance (Kelliher et al. 1995; Baldocchi and Meyers 1998).

By comparing measured evaporation rates, normalized by the rate of equilibrium evaporation (Eq. 7.8),

$$\lambda E_{eq} = \frac{s}{s+\gamma}(R_n - S), \tag{7.8}$$

against estimates of surface resistance (R_c=1/G_c) one can investigate and quantify how biodiversity affects evaporation rates, at the functional level. Figure 7.7 shows that normalized forest evaporation rates vary markedly among forest stands. In general, normalized evaporation rates decrease with increasing surface resistance. Alternatively, we observe that $\lambda E/\lambda E_{eq}$ approaches the value of the Priestly-Taylor coefficient (1.26) as surface resistance drops below 40 s m^{-1}, a value representative of the evaporation ratio for extended, healthy, and well-watered crop surfaces. It is also an observation that is consistent with the theoretical predictions of McNaughton and Spriggs (1986).

Variations in the surface resistance of forest canopies are due to variations in leaf area index, maximum stomatal conductance, and drought (Kelliher et

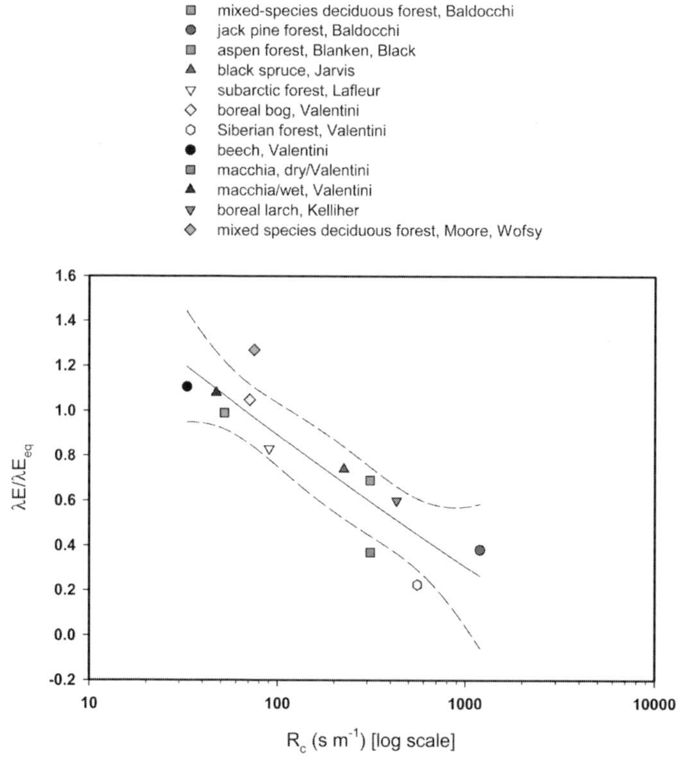

Fig. 7.7. Relationship between latent heat exchange rates of forests, λE, (normalized by the equilibrium rate, λE_{eq}) and their canopy surface resistance, R_c. This figure is adapted from Valentini et al. (1999) and Baldocchi et al. (2000). Listed with each forest stand is the principal investigator of that site. Other sources of data and relevant literature citations are listed in Table 7.2

Table 7.2. List of field sites, species, and citations of data used in Figs. 7.8 and 7.9

Site	Key species, >5 % of number or area	Reference
Prince Albert, Sask.	*Populus tremuloides, Corylus cornuta*	Blanken et al. (2001)
Borden, Ont.	*Populus grandidentata, Acer rubrum, Populus tremuloides, Fraxinus americana, Pinus strobus*	Lee et al. (1999); Staebler et al. (2000); AmeriFlux data of Lee and Fuentes
Petersham, MA	*Quercus rubra, Acer rubrum, Tsuga canadensis, Pinus strobus, Pinus resinosa*	Moore et al. (1996); Barford et al. (2001); Ameriflux data of Wofsy and Munger
Hesse, France	*Fagus sylvatica, Betula pendula, Quercus petraea, Larix decidua, Prunus avium, Fraxinus excelsior, Carpinus betulus*	Granier et al. (2000)
Oak Ridge, TN	*Quercus alba, Q. prinus, Q. rubra, Q. velutina, Acer rubrum, Liriodendron tulipifera, Pinus echinata*	Hutchison et al. (1986); Wilson and Baldocchi (2000); Falge and Schindler (unpubl.)
	P. strobus	
	P. virginia, Carya, Fagus, Juglans, Cornus, Fraxinus	
Pellston, MI	*Populus grandidentata, Pinus strobus, Quercus rubra, Acer rubrum, Betula papyrifera*	Curtis et al. (2002); AmeriFlux data of Curtis et al.

al. 1995; Baldocchi and Meyers 1998). Since these factors are modulated by species (but can also be influenced by ecological and biogeochemical factors), the begging question is whether or not forest evaporation correlates with species diversity?

We address this question next by using the FLUXNET database (http://www-eosdis.ornl.gov/FLUXNET/; Baldocchi et al. 2001) and data published in the literature (Table 7.2). We narrow the scope of this analysis and its susceptibility to confounding factors by focusing only on data from broadleaved deciduous forests. In addition, we confine the analysis to the summer growing season, when the forests have adequate soil moisture and form closed canopies.

We submit a priori that all species do not contribute equally to canopy evaporation. Instead we intend to count the number of species that affect ecosystem functioning most, through their ability to capture and use soil resources (Chapin et al. 1997; Roy 2001). As a first guess, we consider the number of species that make up more the 5 % of the stand on a stem- or leaf-area basis. Using this metric, we observe that normalized evaporation rates decrease as the number of species increase. We also observe that 93 % of the variance in the dependent variable is explained with this independent variable (Fig. 7.8).

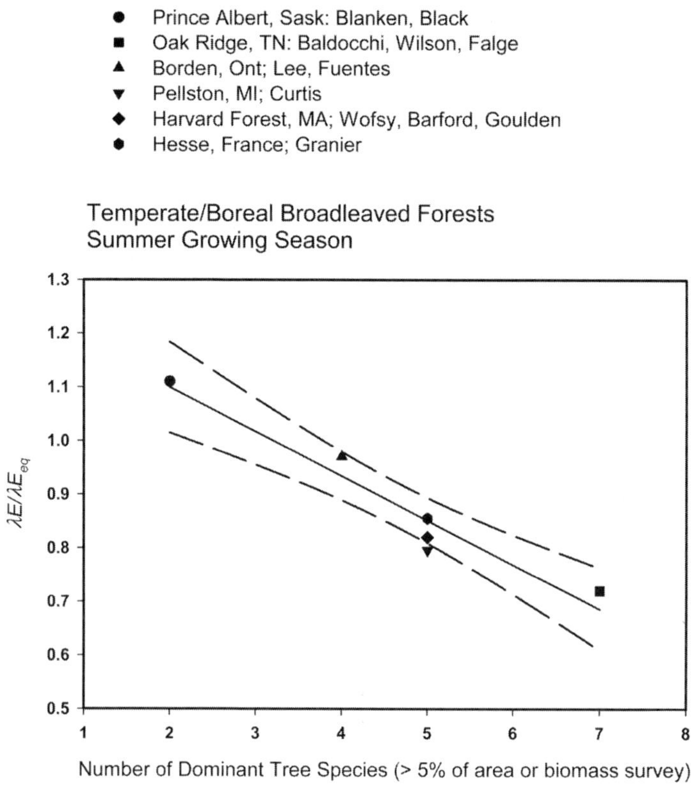

Fig. 7.8. The relationship between normalized evaporation and dominant species number of temperate and boreal deciduous broadleaved forest stands. The coefficients of the linear regression are: 1.26 for the intercept, –0.082 for the slope, and 0.932 for the coefficient of determination. Sources of data are listed in Table 7.2

The choice of 5% of the total stem or leaf area, as the cutoff value for counting the number of key species, is arbitrary and merits further inquiry. If we consider the total number of species in a stand instead, we find that the coefficient of determination of the linear regression with the dependent variable ($\lambda E/\lambda E_{eq}$) is reduced markedly, decreasing from 0.93 to 0.53 (Fig. 7.9). In future, using more sophisticated analyses, one will need to consider wind direction and the composition of vegetation within the defined flux footprint.

Why does normalized evaporation decrease with species diversity? One can hypothesize that deciduous broadleaved stands, composed of greater species diversity, have a greater mix of ring-porous and diffuse-porous trees that possess different abilities to transfer water and transpire (Wullschleger et al. 2001). There may also be an effect of the nitrogen economy of the stand on the surface conductance, as a limited pool of nitrogen must be distributed

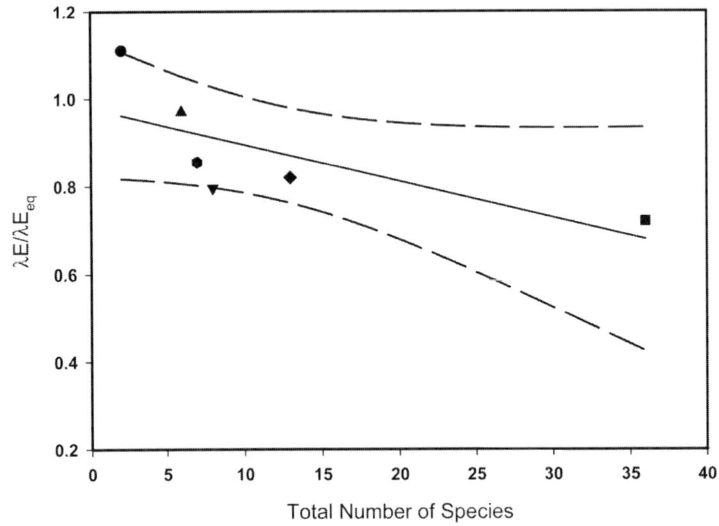

Fig. 7.9. Relationship between normalized latent heat exchange and total number of species in temperate broadleaved forests. Coefficient of variation, r^2, is 0.53. The evaporation data come from summer periods and are averaged over the course of days

among more species and functional types, thereby reducing g_s and G_c. This second hypothesis is proffered on the basis of work by Hooper and Vitousek (1997). They reported that a prevailing functional type or species may dominate the pools of available N and lower the amount available to other species. If this effect occurs in forests, stomatal conductance, and leaf area index will be lower for the canopies with higher numbers of species, and so will their canopy conductance.

It is doubtful that the results shown in Fig. 7.8 are an artifact of different leaf area, since all the stands are closed, and it is known that normalized evaporation rates become saturated at high leaf-area indices. Weather effects, radiation, and temperature, are discounted by the normalization with λE_{eq} and by considering ecosystems in relatively temperate and humid climates. In any event, these results are preliminary and merit further scrutiny as more data are entered into the FLUXNET database.

7.5 Conclusions

In this paper we have discussed the impact of biodiversity on forest evaporation at three scales: leaf, tree, and canopy. Whether or not biodiversity affects evaporation may come down to an argument on semantics. By applying biophysical theory, we show that it is the functional plant variables, e.g., stomatal and boundary layer conductance, hydraulic conductivity, rooting depth, leaf reflectivity, that affect evaporation. So in this context one may argue that functional diversity affects evaporation more than species diversity. On the other hand, evolutionary pressures have forced different species to adopt different functional features. While many different species may transpire at different rates, issues relating to functional convergence can cause many different species to transpire at the same rate, given similar environmental forcings.

At the landscape scale, an evaluation of eddy flux data suggests that increasing biodiversity may result in lower rates of normalized evaporation, but the pool size of the data is small and this hypothesis needs revisiting as more data become available. It also is in contrast with the scale invariant hypothesis of Enquist et al. (1998) and a correlative analysis by Currie and Paquin (1987). However, the variance of information shown at particular class sizes by Enquist et al. (1998) is huge, ranging two orders of magnitude, and the analysis of Currie and Paquin (1987) may be an artifact of using indirect estimates of evaporation, leaving plenty of room for observations reported here.

Acknowledgments: The author expresses thanks to the many sources of eddy covariance and forest biometry data: Oak Ridge: Kell Wilson, Eva Falge; Hesse, France: Andre Granier; Borden, Ontario: Xuhui Lee, Jose Fuentes; Harvard Forest: Steve Wofsy, Mike Goulden, Carol Barford, Bill Munger; and Pellston Michigan: Peter Curtis. The author is very grateful to Stan Wullschleger and Eva Falge for providing very helpful internal reviews of this document. The author also thanks Michael Scherer-Lorenzen and Detlef Schulze for inviting him to the workshop to discuss this topic and for providing expert editorial comments.

The author is supported by the NASA Carbon Science Program and is a contributor to the FLUXNET Project and FLUXNET's participation in the Diversitas, International Biodiversity Observation Year (IBOY).

References

Allen AP, Brown JH, Gillooly JF (2002) Global biodiversity, biochemical kinetics and energetic-equivalence rule. Science 297:1545–1548

Baldocchi DD, Meyers TP (1998) On using eco-physiological, micrometeorological and biogeochemical theory to evaluate carbon dioxide, water vapor and gaseous deposition fluxes over vegetation. Agric For Meteor 90:1–26

Baldocchi DD, Kelliher FM, Black TA, Jarvis PG (2000) Climate and vegetation controls on boreal zone energy exchange. Global Change Biol 6 (Suppl 1):69–83

Baldocchi DD, Falge E, Gu L, Olson R, Hollinger D, Running S, Anthoni P, Bernhofern C, Davis K, Fuentes J, Goldstein A, Katul G, Law B, Lee X, Malhi Y, Meyers T, Munger JW, Oechel W, Pilegaard K, Schmid HP, Valentini R, Verma S, Vesala T, Wilson K, Wofsy S (2001) FLUXNET: a new tool to study the temporal and spatial variability of ecosystem-scale carbon dioxide, water vapor and energy flux densities. Bull Am Meteor Soc 82:2415–2434

Barford CC, Wofsy SC, Goulden ML, Munger JW, Pyle EH, Urbanski SP, Hutyra L, Saleska SR, Fitzjarrald D, Moore K (2001) Factors controlling long- and short-term sequestration of atmospheric CO_2 in a mid-latitude forest. Science 294:1688–1691

Beerling DJ, Osborne CP, Chaloner WG (2001) Evolution of leaf form in land plants linked to atmospheric CO_2 decline in the late Palaeozoic era. Nature 410:352–354

Blanken PD, Black TA, Neumann HH, den Hartog G, Yang PC, Nesic Z, Lee X (2001) The seasonal water and energy exchange above and within a boreal aspen forest. J Hydrol 245:118–136

Campbell GS, Norman JM (1998) An introduction to environmental biophysics. Springer, Berlin Heidelberg New York

Catovsky S, Holbrook NM, Bazzaz FA (2002) Coupling whole-tree transpiration and canopy photosynthesis in coniferous and broadleaved tree species. Can J For Res 32:295–309

Chapin FS, Walker BH, Hobbs RJ, Hooper DU, Lawton JH, Sala OE, Tilman D (1997) Biotic control over the functioning of ecosystems. Science 277:500–504

Collatz GJ, Ball JT, Grivet C, Berry JA (1991) Regulation of stomatal conductance and transpiration: a physiological model of canopy processes. Agric For Meteor 54:107–136

Currie DJ, Paquin V (1987) Large scale biogeographical patterns of species richness of trees. Nature 329:326–327

Curtis PS, Hanson PJ, Bolstad P, Barford C, Randolph JC, Schmid HP, Wilson KB (2002) Biometric and eddy-covariance based estimates of net ecosystem carbon exchange in five eastern North American deciduous forests. Agric For Meteor 113:3–20

Eamus D, Hutley LB, O'Grady AP (2001) Daily and seasonal patterns of carbon and water fluxes above a north Australian savanna. Tree Physiol 21:977–988

Ehleringer JR, Dawson TE (1992) Water uptake by plants: perspectives from stable isotope composition. Plant Cell Environ 15:1073–1082

Enquist BJ (2002) Universal scaling in tree and vascular plant allometry: toward a general quantitative theory linking plant form and function from cells to ecosystems. Tree Physiol 22:1045–1064

Enquist BJ, Brown JH, West GB (1998) Allometric scaling of plant energetics and population density. Nature 395:163–165

Granier A, Biron P, Breda N, Pontailler J-Y, Saugier B (1996) Transpiration of trees and forest stands: short and long term monitoring using sapflow methods. Global Change Biol 2:265–274

Granier A, Ceschia E, Damesin C, Dufrene E, Epron D, Gross P, Lebaube S, Le Dantec V, Le Goff N, Lemoine D, Lucot E, Ottorini JM, Pontailler JY, Saugier B (2000) The carbon balance of a young beech forest. Funct Ecol 14:312–325

Hector A, Schmid B, Beierkuhnlein C, Caldeira MC, Diemer M, Dimitrakopoulus PG, Finn JA, Freitas H, Giller PS, Good J, Harris R, Högberg P, Huss-Danell K, Joshi J, Jumpponen A, Körner C, Leadley PW, Loreau M, Minns A, Mulder CPH, O'Donovan G, Otway SJ, Pereira JS, Prinz A, Read DJ, Scherer-Lorenzen M, Schulze ED, Siamantziouras ASD, Spehn EM, Terry AC, Troumbis AY, Woodward FI, Yachi S, Lawton JH (1999) Plant diversity and productivity experiments in European grasslands. Science 286:1123–1127

Hooper DU, Vitousek PM (1997) The effects of plant composition and diversity on ecosystem processes. Science 277:1302–1304

Hutchison BA, Matt DR, McMillen RT, Gross LJ, Tajchman SJ, Norman JM (1986) The architecture of a deciduous forest canopy in eastern Tennessee, USA. J Ecol 74:635–646

Jackson RB, Canadell J, Ehleringer JR, Mooney HA, Sala OE, Schulze ED (1996) A global analysis of root distributions for terrestrial biomes. Oecologia 108:389–411

Jarvis PG, McNaughton KG (1986) Stomatal control of transpiration. Adv Ecol Res 15:1–49

Kelliher FM, Leuning R, Raupach MR, Schulze ED (1995) Maximum conductances for evaporation from global vegetation types. Agric For Meteor 73:1–16

Kelliher FM, Lloyd J, Rebmann C, Wirth C, Schulze ED, Baldocchi DD (2001) Evaporation in the boreal zone during summer: physics and vegetation. In: Schulze ED, Heimann M, Harrison S, Holland E, Lloyd J, Prentice IC, Schimel D (eds) Global biogeochemical cycles in the climate system. Academic Press, San Diego, pp 151–165

Law BE, Falge E, Gu L, Baldocchi DD, Bakwin P, Berbigier P, Davis K, Dolman AJ, Falk M, Fuentes JD (2002) Environmental controls over carbon dioxide and water vapor exchange of terrestrial vegetation. Agric For Meteor 113:97–120

Lee X, Fuentes JD, Staebler RM, Neumann HH (1999) Long-term observation of the atmospheric exchange of CO_2 with a temperate deciduous forest. J Geophys Res 104:15975–15984

Lewis DC, Burgy RH (1964) The relationship between oak tree roots and ground water in fractured rock as determined by tritium tracing. J Geophys Res 69:2579–2588

McNaughton KG, Spriggs TW (1986) A mixed layer model for regional evaporation. Boundary Layer Meteor 34:243–262

Mencuccini M, Bonosi L (2001) Leaf/sapwood ratios in Scots pine show acclimation across Europe. Can J For Res 31:442–456

Mooney HA (2001) Biological diversity, evolution and biogeochemistry. In: Schulze ED, Heimann M, Harrison S, Holland E, Lloyd J, Prentice IC, Schimel D (eds) Global biogeochemical cycles in the climate system. Academic Press, San Diego, pp 279–284

Moore KE, Fitzjarrald DR, Sakai RK, Goulden ML, Munger JW, Wofsy SC (1996) Seasonal variation in radiative and turbulent exchange at a deciduous forest in central Massachusetts. J Appl Meteor 35:122–134

Niinemets U (2001) Global-scale climatic controls of leaf dry mass per area, density and thickness in trees and shrubs. Ecology 82:453–469

Pataki DE, Oren R, Katul G, Sigmon J. (1998) Canopy conductance of *Pinus taeda*, *Liquidambar styraciflua* and *Quercus phellos* under varying atmospheric and soil moisture conditions. Tree Physiol 18:307–315

Pataki DE, Oren R, Smith WK (2000) Sap flux of co-occurring species in a western subalpine forest during seasonal soil drought. Ecology 81:2557–2566

Paw U (1987) Mathematical analysis of the operative temperature and energy budget. J Therm Biol 12:227–233

Reich PB, Walters MB, Ellsworth DS (1997) From tropics to tundra: global convergence in plant functioning. Proc Natl Acad Sci USA 94:13730–13734

Roy J (2001) How does biodiversity control primary productivity. In: Roy J, Saugier B, Mooney HA (eds) Terrestrial global productivity. Academic Press, San Diego, pp 169–186

Ryan MG, Bond BJ, Law BE, Hubbard RM, Woodruff D, Cienciala E, Kucera J. (2000) Transpiration and whole tree conductance in ponderosa pine trees of different height. Oecologia 124:553–560

Saugier B, Katerji N (1991) Some plant factors controlling evaporation. Agric For Meteor 54:263–277

Schulze ED, Kelliher FM, Körner C, Lloyd J, Leuning R (1994) Relationships between maximum stomatal conductance, ecosystem surface conductance, carbon assimilation rate and plant nitrogen nutrition: a global exercise. Annu Rev Ecol Syst 25:629–660

Staebler RM, Fuentes JD, Lee X, Puckett KJ, Neumann HH, Deary MJ, Arnold JA (2000) Long term flux measurements at the Borden forest. Can Meteor Oceanogr Bull 28:9–16

Stephenson N (1998) Actual evapotranspiration and deficit: biologically meaningful correlates of vegetation distribution across spatial scales. J Biogeogr 25:855–870
Tilman D, Knops J, Wedin D, Reich P, Ritchie M, Seimann E. (1997) The influence of functional diversity and composition on ecosystem processes. Science 277:1300–1302
Valentini R, Baldocchi DD, Tenhunen J (1999) Ecological controls on land surface atmospheric interactions. Integrating hydrology, ecosystem dynamics and biogeochemistry in complex landscapes. Dahlem Konferenzen, pp 117–146
Vertessy RA, Benyon RG, O'Sullivan SK, Gribben PR (1995) Relationships between stem diameter, sapwood area and transpiration in a young mountain ash forest. Tree Physiol 15:559–567
Waide RB, Willig MR, Steiner CF, Mittelbach G, Gough L, Dodson SI, Juday GP, Parmenter R (1999) The relationship between productivity and species richness. Annu Rev Ecol Syst 30:257–300
Waring RH, Coops NC, Ohmann JL, Sarr DA. (2002) Interpreting woody plant richness from seasonal ratios of photosynthesis. Ecology 83:2964–2970
Wilson KB, Baldocchi DD (2000) Seasonal and interannual variability of energy fluxes over a broadleaved temperate deciduous forest. Agric For Meteor 100:1–18
Wullschleger SD, Meinzer FC, Vertessy RA (1998) A review of whole-plant water use studies in trees. Tree Physiol 18:499–512
Wullschleger SD, Hanson PJ, Todd DE (2001) Transpiration from a multi-species deciduous forest as estimated by xylem sap flow techniques. For Ecol Manage 143:205–213
Xu L, Baldocchi DD (2003) Seasonal trend of photosynthetic parameters and stomatal conductance of blue oak (*Quercus douglasii*) under prolonged summer drought and high temperature. Tree Physiol 23, 865–877
Zimmerman R, Schulze ED, Wirth C, Schulze EE, McDonald KC, Vygodskaya NN, Ziegler W (2000) Canopy transpiration in a chronosequence of central Siberia pine forests. Global Change Biol 6:25–37

8 Effects of Tree Species Diversity on Litter Quality and Decomposition

S. Hättenschwiler

8.1 Introduction

During the process of decomposition, dead organic matter is physically and chemically broken down. The conversion from dead organic matter into CO_2 and inorganic nutrients available for plant and microbial uptake is a fundamentally important ecosystem process (Swift et al. 1979; Cadish and Giller 1997). Decomposition exhibits a major control over the carbon cycle, and thus over the atmospheric CO_2 concentration and the global climate. Aging and weathering of soils and the underlying bedrock lead to the depletion of mineral sources of nutrients for plant growth in late successional ecosystems and, therefore, ecosystem functioning increasingly depends on recycled nutrients from decomposition. Climate, soil organisms, and the amount and kind of plant litter input are widely accepted as the three main controlling factors over decomposition (Swift et al. 1979; Coûteaux et al. 1995; Cadish and Giller 1997).

The chemical and physical properties of plant litter have a major influence on nutrient cycling and accumulation of soil organic matter (SOM) within a particular ecosystem, and hence on the properties and functioning of that ecosystem. The high correlation between litter quality and decomposition has convincingly been demonstrated in many detailed studies on litter decay and nutrient release from predominantly monospecific litter material. Since most terrestrial ecosystems are composed of a variety of different plant species, each contributing to the annual litter input, it is evident that the composition of the litter pool has a strong impact on overall litter decomposition. The role of diversity and identity of litter species for decomposition and nutrient cycling, however, is not well known and is not normally taken into account in biogeochemical models.

The aim of this chapter is to summarize the current knowledge on the functional significance of tree species diversity for litter quality and decomposition in forest ecosystems. In the first part, I will give a brief account of

Ecological Studies, Vol. 176
M. Scherer-Lorenzen, Ch. Körner, and E.-D. Schulze (Eds.)
Forest Diversity and Function: Temperate and Boreal Systems

inter- and intraspecific variability in litter traits and rates of decomposition. I will then summarize the results from decomposition experiments specifically designed to address the question whether or not decay rate and nutrient mineralization are influenced by species diversity of tree litter. The final part provides conclusions and prospects for future research.

8.2 Variation in Litter Traits and Decomposition Rates

In terrestrial ecosystems 10–30 % of net primary production (NPP) enters the aboveground litter layer annually (fine litterfall, i.e., leaves and reproductive parts) with approximately the same amount entering the soil as fine-root litter (Chapin et al. 2002). Together this may sum up to the total input of dead plant dry mass of about 800 g m^{-2} $year^{-1}$ (8 tons ha^{-1} $year^{-1}$) in temperate forests and 1,200 g m^{-2} $year^{-1}$ (12 tons ha^{-1} $year^{-1}$) in tropical forests, excluding the contribution of dead woody tissue. For an assessment of carbon and nutrient fluxes at the ecosystem and landscape level, the litter input is usually treated as a single and homogeneous pool characterized by its mean chemical composition. This appears to be appropriate for estimates of decomposition and elemental budgets and cycling at large scales. At smaller scales, however, spatial and temporal variation in litter composition and its quality might be of importance for decomposition processes and nutrient cycling. How variable is litter quality and decomposition within a forest site? What is the functional significance of this variation? These kinds of questions are of immediate relevance for the evaluation of how biodiversity affects ecosystem functioning.

8.2.1 Inter- and Intraspecific Variation in Litter Quality

Leaf chemical composition and physical quality vary tremendously among plant functional types and species (e.g., Perez-Harguindeguy et al. 2000). Part of the variation is related to differences in leaf life span. Nutrient concentrations and specific leaf area (often related to leaf toughness) are typically low in leaves of a long life span, such as conifer needles. In addition, leaf production and accumulation of secondary metabolites, such as lignin and tannins, generally increase as the life span increases. A surprisingly wide variation in leaf quality traits was found even within functionally narrow groups of species, such as broadleaf deciduous trees from temperate forests having essentially the same leaf life span and occurring at the same site (Ricklefs and Matthew 1982). For example, nitrogen concentration varies by a factor of more than two, lignin concentration by a factor of more than five, and calcium concentration by a factor of more than six in leaves among 34 North Ameri-

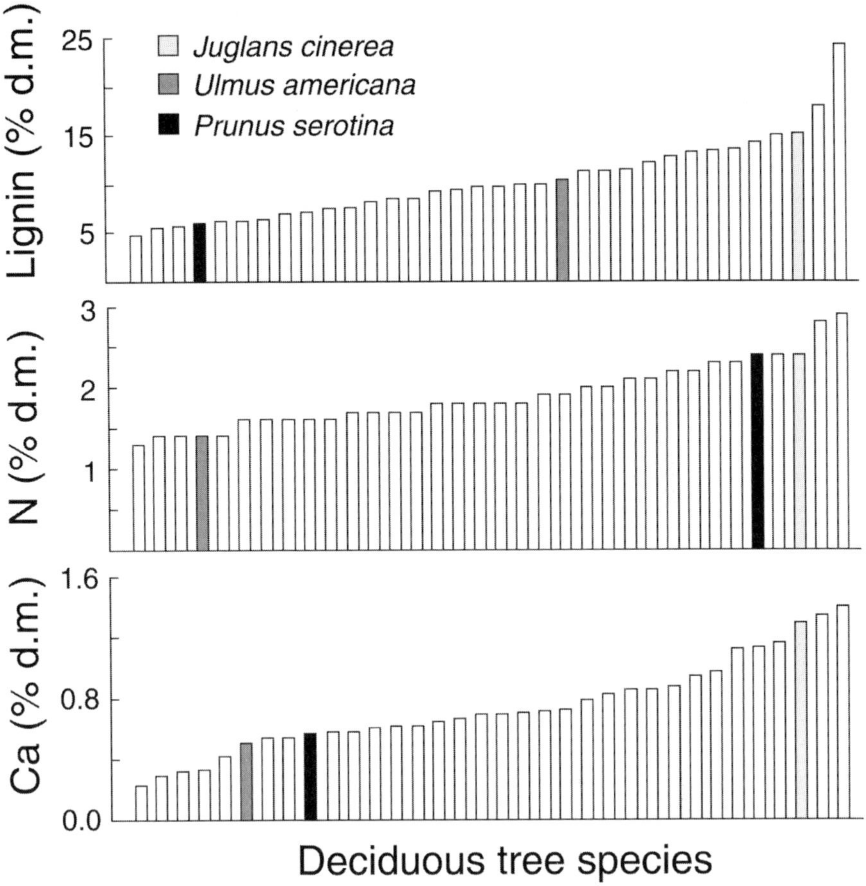

Fig. 8.1. Variation in leaf chemical composition among 34 temperate deciduous forest tree species. Trees were sampled within 24 km in southeastern Ontario. Data reproduced with permission from Ricklefs and Matthew (1982)

can tree species (Fig. 8.1). Moreover, the different compounds and nutrients do not necessarily correlate with each other, creating a variety of elemental ratios. *Prunus serotina*, for example, has a low lignin concentration but a high nitrogen concentration. *Juglans cinerea*, in contrast, has both a high lignin and a high nitrogen concentration, and finally *Ulmus americana* has an intermediate lignin and a low nitrogen concentration. The overall leaf quality and ultimately the litter quality after senescence within a tree species is defined by the relative amounts of many different compounds and nutrients, rather than by the absolute content of a single nutrient.

While variation in litter traits among species is usually well appreciated, variation within species is less frequently considered, but actually may exceed

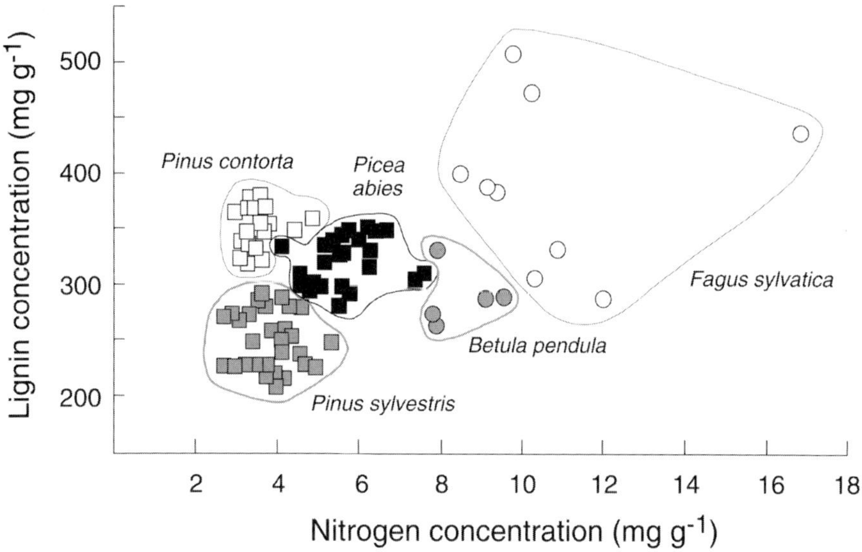

Fig. 8.2. Lignin and nitrogen concentrations in fresh fallen leaf litter of five different tree species sampled across large geographical areas. Reproduced with permission from Berg (2000)

that among species (Fig. 8.2; Berg 2000). *Fagus sylvatica*, for example, exhibits differences in lignin concentration from 25–50 % and nitrogen concentration ranges from 0.8–1.7 % among different sites sampled over large geographical regions. As Berg (2000) pointed out, the different species graphically displayed in Fig. 8.2 form distinct clusters with hardly any overlap along the two axes of lignin and nitrogen concentrations, a fact relevant for the characterization of among-species variation. The differences in litter quality within species (Fig. 8.2) may largely reflect phenotypic variation as a result of the variability in environmental factors and/or biotic interactions across different sites over large areas. A considerable genotypic variation in litter traits, however, has also been observed. For example, polyphenol concentrations in leaf and fine-root litter vary inherently among four distinct populations of *Metrosideros polymorpha*, the dominant tree species in Hawaiian montane forests (Hättenschwiler et al. 2003). Because these forests grew under virtually the same climatic conditions, genotypic variation in polyphenol concentration is most likely related to the large differences in soil age, ranging from 0.3 to 4,100 ka, and fertility. Long-term experimental fertilization, however, did not affect polyphenol concentration at either site, excluding immediate responses to nutrient availability at the level of the phenotype. Interestingly, polyphenol concentrations in leaf litter and fine-root litter varied among populations, suggesting different selection forces and functional roles of leaf-derived versus root-derived polyphenols. Even within a single forest stand, genotypic variation in litter traits can affect decomposition and C and N cycling, as has recently been shown for *Quercus* trees in the southeastern USA (Madritch and Hunter 2002).

It can be concluded that litter quality varies widely among tree species, and also within species due to phenotypic and genotypic differentiation. Thus, biological diversity at both levels, species and genotype, is likely to have significant implications for decomposition and nutrient cycling in forest ecosystems.

8.2.2 Variation in Decomposition Rates

Climate exhibits a major control over rates of decomposition at large geographical scales. Actual evapotranspiration, integrating the effects of precipitation and temperature, is most widely used as an index of climate and can explain a variation of about 49 to 60 % in observed decomposition of a single uniform litter type among sites at regional (Vitousek et al. 1994), continental (Berg et al. 1993), and global scales (Gholz et al. 2000). Testing two different litter species (broadleaf vs. conifer) of contrasting initial quality across a broad global gradient, Gholz et al. (2000) found considerably slower decomposition of the poor-quality conifer litter compared to the high-quality broadleaf litter, indicating an additional strong control of substrate quality over decomposition along the entire gradient of actual evapotranspiration. Litter decay rates differ widely among species decomposing under identical environmental conditions (Cornelissen 1996; Wardle et al. 1997). For example, leaf litter from 16 different European tree species at the same site showed a fivefold difference in decomposition rate (Fig. 8.3.). These differences in decomposition are attributed to variation in litter quality, such as the ratio of carbon to nitrogen and its consequences for microbial activity and substrate utilization. The C/N ratio is widely used as an indicator for tissue quality and often correlates well with decomposition rates (Taylor et al. 1989a; Fig. 8.3). Several other litter properties have also been identified that correlate with mass loss, such as lignin concentration (McClaugherty and Berg 1987), lignin/N ratio (Melillo et al. 1982), and leaf toughness (Perez-Harguindeguy et al. 2000), exhibiting a negative relationship with rates of decomposition.

Based on empirical studies relating decomposition with initial litter quality, litter traits can be used to predict rates of decomposition across species (Aber et al. 1990) and also to serve as input variables in ecosystem C models. However, since single litter quality characteristics do not accurately represent all the different processes involved in litter decay, considerable unexplained variation in decomposition rates among species remains. Moreover, there is good evidence that the correlation between initial litter quality and decay rate changes during the course of decomposition. While high initial litter N concentration correlates positively with litter mass loss rate in the early stage of

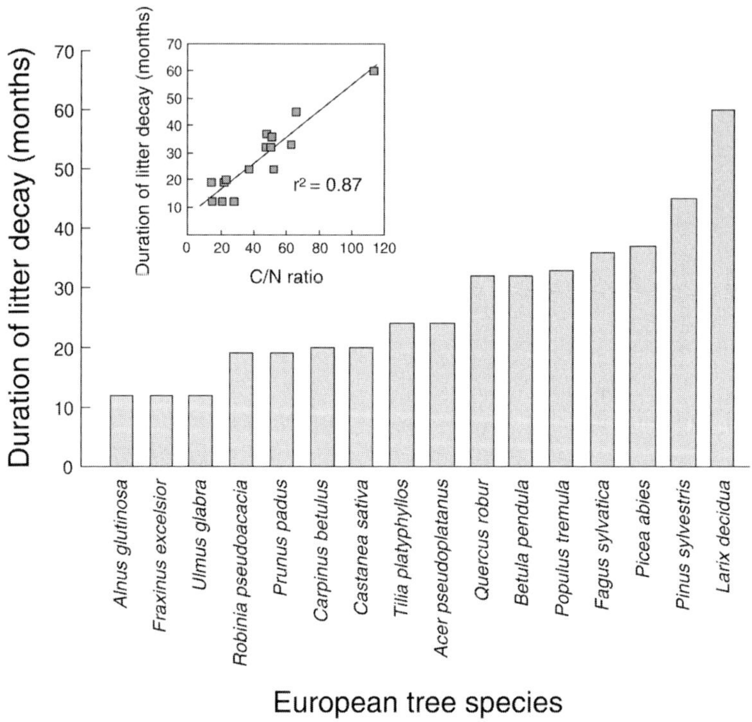

Fig. 8.3. Average duration of decomposition of leaf litter from different European tree species exposed in the field under identical environmental conditions. The duration of litter decomposition correlates positively with the C/N ratio, i.e., the wider the C/N ratio, the slower decomposition proceeds (*insert*). Data from Ellenberg (1986) and modified in presentation

decomposition, this correlation later becomes negative (Berg 2000). In his "limit-value concept" Berg (2000) provides evidence for a higher proportion of litter mass converted to the pool of stable organic matter from litter material rich in N, despite its faster initial decomposition, compared to poor-quality litter of low initial N concentration. Deviations from fitted regressions between litter quality and decomposition may be of minor importance in global production/decomposition models using litter quality parameters as key variables, but are likely to increase in significance at decreasing scales of observation. The good correlation between decomposition rate and litter quality across species (Melillo et al. 1982; Aber et al. 1990; Fig. 8.3) suggests that decomposition at the community level can also be sufficiently well described using the initial quality of the bulk litter fall. This assumption is an important prerequisite for modeling and predicting decomposition processes and C cycling over large scales. However, it is currently unknown whether decomposition of a mixture of different litter species with a given average litter quality is indeed the same as that of a monospecific litter of identical initial litter quality.

In conclusion, it is clear that litter decomposition differs widely among tree species occurring within a forest stand. It is also clear that a large part of observed variation in decomposition is related to differences in initial litter quality, even though the relative importance of various chemical compounds and the specific mechanisms of control are not well understood. Presently, it is unclear, however, whether the average litter quality of a site, e.g., the site-specific litter lignin/N ratio, sufficiently describes decomposition at the community level, or whether the spatial and temporal heterogeneity of inputs of vastly different litter types are important factors for community-level decomposition. This is important to know for an evaluation of the functional significance of biodiversity for decomposition processes, as well as for a validation of modeling approaches of decomposition and C and nutrient cycles, at least at a local scale.

8.3 Litter Diversity Effects on Decomposition

Litter mass loss and mineralization rates can vary substantially among forest sites differing in tree species richness or composition (Chapman et al. 1988; Prescott et al. 2000; Zimmer 2002). Tree-canopy characteristics have large influence on decomposition and nutrient cycling through alterations in hydrological and temperature conditions, changes in the chemical composition of precipitation, and most importantly, through being a source of leaf litter (Prescott 2002). However, species composition of the tree canopy is only one factor among many others, such as soil physical and chemical properties, soil fauna community structure, and understory vegetation composition, each of which may differ among sites having even the same climate. All these factors exhibit important controls over decomposition and can interact with each other, making an assessment of the importance of tree species diversity per se very difficult. Moreover, tree species composition can influence ecosystem nutrient cycling not only through the production of diverse litter, but also through plant nutrient uptake and use, rhizosphere interactions, and changes in the decomposition microenvironment (Hobbie 1992). Litter composition is certainly of major importance, as was outlined in the previous paragraphs, and possible interactions among litter species could have further significant implications for ecosystem functioning. Litter species may interact in essentially two different ways, synergistically leading to enhanced decomposition of litter mixtures, and antagonistically causing decreased decomposition rates in mixtures compared to single-species decomposition. If no interactions occur, decomposition of litter mixtures can be straightforwardly calculated from decay rates of individual litter species and their relative contribution to the mixture, i.e., purely additive effects. To properly assess and separate the importance of litter mixture effects on decomposition from other site-specific properties, the different litter combinations would have to decompose under standardized conditions. A literature survey, however, showed that such studies are comparatively rare and seldom include more than two tree litter species.

8.3.1 Decomposition of Litter Mixtures

In support of the hypothesis of synergistic interactions among litter types, a number of studies found accelerated litter mass loss rates with increasing litter diversity (Gustafson 1943; Johnston 1953; Rustad and Cronan 1988; Taylor et al. 1989b; Briones and Ineson 1996; McTiernan et al. 1997; Salamanca et al. 1998; Kaneko and Salamanca 1999; see Fig. 8.4). In many other cases, there were no effects of litter species mixtures on litter mass loss (Johnston 1953; Blair et al. 1990; Klemmedson 1992; McTiernan et al. 1997; Hansen and Coleman 1998; Nilsson et al. 1999; Prescott et al. 2000; see Fig. 8.4), supporting the hypothesis of purely additive effects. In only two cases was a reduced decay

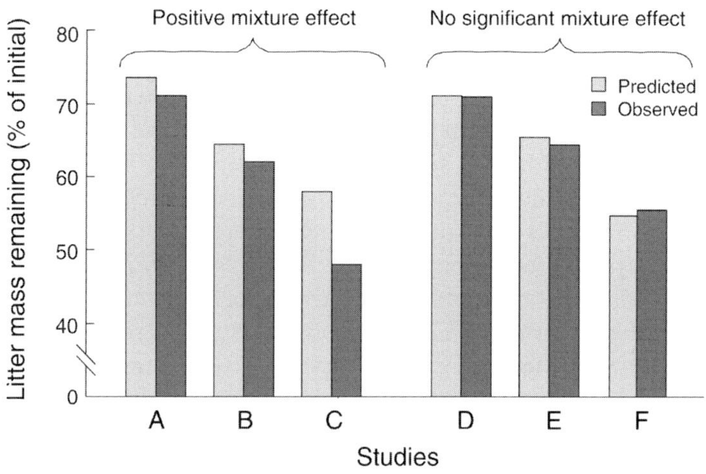

Fig. 8.4. Comparisons of observed litter mass remaining of mixed-species leaf litter with predicted values calculated from measurements of component species from different studies. All studies used litterbags exposed in the field. Species composition and duration of exposure are as follows: **A** *Pinus densiflora/Quercus serrata* for 365 days (Salamanca et al. 1998), **B** *Populus tremuloides/Alnus crispa* for 730 days (Taylor et al. 1989b), **C** *Pinus strobus/Picea rubens/Acer rubrum* for 730 days (Rustad and Cronan 1988), **D** *Quercus rubra/Acer saccharum/Betula alleghaniensis* for 251 days (Hansen and Coleman 1998), **E** *Pinus ponderosa/Quercus gambilii* for 728 days (Klemmedson 1992), **F** *Quercus prinus/Acer rubrum/Cornus florida* for 378 days. (Blair et al. 1990)

rate observed in mixtures compared to single-species litters (Fyles and Fyles 1993; McTiernan et al. 1997), indicating antagonistic effects. Excepting the studies by Rustad and Cronan (1988), Blair et al. (1990), Hansen and Coleman (1998), and Kaneko and Salamanca (1999), who included three-species mixtures in their tests, all other studies compared combinations of only two litter species with their respective monocultures. This strongly limits a thorough assessment of diversity effects, and a more general description of litter mass loss as a function of tree litter diversity. A few investigations with mainly herbaceous species included many species and several levels of diversity (e.g., Wardle et al. 1997; Hector et al. 2000), and found largely unpredictable, idiosyncratic responses of litter mass loss to increasing species richness. Likewise, litter mixing had non-additive, idiosyncratic effects on N loss from litter material, possibly correlating with changes in net N mineralization rates (Wardle et al. 1997). Those tree litter mixture studies mentioned above that actually measured N fluxes in addition to mass loss reported variable mixture effects on N loss that were quite independent from those on mass loss. Out of a total of 28 different tree litter combinations tested (a large number in the laboratory experiment by McTiernan et al. 1997), there was no change in N fluxes in 17 cases, a decreased N loss from mixtures in 7 cases, and a higher N loss in mixtures in 4 cases. Even small changes in net N mineralization due to non-additive litter mixture effects might significantly change plant N availability over larger spatial and temporal scales (Finzi and Canham 1998), with considerable consequences for community dynamics and ecosystem functioning.

Changes in decomposition of individual species within litter mixtures may not be detected when the mixture is measured as a whole, particularly if the mixture is composed of several different species and if there are opposite diversity effects within the same mixture. Such diversity effects, however, result in altered turnover rates of specific litter types and might have important implications for nutrient dynamics and soil organic matter formation. Species-specific differences have been observed in some species combinations in the few studies separating decomposition among individual species within mixtures (Briones and Ineson 1996; Conn and Dighton 2000; Prescott et al. 2000; Gasser and Hättenschwiler, unpubl.), not commonly done in most experiments. For example, Gasser and Hättenschwiler (unpubl.), studying mixtures of leaf litter from six temperate-forest tree species, observed that mass loss during the initial phase of decomposition was either not changed in three of the species, significantly increased in one species, or decreased in two species compared to monocultures of the same species (Fig. 8.5). The changes in decay rates of certain litter species in the mixture leads to altered temporal dynamics of the litter layer composition, and possibly to changes in nutrient turnover. This example demonstrates that synergistic, antagonistic, and additive effects are not exclusive interactions of litter species, but can actually occur together at the same time. Most importantly, if only litter mass loss of

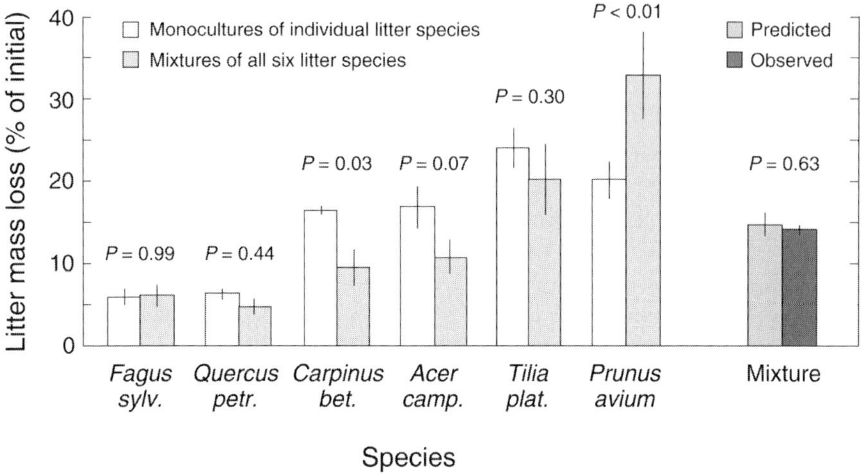

Fig. 8.5. Leaf litter mass loss of six European temperate forest tree species in monocultures and in mixtures of all six species after 92 days of decomposition in a mixed deciduous forest near Basel, Switzerland. Litter was exposed in microcosms inserted into the ground covering an area of 0.0177 m^2 each. Microcosms were covered with 0.5-mm nylon mesh and equally distributed among three patches within a single forest stand (n=3 microcosms per treatment). In addition to mass loss of each individual species (determined after identification and separation of all remaining leaf parts), mass loss of the entire mixture is shown and compared to the predicted value based on measured mass losses in monocultures of each component species (*right end of the figure*; Gasser and Hättenschwiler, unpubl. data)

the entire mixture of all six species had been measured and compared to the predicted value based on measurements of monocultures, we would wrongly conclude that decomposition is not affected by diversity (Fig. 8.5). Such mixture effects on particular litter species might be more common than is usually believed, but have remained undetected because individual litter species have not been measured separately.

The currently available data suggest that non-additive litter mixture effects on mass loss and nutrient mineralization likely occur in many litter species combinations and need to be taken into account to accurately predict decomposition processes at the ecosystem level. The question why some litter combinations exhibit non-additive effects upon decomposition and why others do not, however, is difficult to answer. Even within the most often studied two-species mixtures of the same genera, *Pinus* and *Quercus,* no consistent mixture effects were observed (Table 8.1), and the underlying mechanisms of these contrasting mixture effects have not yet been identified. The fact that individual species are affected differently even within the same mixture suggests that various mechanisms are involved at the same time.

Table 8.1. Conclusions from experiments testing the effects of mixtures between conifer (*Pinus* spp.) and broadleaf deciduous tree litter (mostly *Quercus* spp.) over the last 60 years of research

Conclusion	Species mixed	Author(s)
A mixture of pine and hardwood leaves increases the rate of decay of both kinds of leaves	*Pinus resinosa* *Quercus velutina*	Gustafson (1943)
Hardwood leaves do not accelerate decomposition of conifer litter	*Pinus taeda* *Cornus florida*	Thomas (1968)
Clearly, there was no synergistic effect between pine and oak leaf litter	*Pinus ponderosa* *Quercus gambilii*	Klemmedson (1992)
Mixing litter increased decomposition rates relative to those expected from single litter treatments	*Pinus densiflora* *Quercus serrata*	Kaneko and Salamanca (1999)
The presence of pine needles increased the rate of oak leaf decomposition. The presence of oak leaves had no effect on the decay of pine needles	*Pinus rigida* *Quercus* spp.	Conn and Dighton (2000)

8.3.2 The Nature of Litter Species Interactions

A mechanistic approach is ultimately needed to understand interactive effects among litter species on decomposition and to provide general predictive tools for the functional significance of litter diversity for ecosystem processes. One possible suite of mechanisms is litter diversity effects on the composition and activity of higher trophic levels and their feedbacks on decomposition.

In a litterbag decomposition experiment, Hansen and Coleman (1998) quantified microhabitat variety and associated species richness of oribatid mites. They found a significantly greater variety of microhabitats in three-species mixtures than in the three monocultures, and also, yet not so clearly expressed, a higher species richness of mites. However, these results did not correlate with higher litter mass loss rates from mixed litter. It could well be, however, that a different composition of the mite community has an effect on decomposition in the longer term or on nutrient fluxes, similar to that found in the study by Blair et al. (1990). They observed that significantly more nitrogen was lost initially and less nitrogen immobilized in later stages from litter mixtures than would be predicted from component species. This observation was accompanied by significantly fewer fungal hyphae and more nematodes, apparently including fungal feeders, in litter mixtures. A different structure and abundance of microhabitats appear to be likely mechanisms for altered decomposition in mixed-species litter via changes in diversity and/or abundance of soil fauna. In addition to non-trophic litter diversity effects on soil animals, the more direct influence of food diversity for saprophagous animals can be important as well. The litter-feeding macrofauna (gastropods, earthworms, millipedes, isopods, insect larvae) prefers certain litter types and is quite sensitive to small differences in litter quality (Hassall et al. 1987; Hättenschwiler and Bretscher 2001). The overall performance, feeding behavior, and abundance of macrofauna, thus, are influenced by the temporal variation in the amount and composition of litter. Because the saprophagous macrofauna processes large amounts of litter and has a tremendous impact on the smaller litter and soil fauna (Scheu 1987; Anderson 1988; Càrcamo et al. 2000), litter diversity effects on the macrofauna may determine the overall decomposition process to a large extent. The interactions between litter diversity and macrofauna, however, are very poorly studied, mainly because the traditional approach of using litterbags to study decomposition excludes these larger animals. An important issue in the context of litter interactions with decomposers is the functional significance of the diversity itself of soil organisms, not covered here but summarized later in this volume (Scheu, Chap. 11).

Another group of mechanisms for litter diversity effects may be changes in processes during decomposition of component species by specific litter compounds, such as polyphenols. Phenolic compounds, as an important and abundant group of plant secondary metabolites, are believed to play many different roles in soil processes (Hättenschwiler and Vitousek 2000), ranging from control over feeding behavior of macrofauna to specific interactions with microbes and the formation of complexes with proteins. The role of polyphenols within litter and soil is only beginning to be studied more thoroughly, and so far has not been discussed explicitly in the context of litter diversity effects on decomposition. Schimel et al. (1998), for example, observed diverse effects on soil processes of polyphenols from *Populus balsamifera* leaf litter in Alaskan taiga forest ecosystems. Phenolic acids provide a microbial growth substrate leading to increasing microbial N immobilization, while specific tannins inhibit microbial activity and suppress N_2 fixation in the early successional *Alnus tenuifolia*. These diverse effects of polyphenols from a specific litter species may ultimately enhance successional dynamics and change the nitrogen availability in these ecosystems. It seems likely that at least some of the polyphenols present in virtually all litter species may be involved in observed antagonistic or synergistic effects on decomposition of litter mixtures.

One of the most prominently discussed mechanisms for synergistic mixture effects is the stimulation of decomposition of a low-quality litter type by the presence of a fast-decomposing high-quality litter (Seastedt 1984; Chapman et al. 1988; Wardle et al. 1997). The high-quality litter is preferentially exploited by decomposer organisms, eventually leading to a high nutrient availability allowing nutrient transfer to the low-quality litter. This in turn could lead to a more rapid utilization of carbon substrates of the low-quality litter and consequently to an overall faster decomposition of the litter mixture. From the studies reviewed here, only the results by Salamanca et al. (1998), and in one of the species pair tested by Briones and Ineson (1996), support this mechanism for synergistic effects. In both cases, an enhanced mass loss of the low-quality litter correlated with increased microbial activity and with an apparent net N transfer from the high- to the low-quality litter type. Nutrient transfer, however, might be more commonly involved in positive interactions among different litter types, but has rarely been evaluated comprehensively.

8.4 Conclusions

Inter- and intraspecific variation in leaf litter quality is substantial and substrate quality is of overriding importance for rates of litter decay and mineralization within a forest site. Although this has long been recognized, the functional significance of litter mixtures for decomposition processes and ecosystem functioning has surprisingly been little explored. The currently available literature summarized here shows that studies specifically addressing litter diversity effects on decomposition rarely included more than two species and were carried out predominantly in temperate forests with other forest ecosystems outside the temperate zone (most importantly tropical forests) critically underrepresented.

Nevertheless, non-additive litter mixture effects on mass loss and/or nutrient mineralization have been observed in about half of all studies, suggesting that litter species richness and/or composition can have important implications for decomposition processes and ecosystem functioning. The relationship between litter species richness and process rate, however, does not yet appear to be predictable, and the currently available data suggest that the identity of species within a mixture is more important than the number of species. A comprehensive mechanistic approach in the analysis of diversity effects would substantially improve our understanding of the functional significance of litter diversity for decomposition and should be of high priority for future research. To get there, I think we first need to abandon the idea of a single, true correlation between diversity and process rate. The few studies separating diversity effects on decomposition of individual component species strongly suggest that we have to focus beyond such simple correlations based on litter mixtures as a whole. Unraveling interactions across trophic levels, identifying specific compounds such as polyphenols acting as inhibitors or stimulators in the process of decomposition, and the investigation of nutrient transfer among litter species are three promising areas of future, mechanistically oriented research.

References

Aber JD, Melillo JM, McClaugherty CA (1990) Predicting long-term patterns of mass loss, nitrogen dynamics, and soil organic matter formation from initial fine litter chemistry in temperate forest ecosystems. Can J Bot 68:2201–2208

Anderson JM (1988) Spatiotemporal effects of invertebrates on soil processes. Biol Fertil Soils 6:216–227

Berg B (2000) Litter decomposition and organic matter turnover in northern forest soils. For Ecol Manage 133:13–22

Berg B, Berg MP, Bottner P et al (1993) Litter mass loss rates in pine forests of Europe and Eastern United States: some relationships with climate and litter quality. Biogeochemistry 20:127–159

Blair JM, Parmelee RW, Beare MH (1990) Decay rates, nitrogen fluxes, and decomposer communities of single- and mixed-species foliar litter. Ecology 71:1976–1985

Briones MJI, Ineson P (1996) Decomposition of eucalyptus leaves in litter mixtures. Soil Biol Biochem 28:1381–1388

Cadish G, Giller KE (1997) Driven by nature: plant litter quality and decomposition. CAB International, Wallingford

Càrcamo HA, Abe TA, Prescott CE, Holl FB, Chanway CP (2000) Influence of millipedes on litter decomposition, N mineralization, and microbial communities in a coastal forest in British Columbia, Canada. Can J For Res 30:817–826

Chapin FS III., Matson PA, Mooney HA (2002) Principles of terrestrial ecosystem ecology. Springer, Berlin Heidelberg New York

Chapman K, Whittaker JB, Heal OW (1988) Metabolic and faunal activity in litters of tree mixtures compared with pure stands. Agric Ecosyst Environ 24:33–40

Conn C, Dighton J (2000) Litter quality influences on decomposition, ectomycorrhizal community structure and mycorrhizal root surface acid phosphatase activity. Soil Biol Biochem 32:489–496

Cornelissen JHC (1996) An experimental comparison of leaf decomposition rates in a wide range of temperate plant species and types. J Ecol 84:573–582

Coûteaux M-M, Bottner P, Berg B (1995) Litter decomposition, climate and litter quality. Trends Ecol Evol 10:63–66

Ellenberg H (1986) Vegetation Mitteleuropas mit den Alpen in ökologischer Sicht. Ulmer, Stuttgart

Finzi AC, Canham CD (1998) Non-additive effects of litter mixtures on net N mineralization in a southern New England forest. For Ecol Manage 105:129–136

Fyles JW, Fyles IH (1993) Interaction of Douglas-fir with red alder and salal foliage litter during decomposition. Can J For Res 23:358–361

Gholz HL, Wedin DA, Smitherman SM, Harmon ME, Parton WJ (2000) Long-term dynamics of pine and hardwood litter in contrasting environments: toward a global model of decomposition. Global Change Biol 6:751–765

Gustafson FG (1943) Decomposition of the leaves of some forest trees under field conditions. Plant Physiol 18:704–707

Hansen RA, Coleman DC (1998) Litter complexity and composition are determinants of the diversity and species composition of oribatid mites (Acari: Oribatidae) in litterbags. Appl Soil Ecol 9:17–23

Hassall M, Turner JG, Rands MRW (1987) Effects of terrestrial isopods on the decomposition of different woodland leaf litter. Oecologia 72:597–604

Hättenschwiler S, Vitousek PM (2000) The role of polyphenols in terrestrial ecosystem nutrient cycling. Trends Ecol Evol 15:238–243

Hättenschwiler S, Bretscher D (2001) Isopod effects on decomposition of litter produced under elevated CO_2, N deposition and different soil types. Global Change Biol 7:565–579

Hättenschwiler S, Hagerman AE, Vitousek PM (2003) Polyphenols in litter from tropical montane forests across a wide range in soil fertility. Biogeochemistry 64:129–148

Hector A, Beale AJ, Minns A, Otway SJ, Lawton JH (2000) Consequences of the reduction of plant diversity for litter decomposition: effects through litter quality and microenvironment. Oikos 90:357–371

Hobbie SE (1992) Effects of plant species on nutrient cycling. Trends Ecol Evol 7:336–339

Johnston DR (1953) A laboratory study of the decomposition of vegetable debris in relation to the formation of raw humus. Plant Soil 4:345–369

Kaneko N, Salamanca EF (1999) Mixed leaf litter effects on decomposition rates and soil microarthropod communities in an oak-pine stand in Japan. Ecol Res 14:131–138

Klemmedson JO (1992) Decomposition and nutrient release from mixtures of Gambel oak and ponderosa pine leaf litter. For Ecol Manage 47:349–361

Madritch MD, Hunter MD (2002) Phenotypic diversity influences ecosystem functioning in an oak sandhills community. Ecology 83:2084–2090

McClaugherty C, Berg B (1987) Cellulose, lignin and nitrogen levels as rate regulating factors in late stages of forest litter decomposition. Pedobiologia 30:101–112

McTiernan KB, Ineson P, Coward PA (1997) Respiration and nutrient release from tree leaf litter mixtures. Oikos 78:527–538

Melillo JM, Aber JD, Muratore JF (1982) Nitrogen and lignin control of hardwood leaf litter decomposition dynamics. Ecology 63:621–626

Nilsson M-C, Wardle DA, Dahlberg A (1999) Effects of plant litter species composition and diversity on the boreal forest plant-soil system. Oikos 86:16–26

Perez-Harguindeguy N, Diaz S, Cornelissen JHC, Venramini F, Cabido M, Castellanos A (2000) Chemistry and toughness predict leaf litter decomposition rates over a wide spectrum of functional types and taxa in central Argentina. Plant Soil 218:21–30

Prescott CE (2002) The influence of the forest canopy on nutrient cycling. Tree Physiol 22:1193–1200

Prescott CE, Zabek LM, Staley CL, Kabzerns R (2000) Decomposition of broadleaf and needle litter in forests of British Columbia: influences of litter type, forest type, and litter mixtures. Can J For Res 30:1742–1750

Ricklefs RE, Matthew KK (1982) Chemical characteristics of the foliage of some deciduous trees in southeastern Ontario. Can J Bot 60:2037–2045

Rustad LE, Cronan CS (1988) Element loss and retention during litter decay in a red spruce stand in Maine. Can J For Res 18:947–953

Salamanca EF, Kaneko N, Katagiri S (1998) Effects of leaf litter mixtures on the decomposition of *Quercus serrata* and *Pinus densiflora* using field and laboratory microcosm methods. Ecol Eng 10:53–73

Seastedt TR (1984) The role of arthropods in decomposition and mineralization processes. Annu Rev Entomol 29:25–46

Scheu S (1987) Microbial activity and nutrient dynamics in earthworm casts (Lumbricidae). Biol Fert Soils 5:230–234

Schimel JP, Cates RG, Ruess R (1998) The role of balsam poplar secondary chemicals in controlling soil nutrient dynamics through succession in the Alaskan taiga. Biogeochemistry 42:221–234

Swift MJ, Heal OW, Anderson JM (1979) Decomposition in terrestrial ecosystems. Univ California Press, Berkeley

Taylor BR, Parkinson D, Parsons WFJ (1989a) Nitrogen and lignin content as predictors of litter decay rates: a microcosm test. Ecology 70:97–104
Taylor BR, Parsons WFJ, Parkinson D (1989b) Decomposition of *Populus tremuloides* leaf litter accelerated by addition of *Alnus crispa* litter. Can J For Res 19:674–679
Thomas WA (1968) Decomposition of loblolly pine needles with and without addition of dogwood leaves. Ecology 49:568–571
Vitousek PM, Turner DR, Parton WJ, Sanford RL (1994) Litter decomposition on the Mauna Loa environmental matrix, Hawai'i: patterns, mechanisms, and models. Ecology 75:418–429
Wardle DA, Bonner KI, Nicholson KS (1997) Biodiversity and plant litter: experimental evidence which does not support the view that enhanced species richness improves ecosystem function. Oikos 79:247–258
Zimmer M (2002) Is decomposition of leaf litter influenced by its species richness? Soil Biol Biochem 34:277–284

9 The Effect of Biodiversity on Carbon Storage in Soils

G. Gleixner, C. Kramer, V. Hahn, and D. Sachse

9.1 Introduction

Only limited knowledge is presently available about how carbon is stored in soils and how this process can be influenced by abiotic processes (Schimel et al. 2001). Even less is known about the effects of biodiversity on carbon storage (Catovsky et al. 2002), especially with respect to the role of tree biodiversity. Most experimental investigations on biodiversity deal with grassland experiments focusing on ecosystem functions of biomass productivity or nutrient retention (Kinzig et al. 2001; Loreau et al. 2002). In order to identify possible interactions of biodiversity with carbon storage, this chapter summarizes current knowledge on carbon storage and emphasize the importance of this process for ecosystem functioning. Several key areas will be identified where plant biodiversity might influence carbon storage.

9.2 Formation of Soil Carbon

Generally, the accumulation of carbon in soil is the result of ecosystem development driven by the input and decomposition of plant-derived carbon. In the early stages of ecosystem development during primary succession, i.e., after the retreat of ice in the late Pleistocene, mainly lichens and mosses added carbon to the bare surface. In consequence, surface rocks were biologically weathered (Barker and Banfield 1996; Banfield et al. 1999) and the first soil organic matter (SOM) was formed from decomposing biomass. The increase in temperature, nutrient availability due to weathering and water-holding capacity due to SOM, enabled further progress in ecosystem development, to be seen in the development of soil profiles (Fig. 9.1). Increasing biomass and litter production formed a litter layer (L horizon) of undecomposed dead plant material. Underneath this litter layer organic layers (O horizon)

Ecological Studies, Vol. 176
M. Scherer-Lorenzen, Ch. Körner, and E.-D. Schulze (Eds.)
Forest Diversity and Function: Temperate and Boreal Systems

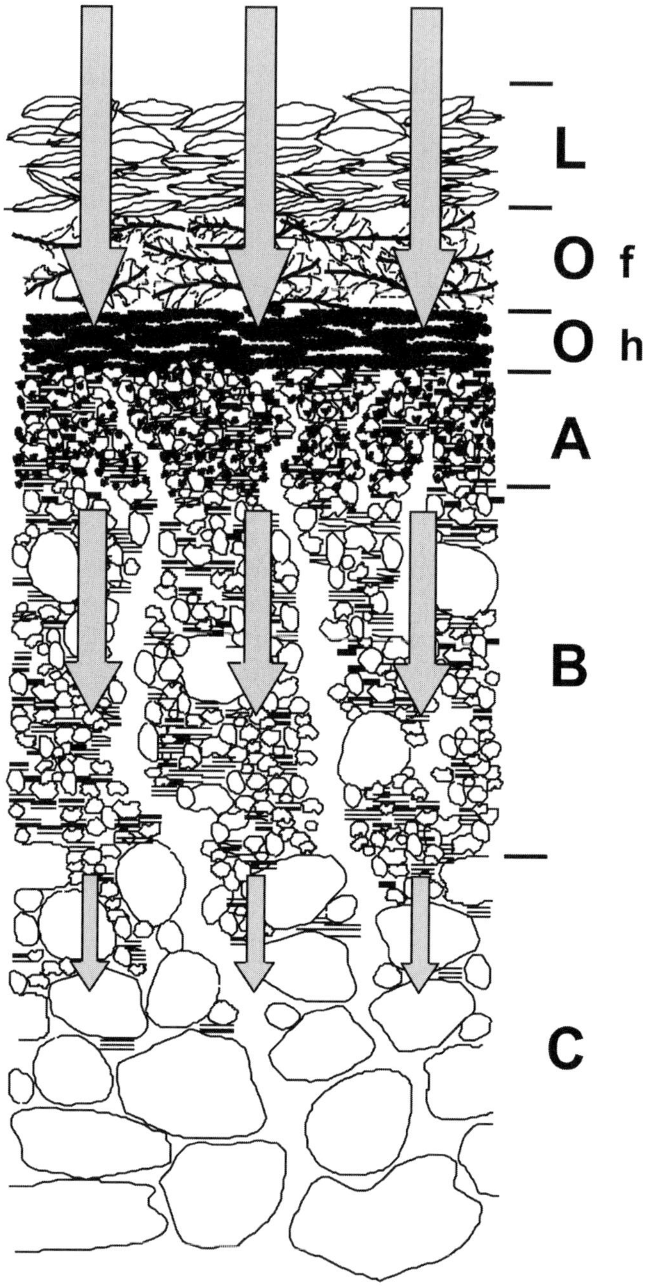

Fig. 9.1. Different horizons of a soil profile. *L* Litter, *O* organic layer, *A* mineral layer with organic carbon and leached minerals, *B* mineral layer with precipitation of oxides/hydroxides and/or carbon, *C* unaltered parent substrate. *Arrows* indicate the decreasing water flow down the soil profile

developed of partially degraded and fermented plant material (Of horizon). In the humic horizon (Oh horizon) above the mineral layer no plant structures can be identified. Organic matter is also transported into deeper mineral soil layers either by digging soil organisms or by percolating rainwater. The latter process is most important for the development of the soil profiles. The transport of carbon from the O horizons to the upper mineral horizon, and the export of minerals and metal oxides from this horizon through percolating soil water, form a mineral-depleted A horizon in the deeper mineral soil. Below the A horizon an often brownish or reddish mineral enriched B horizon is formed due to the precipitation of leached weathering products, i.e., iron oxides/hydroxides and/or humic substances, from the percolating stream of soil water. Underneath the developed soil profile, unaltered parent substrate remains in the C horizon.

Carbon found in soils is thus primarily produced by plants from atmospheric CO_2 and enters the soil as root or leaf litter or as root exudate. Soil organisms decompose this litter and root exudate and they release most of the assimilated carbon again as CO_2 back into the atmosphere. Some of the plant-derived litter may remain untouched above the soil (raw humus), but most of the litter-derived carbon remaining in soil is transformed to soil organic matter by the action of soil organisms (Hättenschwiler, Chap. 8, this Vol.). The complex process of soil organic matter formation is an achievement of the trophic networks in soil and might be influenced by the biodiversity of the soil organisms (Scheu, Chap. 11, this Vol.). In general, shredding organisms such as earthworms or woodlice, break litter into small pieces and extract digestible compounds. This process increases the surface area of litter and inoculates it with decomposing microorganisms that degrade indigestible compounds externally (Meyer 1993). Soil animals such as nematodes, woodlice, collembola, and mites feed on these nutrient-rich microorganisms, and predators hunt these microbe-feeding soil animals in the soil. Finally, decomposers mineralize dead soil animals closing the carbon cycle in the soil. In short, the formation and turnover of soil carbon depend on the interaction of plants and soil organisms. The biodiversity of both plants and soil organisms thus influences the soil C dynamics.

9.3 Consequences of Plant Diversity on the Quality of Carbon Input

The stability of soil carbon and hence the amount of stored carbon depends on the chemical structure (i.e., the intrinsic stability) of molecules, their interaction with mineral surfaces (i.e., their "storage capacity"), and the amount of carbon submitted to decomposition (Lichtfouse et al. 1998; Kaiser and Guggenberger 2003). We will not discuss the latter two factors, as the carbon storage capacity is mainly controlled by parent soil material and it is not influenced by biodiversity. The effect of biodiversity on the productivity of forests and consequently on the litter production has already been discussed (Pretzsch, Chap. 3, this Vol.). The main focus of this chapter will be the chemical and isotopic composition of organic matter.

Most plant-derived carbon belongs to a small number of chemical structures. These are mainly carbohydrates, organic acids, lipids, lignin, and proteins. Some of them, such as carbohydrates, organic acids, and proteins, are preferred energy sources for soil organisms and thus less recalcitrant in soils than lignin or lipids (Gleixner et al. 2001a). In consequence, the decomposition rate of plant litter will change with litter quality (see also Hättenschwiler, Chap. 8, this Vol.) and stable plant-derived structures may accumulate in soil. Wood for example, as the most abundant plant biomass, mainly consists of cellulose and lignin (Fig. 9.2a, b). Cellulose is chemically less stable than lignin and lignin accumulates, i.e., is selectively preserved, in wood decomposition. This is well known for example for brown rot fungi (Gleixner et al. 1993).

Lignin itself is a complex polymer made from three different lignin monomers, coumaryl, coniferyl, and sinapyl alcohols, differing in their methoxyl substitution in the ortho position of the phenolic ring (Fig. 9.2 c). The contribution of the three monomers characterizes the lignin and indicates its origin. Monocotyledons, like grasses, are rich in coumaryl alcohol, whereas dicotyledons are rich in coniferyl and sinapyl alcohol. In conifers coniferyl alcohol is the main lignin monomer, whereas in broadleaf trees sinapyl alcohol dominates. Depending on the biodiversity of the plant community, the composition of lignin biomarkers might differ and the selective preservation of remaining lignin molecules might determine the quality of stored carbon.

The selective preservation of chemically resistant molecules is also known to occur in lipids (Lichtfouse et al. 1998), e.g., alkanes (Fig. 9.2d). The composition of alkanes, which are part of the epicuticular waxes (Eglinton et al. 1962), is characteristic for different plant types and enables reconstruction of the paleoenvironment (Brassell et al. 1986; Eglinton and Hamilton 1967). Other constituents of the epicuticular waxes, such as alkanoic acids, hydroxyalkanoic acids, alcohols, alkanediols, alkanals, or alkyl esters are less stable than alkanes (and have inspired less taxonomic information; Riederer 1989). Green algae, often present in soils, synthesize alkanes mainly with a chain length of 17 carbon atoms, whereas higher plants synthesize alkanes with chain lengths of mainly 27, 29, and 31 carbon atoms (Rieley et al. 1991). The relative composition of each different alkane depends on its origin (Schwark et al. 2002). Grasses of the understory vegetation are dominated by the C31 alkane (Cranwell et al. 1987; van Bergen et al. 1997), whereas deciduous trees consist of mainly C27 and C29 alkanes (Almendros et al. 1996; Spooner et al. 1994). Using a ternary mixing diagram for the C27, C29, and C31 alkanes, the different alkane composition of various trees from southern Italy can be

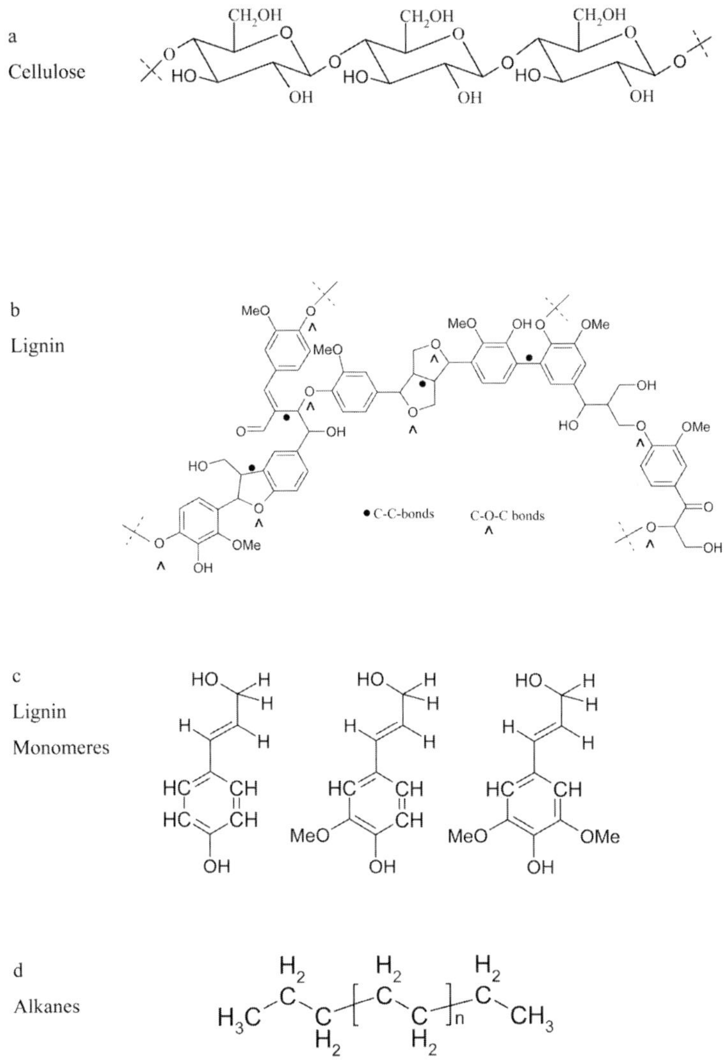

Fig. 9.2. Chemical structure of main biochemical elements of plants. **a** Cellulose, **b** lignin, **c** lignin monomers, **d** alkanes; *n* indicates the number of repeated structures to reach the corresponding total chain length

noted (Fig. 9.3). *Fagus sylvatica* synthesizes mainly the C29 alkane, *Quercus cerris* mainly C31. The relative abundance of these three alkanes is also reflected in lipid extracts from soil. In beech forests the upper horizons of the soil are clearly dominated by the C29 alkane (Fig. 9.3).

These two examples, lignin and alkanes, suggest that plant-derived differences in chemical composition (i.e., the quality) of plant biomass depend on plant species composition, and the occurrence of such biomarkers in soil car-

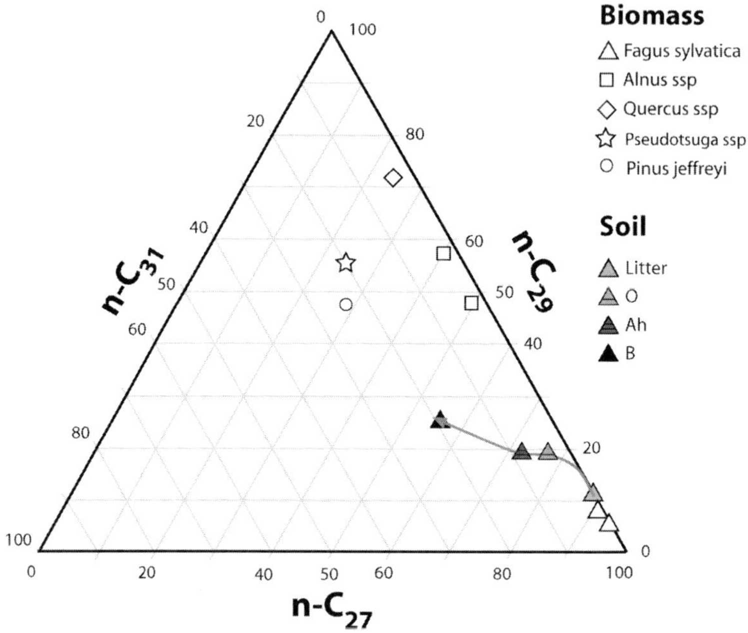

Fig. 9.3. Relative composition of alkanes C27, C29, and C31 in various plant species and in depth horizons of soil under beech vegetation

bon differs accordingly. However, not only the quality of biomass produced depends on the species composition. The amount and composition of carbon flow to roots for exudation are also species-dependent (Grayston et al. 1997). The relative amount of root exudate as fraction of plant carbon assimilation varies thusly: 40 % for *Liriodendron tulipifera*, 60 % for *Pinus sylverstris,* and even 78 % for *Pseudotsuga menziesii*. In general, the exudates consist of various carbohydrates, amino acids, aliphatic and aromatic fatty acids, sterols, and enzyme- and hormone-like substances (Grayston et al. 1997). The composition of the exudates varies greatly between different species. For example, deciduous trees exude preferentially the amino acids cysteine and homoserine, whereas evergreens have no preferential amino acid exudation pattern. Exudation patterns of carbohydrates, like glucose, fructose and sucrose, and organic acids, such as acetic, succinic, and oxalic acid, all of which may be major components in tree root exudates, also differ with tree species. However, no clear pattern is obvious. Even different species of the same genus *Pinus* have different exudate compositions. Moreover, age and developmental stage of the trees and environmental conditions, such as nutrient status, pH, water availability, temperature, light intensity, carbon dioxide concentrations, and presence of microorganisms, affect the quality and quantity of root exudation (Grayston et al. 1997).

Root exudates are the major carbon source for soil microorganisms living in association or symbiosis with tree roots. Soil microorganisms oxidize most root-derived carbon, and microbe-derived compounds, like fucose or rhamnose, are dominant in the dissolved carbon. In consequence, the direct impact of root exudates on carbon storage is small. However, strong feedback can be expected between the tree species composition as expressed in root exudation and soil microbial composition. Evidence exists that changing artificial root exudations can affect the species composition of soil microorganisms (Baudoin et al. 2003). Spore germination (mainly), hyphal elongation, and branching as well as chemotaxis are effected by root exudates. Some feedbacks are rather specific, such as is the symbiosis between N-fixing *Frankia* and *Alnus*; or they may be nonspecific, such as are the ectomycorrhizal fungi *Laccaria laccata* or *Boletus edulis*, which have a broad range of host plants. However, the impact of differences in the below-ground biodiversity on carbon storage is not well understood and controversial (Hooper et al. 2000).

In addition to the importance of chemical structure for the quality of carbon input to soils, all organic compounds have a unique isotopic "fingerprint" characteristic for its origin (O'Leary 1981; Schmidt and Gleixner 1998). This fingerprint can be used to estimate the importance of biomass from different trees for carbon storage. Well known are interspecies differences, as with the greater enrichment of ^{13}C in C4 versus C3 plants of ~12–15 ‰ or the isotopic enrichment of wood and litter from conifers versus broadleaf trees of ~5 ‰. Moreover, N-fixing plants like Fabaceae or *Alnus* species have a unique ^{15}N signal; and, independently of the transpiration rate and leaf anatomy, the D content of trees also varies. Intermolecular isotopic differences are known as well. Lignin, for example, is depleted in ^{13}C relative to cellulose by up to 6 ‰. This isotope information is widely used to trace the origin and turnover of soil carbon (Boutton and Yamasaki 1996). However, most investigations are made using only bulk soil or plant material. Differing decomposition rates of chemical structures introduce isotopic shifts of bulk soil organic matter that mimic isotope effects or source differences. For example, the relative increase of the lignin content in remaining wood will cause an isotopic shift of the remaining wood to more-depleted $\delta^{13}C$ values. Using the isotopic information of individual molecules overcomes this problem; molecules isolated from soil found to have the same isotope content as their plant precursors indicate their selective preservation (Kracht and Gleixner 2000).

[1] The $\delta^{13}C$ value in "per mille" [‰] is the relative difference of the isotopic ratio R of the heavy isotope (^{13}C) to the light isotope (^{12}C) of a sample to a reference material times 1,000. Thus, $\delta^{13}C$ value [‰]=($R_{sample}-R_{standard}/R_{standard}$) × 1,000. International standard for carbon is V-PDB, a carbonate (Coplen 1996).

9.4 Distribution of Carbon and Nitrogen and Their Stable Isotopes in Soil Profiles

The main sources for soil organic matter in natural systems are leaf litter input to the soil surface and root litter and exudate inputs within the soil profile. We evaluated the relative distribution of soil carbon and root biomass with depth, from a global dataset of 2,721 carbon samples and 117 root samples. The samples originated from all major biomes of the earth, i.e., boreal forest, crops, desert, sclerophyllous shrubs, temperate deciduous forest, temperate evergreen forest, temperate grassland, tropical deciduous forest, tropical evergreen forest, tropical grassland/savanna, and tundra (Jobbagy and Jackson 2001). On a global average over 60 % of the root biomass was found in the top 20 cm of soil, and it logarithmically decreases with depth (Fig. 9.4). Only 14 % of root biomass was found below 40 cm. In contrast, only 40 % of the soil carbon was located in the top 20 cm of soil. It also decreases logarithmically; however, 36 % of the soil carbon was found at depth below 40 cm.

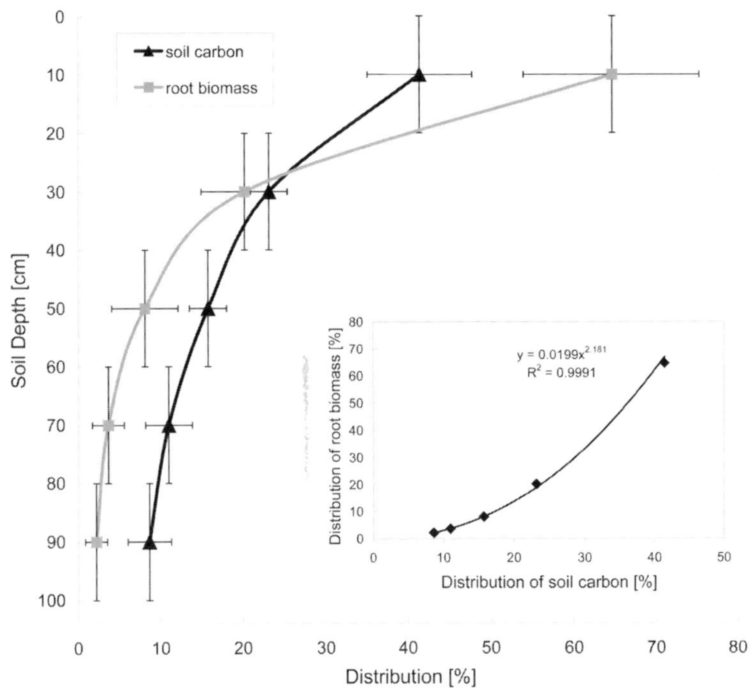

Fig. 9.4. Global summary of the distribution of soil carbon and root biomass in depth profiles of the world's major ecosystems. *Y error bars* indicate sampling interval, *x error bars* indicate standard deviation from 11 biomes summarizing 2,721 soil samples and 117 root biomass samples. (Data from Jobbagy and Jackson 2001)

Thus, soil carbon enriches relative to root biomass with soil depth. The correlation between root biomass distribution and soil carbon distribution suggests an important role of root-derived carbon for the formation of soil carbon ($y=0.0199x^{2.181}$, $R^2=0.9991$). However, in relation to root biomass less carbon is found in the top 20 cm of soils and more carbon in the subsoil. This underlines the importance of (1) microbial degradation of biomass in the upper 20 cm, (2) water for the downward transport of dissolved organic carbon, and (3) the sorption of carbon to the inorganic phase in deeper soil horizons. The distribution of root carbon to the soil is known to be influenced by the species composition and therefore by the biodiversity of the trees (Rothe and Binkley 2001). Such diversity might thus be a factor in controlling carbon storage. However, in the upper 20 cm of soil profiles, the decomposition and hence the diversity of soil organisms, appear to exert a stronger control on carbon storage, whereas in deeper soil horizons intrinsic inorganic soil factors might be more dominant for carbon storage.

Coinciding with the decrease in carbon concentration with soil depth, the concentration of nitrogen decreases, too, although to a lesser extent. Consequently, the C/N ratio of soil organic matter generally decreases with depth from values of above 30±15 characteristic for plant litter to values of 10±2 characteristic for microbial biomass (Fig. 9.5). This change can be observed

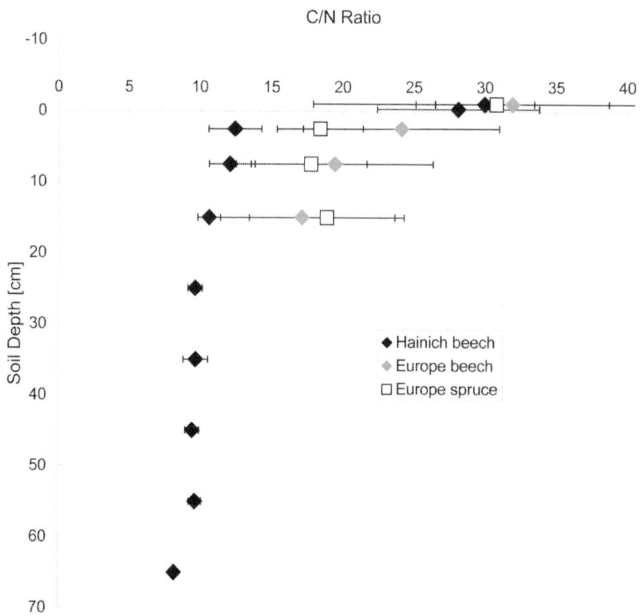

Fig. 9.5. C/N ratio of soil organic matter from different depth intervals of 100 independent replicates of an old-growth beech stand in the National Park Hainich, Germany, (unpubl.), and of four beech stands and six spruce stands from a latitudinal gradient from Europe (Schulze 2000)

for small-scale variations of independent replicates from an old growth beech forest in the National Park Hainich, Germany, as is true for large-scale variations of different beech and spruce stands in the European latitudinal gradient (Schulze 2000). However, no clear trends for different tree species were observed. For the interpretation of C and N distributions in soils, site history and site management have to be considered as additional important factors responsible for variability. In general, the C/N ratios above 20 cm suggest that in the upper layer of the soil profile litter-derived carbon may be part of the SOM pool. Consequently undecomposed remaining chemical structures or carbon with identical isotopic signals as in litter should be found in this layer. Moreover, the composition of carbon in the organic layer and in the Ah layer may vary with the quality of litter as influenced by tree biodiversity. In contrast, the narrow C/N ratio of 10 in deeper horizons is indicative for microbial biomass and independent of root input. Consequently, SOM found in deeper soil horizons might originate from soil microorganisms and be primarily realized by belowground diversity.

In order to demonstrate the microbial origin of carbon in deeper horizons we used the enrichment of ^{13}C and ^{15}N values in trophic networks. This enrichment is known to be between 0–1‰ for C and between 3–4 ‰ for N (Rothe and Gleixner 2000 and references therein). We used $\delta^{13}C$ and $\delta^{15}N$ values of 10 different beech and spruce stands over a latitudinal gradient in Europe (Schulze 2000) and 100 independent depth profiles from the National Park Hainich, Germany. Interestingly, for both cases the $\delta^{13}C$ and $\delta^{15}N$ values were highly correlated (Fig. 9.6), indicating that both the ^{13}C and the ^{15}N values of soil organic matter were increasing with depth at a slope between 3.7 and 4.6. This value is in nice agreement with the trophic level shift suspected from food chains and suggests that soil carbon is continuously recycled in the trophic network of soil organisms. However, this idea contrasts with current theory of soil organic matter formation, which suggests that soil carbon is stabilized mainly by physical (Christensen 1992) and chemical mechanisms (Lichtfouse et al. 1998). We may only speculate at this point that changes in soil microbial diversity may also influence the stabilization and storage of organic matter in deeper horizons of mineral soils.

9.5 Dynamic of Soil Organic Matter

In order to understand the contradicting processes that appear to affect soil organic matter formation it is necessary to evaluate the dynamics, i.e., the turnover, of soil organic matter. Current models for the turnover of SOM suggest that there are three pools of organic matter. These exhibit turnover times of below 5 years for the fast pool, of 50 to 100 years for the slow pool, and of more than 1,000 years for the passive or inert pool (Jenkinson et al. 1987; Par-

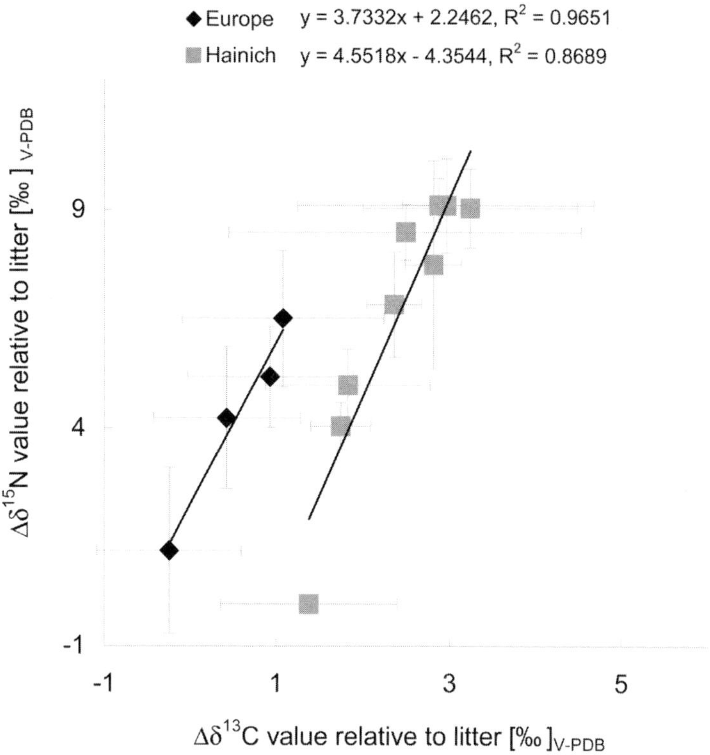

Fig. 9.6. Difference in the $\delta^{13}C$ and $\delta^{15}N$ values of soil organic matter at various depths from the $\delta^{13}C$ and $\delta^{15}N$ values of litter from 100 independent samples from an old-growth beech stand in the national park Hainich, Germany (unpubl.), and of four beech stands and six spruce stands from a latitudinal gradient from Europe. (Schulze 2000)

ton et al. 1987). The latter pool would consist of selectively preserved organic matter, whereas physically protected (i.e., adsorbed) carbon would be found depending on the sorption constants in all three pools.

It is state-of-the-art to experimentally determine the dynamics of soil organic matter turnover using natural labeling experiments (Fig. 9.7). The existing vegetation is replaced by structural similar but isotopically different vegetation, such as by planting conifers having a $\delta^{13}C$ value of –26‰ on deciduous tree habitats that have a $\delta^{13}C$ values of –29 ‰, or by replacing C3 plants, like wheat or rye having a $\delta^{13}C$ value of ~ -25 ‰, by C4 plants, like maize, with a $\delta^{13}C$ value of ~ –12 ‰, in agricultural systems. The change in $\delta^{13}C$ values in soil organic matter can be used to calculate the fraction of remaining C3-derived carbon (Balesdent and Mariotti 1996). Assuming exponential decay of carbon in soils at steady state, the apparent residence time of total soil carbon or of individual compounds of soil organic matter can be determined (Gleixner et al. 1999).

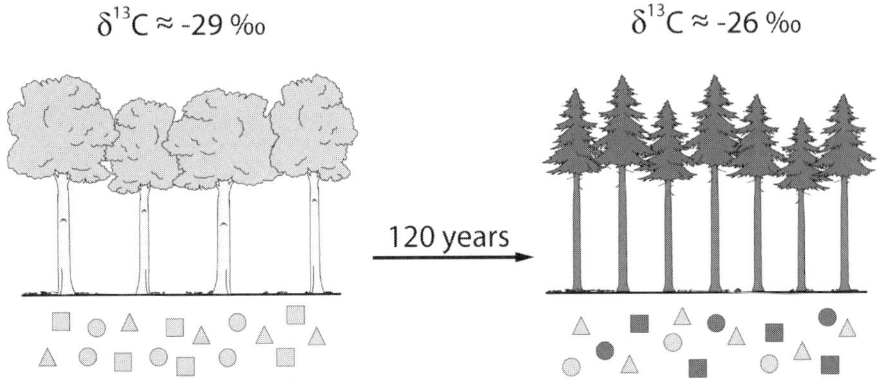

Fig. 9.7. Scheme of a natural labeling experiment. Existing vegetation, such as beech in forest or wheat in agricultural fields (not shown), are replaced by structurally similar but isotopically different plants, e.g., conifers in forest or maize in agriculture (not shown). Individual molecules from soil organic matter, represented by *circles, triangles* and *squares*, are labeled with different speeds as dependent on turnover time

Corresponding turnover times are, for bulk soil organic matter in the upper 25 cm, between 10 and 100 years (Balesdent and Mariotti 1996; Collins et al. 2000; Paul et al. 2001). In forest ecosystems the turnover was also estimated in free air carbon enhancement (FACE) experiments, producing ^{13}C-labeled litter. Interestingly, only very little plant-derived labeled carbon entered the organic layer of the corresponding forest soils (Schlesinger and Lichter 2001), indicating that most litter-derived carbon was respired and did not enter into carbon storage. Similarly, in a 120-year-old vegetation change from an autochthonous beech stand to spruce at the Waldstein, Fichtelgebirge, Germany, this extremely low input of new carbon to soil organic matter was also confirmed (Fig. 9.8). The calculated mean residence time for soil organic matter increases from 60 years in the litter layer to more than 5,000 years at the 10–30 cm depth. However, these turnover rates contrast with calculated turnover rates from ^{14}C ages (Harrison et al. 2000), which hardly reach 500 years in bulk soil organic matter in these profiles. Obviously, carbon input occurs not exclusively through the litter path. We only can speculate that root carbon and root exudates are introducing ^{14}C young carbon with a similar ^{13}C value into deeper soil layers. This would produce low ^{14}C ages and high mean turnover times of soil carbon in the soil profile. In consequence, differences in the root architecture of tree species will inject carbon at different soil depths and, therefore, will influence soil carbon storage. However, more experimental data are needed to verify this assumption.

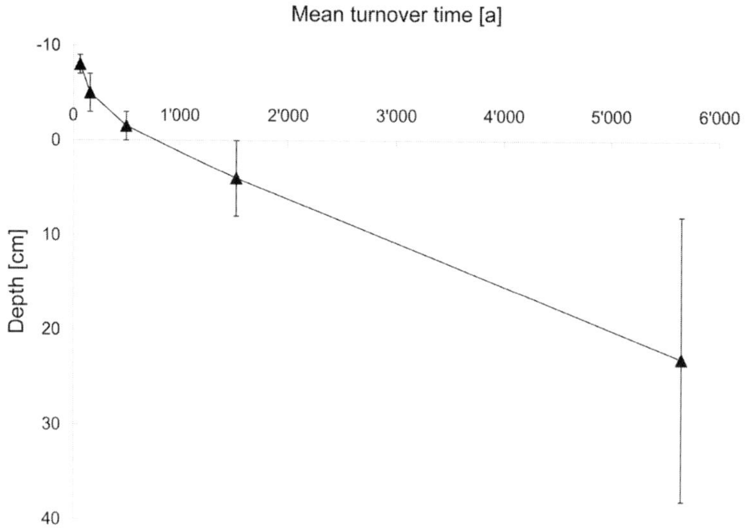

Fig. 9.8. Turnover time of soil organic matter from different depths of an autochthonous beech stand converted to spruce. Litter layers are indicated by negative depth

9.6 Molecular Turnover of Soil Organic Matter

To estimate the mechanisms of carbon storage we have to understand the interactions between above-ground and below-ground diversity (Hooper et al. 2000). The turnover of specific plant markers, such as lignin, or microbial markers, e.g., phospholipid fatty acids, can identify the importance of plant versus microbe-derived chemical structures for carbon storage. Based on these markers, one can estimate the role of microorganisms in this process. To determine the compound-specific isotope ratios of individual compounds and consequently their molecular turnover, two methods appear to be suitable. Firstly, nonpolar solvents can extract soluble compounds, e.g., lipids, and isotope ratios of specific molecules with known origin or stability (such as recalcitrant alkanes from plants or labile phospholipid fatty acids (PLFA) from microbial cell walls) can be determined using isotope ratio mass spectrometry (GC-C-IRMS). Secondly, for structural, insoluble compounds of plants or soil, e.g., cellulose or lignin, thermal extraction of breakdown products and subsequent isotope ratio determination has been developed (Gleixner and Schmidt 1998).

As already mentioned for experiments that changed the vegetation (broadleaf to conifer stands or C3 to C4 plants), the turnover of individual compounds can be determined according to the analytical procedure known for bulk soil organic matter. The squares in Fig. 9.7 indicate already complete labeling of the new vegetation with the isotopic signals, whereas circles show no labeling at all. However, corresponding investigations have thus far not been performed in forest ecosystems. We must therefore demonstrate the potential of the compound specific isotope ratios to estimate the processes of carbon storage in forest ecosystems using agricultural sites.

In contrast to current assumptions on SOM stability, turnover times of under 1 year for the major plant-derived molecules such as "stable" lignin and cellulose molecules were found in agricultural soils (Gleixner et al. 1999, 2001b). This supports the idea that plant-derived carbon skeletons are neither chemically nor physically stabilized in soil in their original structure. Consequently, the influence of plant diversity on the chemistry of soil carbon will be small. No indication for specific, recoverable molecules with turnover times in the millennium range, as suggested by soil carbon models, could be found in the soils under investigation. The existence of this pool is thus in question, although ^{14}C ages of bulk soil carbon have been found in the millennium range (Wang et al. 1996). In contrast, pyrolysis products of carbohydrates and proteins, which were only present in soil samples, had unexpectedly long turnover times of between 20 and 100 years (Fig. 9.9). These turnover times are in agreement with those of the bulk soil. Carbohydrates and proteins are

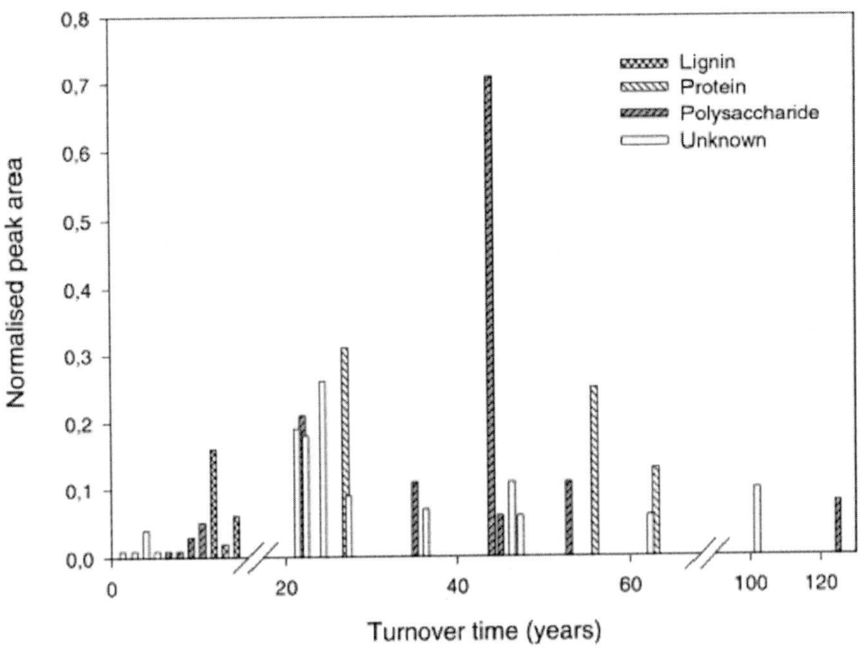

Fig. 9.9. Turnover time and relative peak area of individual pyrolysis products for bulk soil submitted to vegetation change from C3 plants to C4 plants. (Gleixner et al. 2002)

thought to be unstable in soil (Trojanowski et al. 1984), but they are also known to be major parts of soil microorganisms. Consequently, carbon turnover and storage might be controlled by soil-organism derived C. In fact, the below-ground biodiversity might be of higher importance for carbon storage then plant biodiversity. Although our knowledge on the relation of soil microorganisms and carbon storage in forest ecosystems is very limited, the role of soil microbiota in carbon storage has been estimated in agricultural systems using labile PLFAs to trace the flow of labeled carbon into the microbial carbon pool. PLFAs were extracted from soils in 40-year-old C3/C4 vegetation change experiments in Halle, Germany. In contrast to previous understanding, it has been found that only some organisms feed on the new plant-derived carbon being labeled with the new isotopic signal. Most organisms were using partially or completely "humified" soil organic matter as a carbon source, which was only partially or not at all labeled by the new vegetation (Fig. 9.10). Obviously, the flow of carbon in agricultural soils is mainly controlled by soil organisms. This suggests that soil carbon is not "stable" in soils but is continuously reused, and that the observed SOM is remaining microbial biomass. Consequently, every process that keeps the individual carbon atoms in this recycling process possibly increases the carbon storage in soils. Belowground biodiversity or the size of belowground food webs might be of major importance. However, so far no investigation linking below ground biodiversity and carbon storage exists.

The assumption that soil microbes control carbon storage is strongly supported by results from the Long Term Ecological Research site at the Niwot

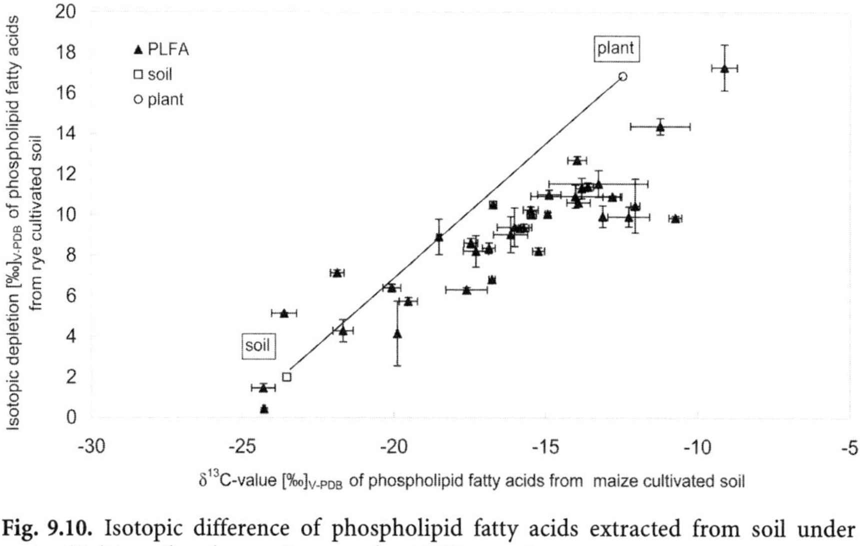

Fig. 9.10. Isotopic difference of phospholipid fatty acids extracted from soil under maize (C4) to soil under continuous wheat (C3) cropping

Ridge, Colorado (Neff et al. 2002). In this extremely N-limited environment, the addition of nitrogen increased both the primary production and the plant species richness and composition. However, neither carbon storage nor ^{14}C content of soil organic matter was significantly affected over a period of 13 years by this change. Compound-specific isotope ratios have demonstrated that "young" plant-derived carbon structures of cellulose and lignin are completely degraded through the addition of nitrogen. At the same time, the turnover of the mineral-associated carbon accelerated and new carbon from the decomposition process entered this pool. Both effects are related to enhanced microbial activity and can only be understood at the molecular level of soil carbon.

9.7 Conclusion

Recent advances in knowledge of molecular turnover rates of soil organic matter suggest that our understanding of carbon storage in soil is rather limited. Moreover, new findings suggest that a direct link between plant diversity and carbon storage may not exist. Carbon storage, however, appears to be indirectly influenced by plant biodiversity through the change of nutrients or water availability, and by quality, quantity, and distribution of leaf and root litter (Langley and Hungate 2003), or by the feedback between above- and belowground diversity, the latter of which is strongly influenced by root exudates of plants. Our results highlight the major importance of the composition of the whole belowground food web in carbon storage. This factor is often inherited and changes only slightly with aboveground biodiversity and/or time (Kowalchuk et al. 2002; Wardle et al. 2003). Interestingly, our results suggest that carbon turnover, perhaps controlled by the biodiversity of soil organisms, is of greater importance for carbon storage than storage capacity, which depends on soil mineralogy.

References

Almendros G, Sanz J, Velasco F (1996) Signatures of lipid assemblages in soils under continental Mediterranean forests. Eur J Soil Sci 47:183–196

Balesdent J, Mariotti A (eds) (1996) Measurement of soil organic matter turnover using ^{13}C natural abundance. Dekker, New York

Banfield JF, Barker WW, Welch SA, Taunton A (1999) Biological impact on mineral dissolution: Application of the lichen model to understanding mineral weathering in the rhizosphere. Proc Natl Acad Sci USA 96:3404–3411

Barker WW, Banfield JF (1996) Biologically versus inorganically mediated weathering reactions – relationships between minerals and extracellular microbial polymers in lithobiontic communities. Chem Geol 132:55–69

Baudoin E, Benizri E, Guckert A (2003) Impact of artificial root exudates on the bacterial community structure in bulk soil and maize rhizosphere. Soil Biol Biochem 35:1183–1192
Boutton TW, Yamasaki S (eds) (1996) Mass spectrometry of soils. Dekker, New York
Brassell SC, Eglinton G, Marlowe IT, Pflaumann U, Sarnthein M (1986) Molecular stratigraphy – a new tool for climatic assessment. Nature 320:129–133
Catovsky S, Bradford MA, Hector A (2002) Biodiversity and ecosystem productivity: implications for carbon storage. Oikos 97:443–448
Christensen BT (1992) Physical fractionation of soil and organic matter in primary particle size and density separates. Adv Soil Sci 20:1–90
Collins HP, Elliott ET, Paustian K, Bundy LC, Dick WA, Huggins DR, Smucker AJM, Paul EA (2000) Soil carbon pools and fluxes in long-term corn belt agroecosystems. Soil Biol Biochem 32:157–168
Coplen TB (1996) New guidelines for reporting stable hydrogen, carbon, and oxygen isotope-ratio data. Geochim Cosmochim Acta 60:3359–3360
Cranwell PA, Eglinton G, Robinson N (1987) Lipids of aquatic organisms as potential contributors to lacustrine sediments 2. Org Geochem 11:513–527
Eglinton G, Hamilton RJ (1967) Leaf epicuticular waxes. Science 156:1322–1334
Eglinton G, Raphael RA, Gonzalez AG, Hamilton RJ (1962) Hydrocarbon constituents of wax coatings of plant leaves – a taxonomic survey. Phytochemistry 1:89–102
Gleixner G, Schmidt HL (1998) On-line determination of group-specific isotope ratios in model compounds and aquatic humic substances by coupling pyrolysis to GC-C-IRMS. In: Stankiewicz AB, van Bergen PF (eds) Nitrogen-containing macromolecules in the bio- and geosphere. ACS symposium series, vol 707. American Chemical Society, Washington, DC, pp 34–46
Gleixner G, Danier HJ, Werner RA, Schmidt H-L (1993) Correlations between the ^{13}C content of primary and secondary plant products in different cell compartments and that in decomposing basidomycetes. Plant Physiol 102:1287–1290
Gleixner G, Bol R, Balesdent J (1999) Molecular insight into soil carbon turnover. Rapid Commun Mass Spectr 13:1278–1283
Gleixner G, Czimczik CJ, Kramer C, Lühker B, Schmidt MWI (2001a) Plant compounds and their turnover and stabilization as soil organic matter. In: Schulze ED, Heimann M, Harrison S et al (eds) Global biogeochemical cycles in the climate system. Academic Press, San Diego, pp 201–215
Gleixner G, Kracht O, Schmidt HL (2001b) Group specific isotope ratios of humic substances. In: Swift SS, Spark KM (eds) Understanding and managing organic matter in soils, sediments and waters. International Humic Substance Society, St. Paul, pp 195–201
Gleixner G, Poirier N, Bol R, Balesdent J (2002) Molecular dynamics of organic matter in a cultivated soil. Org Geochem 33:357–366
Grayston SJ, Vaughan D, Jones D (1997) Rhizosphere carbon flow in trees, in comparison with annual plants – the importance of root exudation and its impact on microbial activity and nutrient availability. Appl Soil Ecol 5:29–56
Harrison AF, Harkeness DD, Rowland AP, Garnett JS, Bacon PJ (2000) Annual carbon and nitrogen fluxes in soils along the European forest transect, determined using the ^{14}C-Bomb. In: Schulze ED (ed) Carbon and nitrogen cycling in European forest ecosystems, Ecological Studies, vol 142. Springer, Berlin Heidelberg New York, pp 237–256
Hooper DU, Bignell DE, Brown VK, Brussaard L, Dangerfield JM, Wall DH, Wardle DA, Coleman DC, Giller KE, Lavelle P, Van der Putten WH, De Ruiter PC, Rusek J, Silver WL, Tiedje JM, Wolters V (2000) Interactions between aboveground and belowground biodiversity in terrestrial ecosystems: Patterns, mechanisms, and feedbacks. Bioscience 50:1049–1061

Jenkinson DS, Hart PBS, Rayner JH, Parry LC (1987) Modelling the turnover of organic matter in long-term experiments at Rothamsted. INTECOL Bull 15:1–8

Jobbagy EG, Jackson RB (2001) The distribution of soil nutrients with depth: Global patterns and the imprint of plants. Biogeochemistry 53:51–77

Kaiser K, Guggenberger G (2003) Mineral surfaces and soil organic matter. Eur J. Soil Sci 54:219–236

Kinzig AP, Pacala SW, Tilman D (eds) (2001) Functional consequences of biodiversity. Empirical progress and theoretical extensions. Princeton Univ Press, Princeton

Kowalchuk GA, Buma DS, de Boer W, Klinkhamer PGL, van Veen JA (2002) Effects of above-ground plant species composition and diversity on the diversity of soil-borne microorganisms. Antonie Van Leeuwenhoek 81:509–520

Kracht O, Gleixner G (2000) Isotope analysis of pyrolysis products from *Sphagnum* peat and dissolved organic matter from bog water. Org Geochem 31:645–654

Langley JA, Hungate BA (2003) Mycorrhizal controls on belowground litter quality. Ecology 84:2302–2312

Lichtfouse E, Chenu C, Baudin F, Leblond C, Da Silva M, Behar F, Derenne S, Largeau C, Wehrung P, Albrecht P (1998) A novel pathway of soil organic matter formation by selective preservation of resistant straight-chain biopolymers – chemical and isotope evidence. Org Geochem 28:411–415

Loreau M, Naeem S, Inchausti P (eds) (2002) Biodiversity and ecosystem functioning. Synthesis and perspectives. Oxford Univ Press, Oxford

Meyer O (1993) Functional groups of microorganisms. In: Schulze ED, Mooney HA (eds) Biodiversity and ecosystem function, Ecological Studies, vol 99. Springer, Berlin Heidelberg New York, pp 67–96

Neff JC, Townsend AR, Gleixner G, Lehman SJ, Turnbull J, Bowman WD (2002) Variable effects of nitrogen additions on the stability and turnover of soil carbon. Nature 419:915–917

O'Leary MH (1981) Carbon isotope fractionation in plants. Phytochemistry 20:553–567

Parton WJ, Schimel DS, Cole CV, Ojima DS (1987) Analysis of factors controlling soil organic matter levels in great plains grassland. Soil Sci Soc Am J 51:1173–1179

Paul EA, Collins HP, Leavitt SW (2001) Dynamics of resistant soil carbon of midwestern agricultural soils measured by naturally occurring ^{14}C abundance. Geoderma 104:239–256

Riederer M (1989) The cuticles of conifers: structure, composition and transport properties. In: Schulze ED, Lange OL, Oren R (eds) Forest decline and air pollution, Ecological Studies, vol 77. Springer, Berlin Heidelberg New York, pp 157–192

Rieley G, Collier RJ, Jones DM, Eglinton G, Eakin PA, Fallick AE (1991) Sources of sedimentary lipids deduced from stable carbon isotope analyses of individual compounds. Nature 352:425–427

Rothe J, Gleixner G (2000) Do stable isotopes reflect the food web development in regenerating ecosystems? Isotopes Environ Health Stud 36:285–301

Rothe A, Binkley D (2001) Nutritional interactions in mixed species forests: a synthesis. Can J For Res 31:1855–1870

Schimel DS, House JI, Hibbard KA, Bousquet P, Ciais P, Peylin P, Braswell BH, Apps MJ, Baker D, Bondeau A, Canadell J, Churkina G, Cramer W, Denning AS, Field CB, Friedlingstein P, Goodale C, Heimann M, Houghton RA, Melillo JM, Moore B, Murdiyarso D, Noble I, Pacala SW, Prentice IC et al (2001) Recent patterns and mechanisms of carbon exchange by terrestrial ecosystems. Nature 414:169–172

Schlesinger WH, Lichter J (2001) Limited carbon storage in soil and litter of experimental forest plots under increased atmospheric CO_2. Nature 411:466–469

Schmidt H-L, Gleixner G (1998) Carbon isotope effects on key reactions in plant metabolism and ^{13}C-patterns in natural compounds. In: Griffiths H (ed) Stable isotopes:

integration in biological, ecological and geochemical processes. BIOS, Oxford, pp 13–26

Schulze ED (2000) Carbon and nitrogen cycling in European forest ecosystems. Ecological Studies, vol 142. Springer, Berlin Heidelberg New York

Schwark L, Zink K, Lechterbeck J (2002) Reconstruction of postglacial to early Holocene vegetation history in terrestrial Central Europe via cuticular lipid biomarkers and pollen records from lake sediments. Geology 30:463–466

Spooner N, Rieley G, Collister JW, Lander M, Cranwell PA, Maxwell JR (1994) Stable carbon isotopic correlation of individual biolipids in aquatic organisms and a lake bottom sediment. Org Geochem 21:823–827

Trojanowski J, Haider K, Huettermann A (1984) Decomposition of carbon-14-labeled lignin holocellulose and lignocellulose by mycorrhizal fungi. Arch Microbiol 139:202–206

Van Bergen PF, Bull ID, Poulton PR, Evershed RP (1997) Organic geochemical studies of soils from the Rothamsted classical experiments. 1. Total lipid extracts, solvent insoluble residues and humic acids from Broadbalk wilderness. Org Geochem 26:117–135

Wang Y, Amundson R, Trumbore S (1996) Radiocarbon dating of soil organic matter. Quat Res 45:282–288

Wardle DA, Yeates GW, Williamson W, Bonner KI (2003) The response of a three trophic level soil food web to the identity and diversity of plant species and functional groups. Oikos 102:45–56

10 Silviculture and Its Interactions with Biodiversity and the Carbon Balance of Forest Soils

M. MUND and E.-D. SCHULZE

10.1 Introduction

It is well known that intensive forest management practices can have significant effects on the biogeochemistry and biodiversity of forest ecosystems. For example, planting and thinning affects the structural biodiversity. Planting of nursery trees also determines the species (including mycorrhizae) and genetic diversity. Fertilization changes the nutrient balance, and thus competitive interactions. Clear-cutting combined with intensive soil preparation causes soil erosion, soil compaction, and losses of soil organic carbon and cations, which in turn affects biodiversity (e.g., Heinsdorf and Krauß 1974; Bormann and Likens 1979; Covington 1981; Heinsdorf 1986; Black and Harden 1995; Apps and Price 1996; Nyland 1996; Jurgensen et al. 1997; Rollinger et al. 1998; Worrell and Hampson 1997; Prescott et al. 2000b; Quesnel and Curran 2000; Johnson and Curtis 2001; Block et al. 2002). However, our knowledge about the interactions of biodiversity with silviculture and site-specific factors and the role of biodiversity in biogeochemical cycles is still very limited.

The Kyoto-Protocol (UN 1997) and the "Bonn agreement" (UN 2001), in particular, raised the question if and which forest management practices influence the carbon balance of forest ecosystems. It is evident that increased decomposition of dead organic matter after clear-cutting results in a net loss or a zero carbon balance of the forest ecosystem over about 5–6 years afterwards, even when successful regeneration occurs (Pypker and Fredeen 2002; Rannik et al. 2002). The time period of net carbon release can be prolonged to 14–20 years if growth of the regenerating stands is reduced or if large amounts of dead wood remain on site (e. g., Cohen et al. 1996; Schulze et al. 1999). Nevertheless, the relative contribution of decomposing dead wood, organic-layer material or soil organic matter (SOM) to the net ecosystem carbon balance is still unclear. Also, the mechanisms that could cause the large discrepancies observed between different case studies investigating the

Ecological Studies, Vol. 176
M. Scherer-Lorenzen, Ch. Körner, and E.-D. Schulze (Eds.)
Forest Diversity and Function: Temperate and Boreal Systems

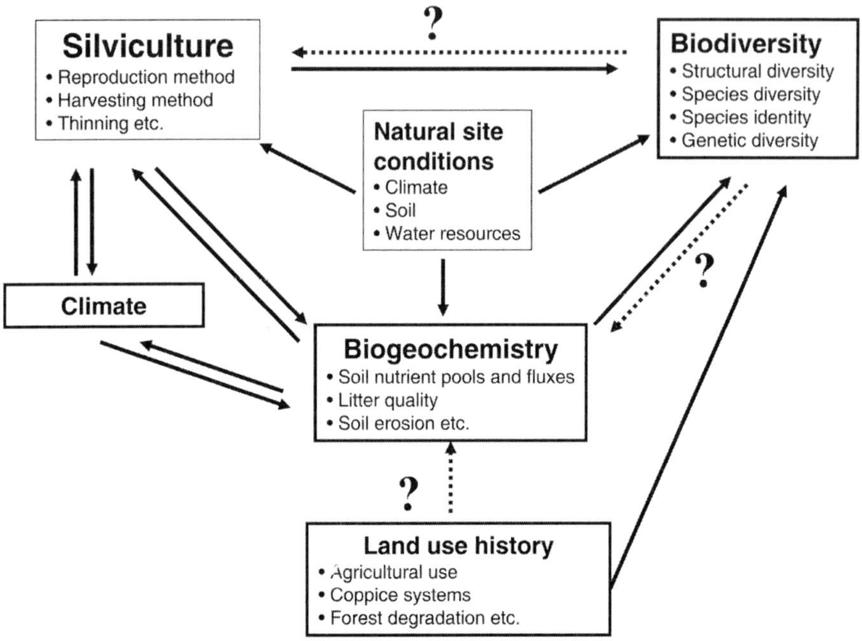

Fig. 10.1. Overview of the general objectives of this chapter. *Questions marks* represent interactions that are still unclear

impacts of forest management on soil organic carbon pools (SOC) are poorly understood.

This paper will present a review of the interactions of silviculture, biodiversity, and the carbon balance of forest ecosystems (Fig. 10.1). We will review published case studies and meta-analyses dealing with (1) different silvicultural systems or harvesting methods in temperate or boreal forests, (2) different tree species, and (3) soil organic carbon pools as indicators of long-term changes in the biogeochemistry of forest ecosystems.

10.2 Overview of Silvicultural Systems, Terms and Definitions

The following terms are sometimes ambiguously or inconsistently defined in the literature, therefore we explicitly define them here, as we use them in the context of this review chapter.

The term "silviculture" includes the art and practice of controlling the regeneration, composition, health, quality, growth, and harvest of forest vegetation. The most important and common silvicultural systems of temperate and boreal forests available for wood supply are characterized in Table 10.1. More comprehensive descriptions of silvicultural systems are given in Burschel and Huss (1987), Matthews (1989), Röhrig and Gussone (1990) and Nyland (1996).

The term "forest management" encompasses all practices of applying scientific, economic, administrative, philosophical, and social principles regarding forested ecosystems. Discussions about effects of forest management on SOC pools are often unclear because the term "forest management" is used to describe the management of already existing forest ecosystems (e.g., thinning, reforestation, and chronosequences after clear-cutting) as well as land-use changes such as afforestation and deforestation. However, land-use changes from forested to non-forested land and vice versa (e.g. Matson et al. 1997; Thuille et al. 2000; Guo and Gifford 2002) represent a very different type of disturbance compared to conversions of primary forests into managed forests or compared to regular silvicultural practices. Therefore in this review we clearly distinguish between "silviculture/management of existing forests" and "land-use changes," and we do not consider afforestation or deforestation except when these processes are helpful for understanding carbon decomposition or accumulation.

In general, it is assumed that the degree of disturbance due to final harvest, regeneration practices, thinning, fertilization, or biocide application

Table 10.1. Brief characterization of silvicultural systems

Reproduction (cutting) method	Clear-cutting	Regular shelterwood	Irregular shelterwood	Selection cutting
Regeneration practice	Planting, seeding, and/or natural regeneration; soil preparation			Mainly natural regeneration
Structure	Even-aged; one or two canopies			Uneven-aged; horizontal and vertical structure
Species diversity	Mono-species stands or plantations			Mono-species or multi-species stands
Tree harvest	Sawlog or whole-tree harvesting [a]			Mainly sawlog harvesting
Tending and further management activities	• Release cutting, thinning • Pruning • Fertilization • Herbicide/pesticide application • etc.			• Release cutting, thinning • Herbicide/pesticide application

[a] "Sawlog harvesting": only bole material is extracted and branches, twigs, and needles or leaves are left on site. "Whole-tree harvesting": entire aboveground biomass, including twigs and needles or leaves, of harvested trees is removed

decreases as follows: clear-cutting >regular shelterwood system >irregular shelterwood system >selection system (see also Marshall 2000). If the monetary investment per unit area or the number of operational actions per rotation is taken to define "intensive" (versus "extensive") silviculture, then the ranking list would be in the opposite direction (Grigal 2000).

"Sawlog harvesting" describes a logging method where only bole material is extracted, and branches, twigs and needles or leaves are left on site as harvest residues (slash). "Whole-tree harvesting" means that the entire aboveground biomass, including twigs and needles or leaves of harvested trees is removed from the forest site. In this review, the term "whole-tree harvesting" includes the immediate removal of whole trees as well as the removal of harvest residues after harvesting the bole (soil preparation, collection of firewood by local people).

The term "soil organic carbon (SOC)" comprises only carbon in dead organic matter in the mineral soil. Organic carbon in the organic layer (L, F, and H horizon) is presented and discussed separately.

The term "pool" is equivalent to the term "stock" and represents a mass per unit area (e.g., t C ha^{-1}) and should be distinguished from fluxes such as net ecosystem exchange (NEP) or decomposition rate (e.g., t C ha^{-1} $year^{-1}$), and concentrations (e.g., g C $g_{soil\ dw}^{-1}$).

10.3 Methodological Restrictions

To interpret results of case studies on silviculture and its interactions with biodiversity and the carbon balance of forest ecosystems the following methodological restrictions should be considered.

10.3.1 Methodological Restrictions with Respect to Biodiversity Effects

Forest ecosystems are characterized by the long lifetimes of trees and, therefore, by processes and mechanisms that affect species or structural diversity only after a delay of several years to decades. The most common ways to deal with this fundamental scientific challenge are by conducting case studies that: (1) compare several forest stands characterized by similar site conditions but different tree species or stand structures (site comparison); (2) investigate single forest stands a few years before and a few years after forest management activities (time series); or (3) analyze stands of different age (forest chronosequences). These approaches are associated with a number of restrictions and assumptions that confine the analysis and interpretation of potential biodiversity effects:

1. In contrast to experimental studies, case studies depend on existing species compositions and management practices, and thus reflect mainly natural site conditions and forestry activities and economical demands of the past.
2. Most temperate and boreal forests that are managed for several decades or centuries are characterized by specific combinations of tree species and silvicultural practices. Many theoretical combinations of species compositions and silviculture are very rare or do not exist at all. For example, in Germany large areas of uneven-aged, mixed-species forests were transformed into pure, even-aged spruce-only stands, and the management scheme continued with specific silvicultural practices such as thinning from below, strip clear-cutting or strip shelterwood method, and a rotation period of 80 to 120 years. Mixed-species (conifers and deciduous trees), even-aged stands, or uneven-aged spruce stands that are older than 50 years are rare.
3. The large spatial heterogeneity within forest ecosystems reduces the ability to detect significant effects of forest management or biodiversity on biogeochemical cycles. Therefore, most case studies initially investigated extreme conditions that promised significant results. Examples are comparisons of clear-cutting and natural forests, or comparisons between pure, even-aged coniferous stands and uneven-aged, mixed deciduous stands.

Considering these many constraints on forest ecosystem case studies, it is not surprising that most studies: (1) deal with species identity effects and not with species diversity or structural diversity effects, and (2) compare the most common and extreme silvicultural practices, neglecting gradual differences and modern, extensive management practices.

10.3.2 Methodological Restrictions with Respect to Soil Organic Carbon Analysis

Intensive soil disturbances due to agricultural cultivation cause significant reductions in the "labile" (fragile or unprotected) and part of the "recalcitrant" (stable or protected with respect to microbial decomposition) SOC pool, resulting in high and long-term reductions of SOC in the bulk soil (Baldock et al. 1992; Bonde et al. 1992; Christensen 1992, 2001; Desjardins et al. 1994; Sollins et al. 1996; Balesdent et al. 1998). If it is assumed that silvicultural practices or changes in biodiversity represent only moderate or at least less-intensive disturbances than the conversion of forests to croplands, then a separation of SOC into a labile pool and a recalcitrant pool would appear to be a reasonable approach to quantify effects of silviculture and biodiversity on SOC (e.g., Guggenberger et al. 1994, 1995; Ellert and Gregorich 1995; Khanna et al. 2001). However, this approach is difficult to apply because the mechanisms involved in the protection of SOC against decomposition (e.g., microbial stabilization or physical protection due to organo-mineral complexes or micro-aggregates, see Gleixner et al., Chap. 9, this Vol.) and the methodology used to determine labile or recalcitrant SOC pools (e.g., density fractionation or a size fractionation) are still under discussion. Therefore, most available case studies focus on impacts of forest management on bulk soil only.

10.4 Effects of Species Composition and Species Identity Effects

The effects of mixed tree species on the nutrient balance and sustainability of forests play a major role in forestry research, and it is expected that species composition influences litter quality, nutrient availability and decomposition of organic matter (e.g., Vesterdal and Raulund-Rasmussen 1998; Prescott et al. 2000 c; Rothe and Binkley 2001; Berger et al. 2002; Prescott 2002; Hättenschwiler, Chap. 8, this Vol.). However, Rothe and Binkley concluded that "general conclusions are limited by the small number of studies that directly address mixed-species effects in forests and the wide variety of observed interactions" (Rothe and Binkley 2001, p. 1855). For instance, in several different studies, litter decomposition rates and nutrient release in mixed-species forests increased, decreased, or did not differ, when compared to monocultures. Only at oligotrophic sites were mixed-species forests found to generally increase forest growth and mineralization compared to mono-species forests, especially when nitrogen-fixing tree species were included (Binkley 1992; Morgan et al. 1992). However, total nutrient pools in the mineral soil seem to be unaffected by tree species diversity during the period of observation (Rothe and Binkley 2001).

It had been expected that the SOC balance of different vegetation types [for example, beech (*Fagus sylvatica*) compared to spruce (*Picea abies*) forest ecosystems] would differ significantly and reveal clear species-identity effects (Grigal and Ohmann 1992; Binkley and Giardina 1998). However, recent studies have shown that site conditions and site history can also influence or superimpose on species-identity effects to such an extent that biodiversity effects could not be detected (Schulze et al. 2000; Wirth et al. 2004). One example can be found in the European research project CANIF (Carbon and Nitrogen Cycling in Forest Ecosystems) that investigated the carbon and nitrogen cycling of mono-species beech (*Fagus sylvatica*) and mono-species spruce (*Picea abies*) forest ecosystems along a north-south transect across Europe (Schulze et al. 2000). In this study, it was evident that carbon pools in the soil organic layer and the mineral soil were higher under spruce stands than under beech stands when the beech or spruce stands existed under the same climatic and edaphic conditions (beech–spruce pairs in Denmark, Czech Republic, Germany, Fig. 10.2A, B). In contrast, the site studies in Italy revealed that the effect of vegetation type can be compensated for by the climatic factors associated with the different elevations of the study sites (the beech stand grew at an elevation of 1,560 m a.s.l., whereas the spruce stand was at 905 m a.s.l.). In study sites in France, SOC pools in the spruce forest were lower than those in the beech forest, probably because of litter raking and grazing at the spruce site. While these results regarding the soil organic layer confirm general expectations (e.g., Matzner 1988; Rehfuess 1990; Ulrich and Puhe 1994), the differences found for the mineral soil were relatively small. Furthermore, Wirth et al. (2004) have reported in a recent survey of the Thuringia state forest a weak trend of higher SOC pools under beech compared to spruce stands, when other factors such as climate and soil texture are excluded.

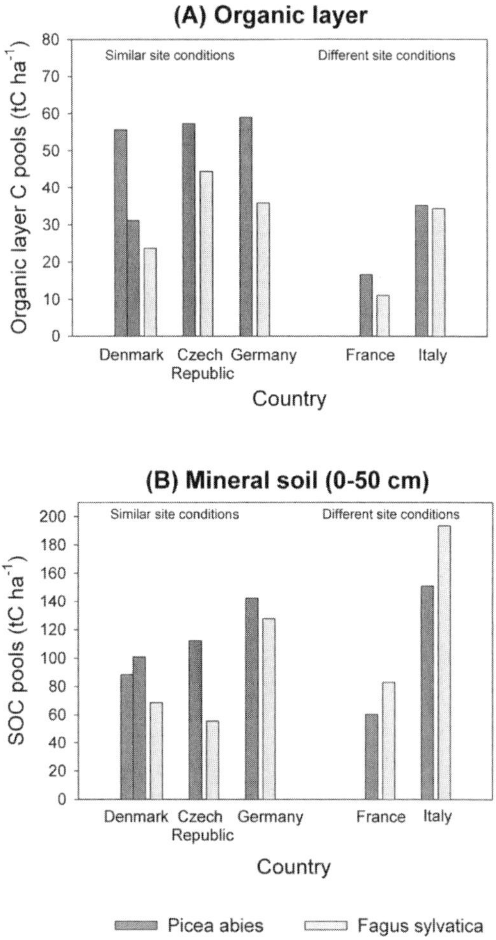

Fig. 10.2A, B. Carbon pools in **A** the forest floor and the **B** the mineral soil (0–50 cm soil depth) of beech forests (*Fagus sylvatica*) compared to spruce forests (*Picea abies*) in different European countries. (Persson et al. 2000)

It seems to be evident that changes in the amount and chemistry of the forest floor and soil cation availability are the most relevant processes affecting soils that were naturally covered by deciduous forests but then planted with conifers (e.g., Matzner 1988; Rehfuess 1990; Ulrich and Puhe 1994; Schulze et al. 1996; Rothe et al. 2002b). For example, it was found that spruce utilizes nitrate at lower rates than beech, and in spite of higher ammonium and nitrate concentrations in beech than in spruce litter, nitrogen leaching is higher under spruce than under beech forests (Gebauer et al. 2000). Furthermore, many case studies in Europe, as summarized by Rothe et al. (2002a), showed a generally higher throughfall deposition and soil leaching of nitrogen and sulfur compounds for spruce forests relative to beech forests. Schulze et al. (2000) suggested that increased nitrogen leaching would also reduce the cation availability in the long term (nitrate and cation cotransport) and this would in turn accelerate podsolization (Nihlgård 1971; Lundström et al. 2000). Podsolization is buffered or accelerated by site-specific factors such as mineral composition of the soil, soil texture, and climate (Lundström et al. 2000). In the beginning, podsolization does not lead to a general reduction of SOC pools but results in a re-translocation of soil carbon pools. Carbon accumulates in the organic layer (Of+Oh horizons) and upper soil horizon (Ah horizon), leaches from the Ae horizon, and accumulates in the Bsh horizon. Only in a very advanced stage is organic carbon remobilized again and released from the Bsh horizon.

10.5 Effects of Conversions of Primary Forests to Managed Forests

A conversion of primary forests or old-growth forests to plantations or managed semi-natural forests leads to a significant reduction of carbon pools in the living and dead aboveground biomass. Depending on thinning regime, rotation period, final harvest, climate, and site productivity, the average living and dead aboveground biomass of managed forests reaches only 20–55 % of the original primary forest biomass (e.g., Houghton et al. 1983; Harmon et al. 1990; Cannell et al. 1992; Burschel et al. 1993; Karjalainen 1996; Fleming and Freedman 1998; Trofymow and Blackwell 1998; Weber 2001; Crow et al. 2002). In particular, the high amount of dead wood (snags and logs) of primary forests, ranging between 30 and 500 t_{dw} ha^{-1}, is reduced to between 2 and 40 t_{dw} ha^{-1} in managed forests (Grier and Logan 1977; Harmon et al. 1986, 1990; Kirby et al. 1998; Duvall and Grigal 1999; Krankina et al. 2002; Pedlar et al. 2002; Mund 2004). Harmon et al. (1990) have calculated that it would take more than 250 years to re-accumulate nearly natural quantities of dead wood following clear-felling of old-growth forests. Duvall and Grigal (1999) even argue that accumulation of dead wood in unmanaged forests does not reach a steady state but continues to increase until either catastrophic disturbances or succession to another vegetation type occur. With respect to structural diversity (stand structure, plant species composition, and landscape patchiness), disturbances due to forest management activities, especially clear-cutting, differ remarkably from natural disturbances (e.g., wildfires, windthrow; Franklin et al. 2002; Lindenmayer and McCarthy 2002; Seymour et al. 2002).

In contrast to the high losses of carbon pools in living and dead wood biomass due to conversion of primary forests to managed forests, there are studies that did not find significant changes in the SOC pools of these converted forests (Fleming and Freedman 1998; Weber 2001). According to a recently published meta-analysis by Guo and Gifford (2002), a conversion of native broadleaved forests into broadleaved plantations in Brazil and Nigeria had on average little effect on SOC pools, whereas the planting of coniferous plantations (*Pinus radiata*) in Australia and New Zealand reduced SOC pools on average by 15 % compared to primary forests (Fig. 10.3). However, the reduction of SOC pools was restricted to sites with annual precipitation exceeding 1,500 mm. Since organic carbon pools in the organic layer and the mineral soil generally increase with increasing precipitation (Ulrich and Puhe 1994;

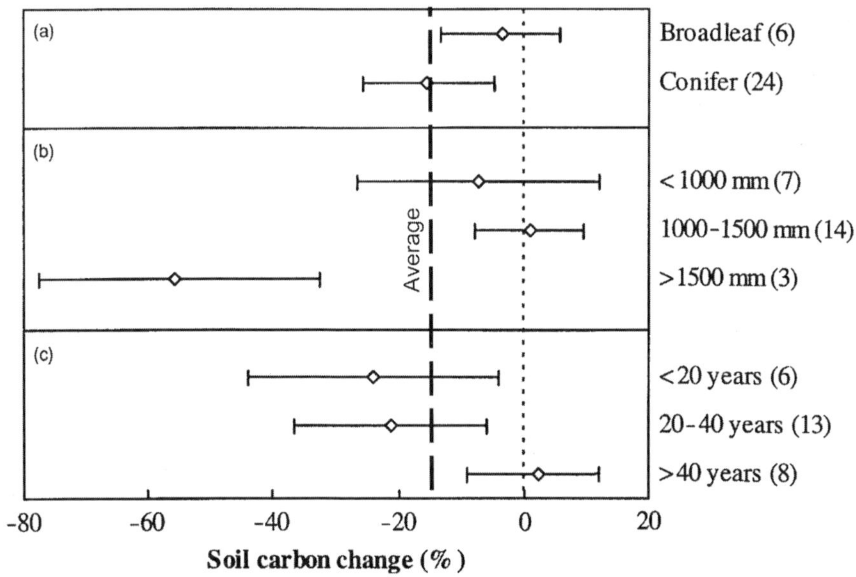

Fig. 10.3a–c. Effects of conversions from native broadleaved forests to broadleaved or coniferous plantations in Brazil, Australia, New Zealand, and Nigeria on soil organic carbon pools. Effects depending on **a** the vegetation type of plantations, **b** precipitation, and **c** stand age of plantations (95 % confidence intervals are shown and numbers of observations are in *parentheses*). (Guo and Gifford 2002)

Wirth et al. 2004), this observation indicates that primary forests with high SOC pools are especially susceptible to carbon losses in the mineral soil when they are converted to managed forests. Under the conditions of plantations in Australia and New Zealand, as reported by Guo and Gifford (2002), a reduction in SOC pools was also restricted to young plantations with stand ages less than 40 years. This time period of 40 years required to accumulate the previous amount of SOC pools is surprisingly short and indicates that most likely only the labile SOC pool was affected by the conversions of primary forests into *Pinus radiata* plantations. (The labile SOC pool of forest soils comprises about 10–60 % of the total organic carbon pool in the bulk soil; Ellert and Gregorich 1995; Entry and Emmingham 1998; Garten et al. 1999).

A distinction between the vegetation types "coniferous forests" and "broadleaved forests" is important for the interpretation of large data sets such as the one analyzed by Guo and Gifford (2002). However, a classification of "coniferous" and "broadleaved" trees represents only "species identity effects" and does not aim at describing interactions of species or structural diversity with land-use change and SOC pools of forests. Because of the annual dynamics of litter-fall, it is very likely that a conversion of deciduous, broadleaved forests into evergreen coniferous plantations (e.g., mixed-hardwood forests of central Europe into pine or spruce plantations) would differ with respect to biogeochemical cycles from a conversion of evergreen, broadleaved forests into pine plantations (e.g., eucalyptus into pine). Under undisturbed, natural conditions of primary forests in a temperate climate, evergreen conifers usually dominate on less favorable sites (infertile, xeric, or wet soils and cool, mainly mountainous climate). Evergreen coniferous forests on such sites are generally characterized by low turnover rates (mineralization) that result in higher accumulation of dead organic matter in the organic layer and A horizon compared to deciduous broadleaved forests (excluding mountainous forests on steep slopes; Schulze 1982; Ellenberg 1996; Landsberg and Gower 1997). Thus, the total amount of carbon in the organic layer and in the upper soil horizon is higher but less stabilized in coniferous forests, and therefore more susceptible to disturbances (Ulrich and Puhe 1994; Ellert and Gregorich 1995; Prescott et al. 2000a). Furthermore, conversions of native broadleaved forests to coniferous plantations are quite common, whereas conversions of native coniferous forests into broadleaved forests are rare. Thus, broadleaved plantations often are more similar to native forests than coniferous plantations.

Despite the fact that SOC decreases in many cases of conversions of primary forests to managed forests, a quantitative conclusion cannot be drawn from available studies. Growth and dominance of conifers or deciduous trees are generally correlated with climatic and edaphic factors, and with different management strategies, and it seems to be impossible to separate species diversity, management, and site-specific effects on SOC pools by observational studies alone. Furthermore, in central Europe there are no primary forests remaining (except for a few sites that have a management history also). Therefore, in central Europe no reference sites exist that could be investigated to clarify the effect of forest use and management on SOC pools.

10.6 Effects of Silvicultural Practices

The most intensive disturbances due to forest management are caused by herbicide treatments, soil preparation (like scalping or bedding), prescribed fires and/or fertilizations, which affect SOC pools (e.g., Heinsdorf and Krauß 1974; Mattson and Smith 1993; Black and Harden 1995; Johnson and Henderson 1995; Johnson and Curtis 2001; Laiho et al. 2003). However, it remains unclear to what extent less disturbing forest management practices, such as "tree harvesting only" instead of "harvesting combined with soil preparation and fertilization" or "single tree cutting" instead of "clear-cutting," affect the biogeochemistry of forest ecosystems, and in particular SOC pools.

10.6.1 Tree Harvesting

In a meta-analysis, Johnson and Curtis (2001) compared the effects of "whole-tree harvesting" (removal of all residues) and "sawlog harvesting" (residues are left on site) on SOC pools in the A horizon, excluding the organic layer. "Whole-tree harvesting" reduced SOC pools in the A horizon by 6 %, while "sawlog harvesting" caused an average increase in SOC of 18 % (Fig. 10.4). However, this positive effect seemed to be restricted to coniferous forest stands only, while hardwoods showed a small negative effect and mixed stands no effect in the case of sawlog harvesting. Furthermore, Johnson and Curtis (2001) consider that the positive effect of residues left on site and incorporated into the soil last only for a few years to decades until the material has been decomposed.

To interpret the results by Johnson and Curtis (2001) with respect to "biodiversity effects" or "management effects," it is important to consider interactions between these factors as well as with site conditions. For example, hardwood forests are more abundant under favorable oceanic climates and on rich soils, conditions that accelerate decomposition and reduce the mean residence time of residues left on site compared to coniferous forests under cold or dry climates and on poor soils. Furthermore, wood of hardwood trees is predominantly infected by white-rot fungi that decompose cellulose and lignin. Coniferous wood is predominantly decomposed by brown-rot fungi that remove cellulose and modify only lignin (Jurgensen et al. 1997; Schwarze et al. 1999). Brown-rotted wood is decomposed slower than white-rotted wood, and can contribute, probably in chemically altered form, to long-term

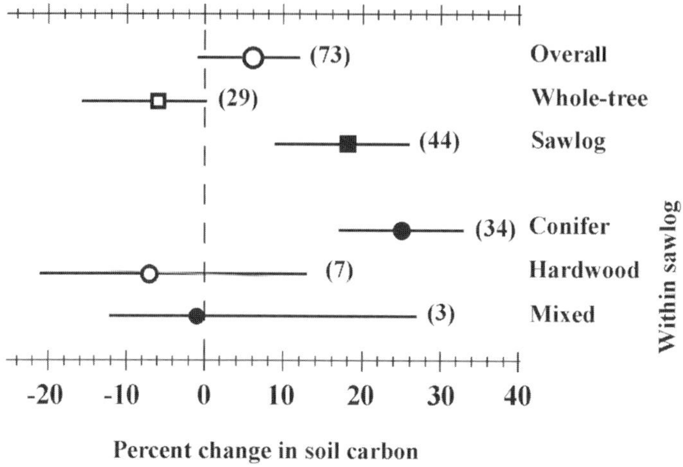

Fig. 10.4. Harvesting effects on soil organic carbon pools (A horizon) presented as percent change after tree harvesting, 99 % confidence intervals and number of studies in *parentheses.* (Johnson and Curtis 2001)

carbon storage in the organic layer (McFee and Stone 1966; Harvey et al. 1981) and in the upper mineral soil.

Some case studies reported a significant reduction of the organic layer after harvesting, but only a small reduction or even an increase of SOC pools, indicating that the organic layer may act as a "buffer" against soil organic carbon losses or as a source of carbon to the upper mineral soil after harvesting (e.g., Heinsdorf 1986; Johnson et al. 1991, 1995; Olsson et al. 1996; Laiho et al. 2003; Mund et al., in prep.). An increased translocation of organic carbon from the organic layer into the mineral soil following tree harvesting can be due either to increased decomposition activities followed by a transport of organic matter (soil fauna or DOC) into the mineral soil, or to mechanical incorporation by the harvesting procedure (harvesting machines, skidding, etc.; Bormann and Likens 1979; Mattson et al. 1987; Huntington and Ryan 1990; Mattson and Smith 1993; Johnson 1995; Johnson et al. 1995; Olsson et al. 1996; Dai et al. 2001; Laiho et al. 2003; Czimczik et al., submitted). If the mineral soil is covered by a thick organic layer, losses of carbon from the organic layer due to harvesting (increased decomposition and DOC export) will be high, but the input of carbon into the mineral soil will also be relatively high, resulting in a net input in the mineral soil. This could be a pattern typical for tree harvesting in evergreen coniferous forests in cold climate and on sandy soils.

Deciduous broadleaved forests on sandy soils and in warm climates are probably most susceptible to losses of SOC, because of a high proportion of unprotected SOC compared to silty/clay soils (Bonde et al. 1992; Sollins et al.

1996; Balesdent et al. 1998; Christensen 2001), and higher decomposition rates and thinner organic layers compared to colder climates and evergreen coniferous forests. Deciduous broadleaved forests on clay soils and in a cold climate may be less susceptible to carbon losses than evergreen coniferous forests under these conditions, because the proportion of unprotected carbon in the mineral soil as well as in the labile organic layer is lower in deciduous broadleaved forests than in evergreen coniferous forests.

10.6.2 Effects of Different Silvicultural Systems and Stand Age

It seems to be evident that small-scaled structural heterogeneity of forests, and therefore also gap cutting of forests, influence nitrogen and cation cycling (Clayton and Kennedy 1985; Vesterdal et al. 1995; Messina et al. 1997; Bradley et al. 2001; Prescott 2002) and the microclimate in forests (Mitscherlich 1981; Liechty et al. 1992; Bauhus and Bartsch 1995; Brumme 1995; Chen et al. 1995; Reynolds et al. 1997; Fleming et al. 1998; Barg and Edmonds 1999; Gray et al. 2002; Laporte et al. 2003). Many studies discuss changes of the microclimate due to timber harvest as an important process for accelerating the decomposition of litter, but we did not find any study that showed a direct effect of changes in the microclimate on SOC pools. Edwards and Ross-Todd (1983), for example, reported significantly higher soil temperatures and soil moisture at a harvested site (5 months after clear-cutting and removal of all woody material) compared to a non-harvested control forest (mixed deciduous forest). However, SOC pools remained constant (33 t C ha^{-1} at 0–45 cm soil depth).

Mund (2004) compared SOC pools (A and B horizon) of two different silvicultural management systems with a mixed deciduous beech forest that has not been managed for about 40 years, representing an advanced stage of nearly natural beech forests in Europe. The regular shelterwood system was represented by two chronosequences of even-aged beech stands (chronosequences "Leinefelde" and "Mühlhausen") and the selection system by three uneven-aged beech stands. All stands grew on nutrient-rich soils on Triassic limestone (partly covered with loess) and in the same climate. The managed stands were regularly thinned but, in contrast to forest plantations, no fertilizers or biocides were applied. The rotation period of the shelterwood system was 120–140 years. Harvest residues were usually collected as firewood after sawlog harvesting. Despite higher species diversity and higher carbon pools in stem biomass (Fig. 10.5) in the unmanaged forests, total SOC pools of the unmanaged forests did not differ significantly from those of the shelterwood systems or the selection systems (one-way ANOVA, $P=0.094$). However, there was a weak trend to higher SOC pools at the unmanaged forests (Fig. 10.6). Small differences among the SOC pools of the managed forests were balanced by carbon pools in stem biomass (Fig. 10.7). Conclusively, total carbon pools

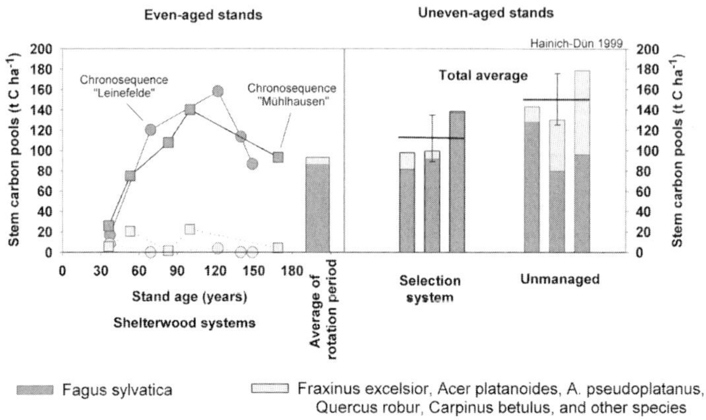

Fig. 10.5. Total carbon pools in stem biomass and contribution of tree species to these pools as a function of age and the silvicultural system (all beech and mixed beech forests grew on silty loam to silty clay soils on limestone) (Mund 2004)

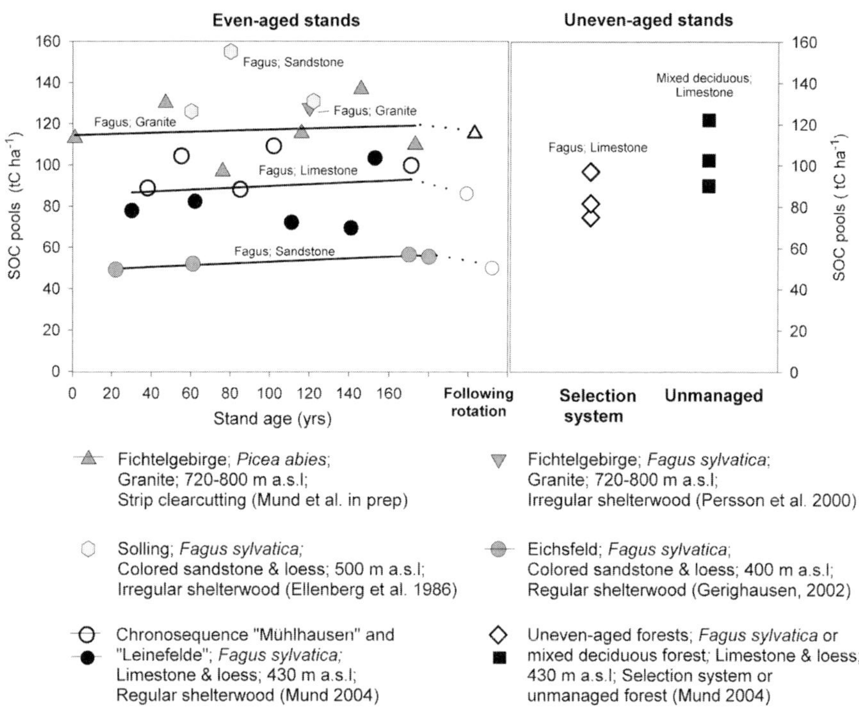

Fig. 10.6. Soil organic carbon pools of different beech or spruce chronosequences representing different silvicultural systems and climatic and edaphic conditions in Germany. Dotted lines show hypothesized SOC pools at the beginning of the following rotation (Ellenberg et al. 1986; Persson et al. 2000; Gerighausen 2002; Mund 2004; Mund et al., in prep.)

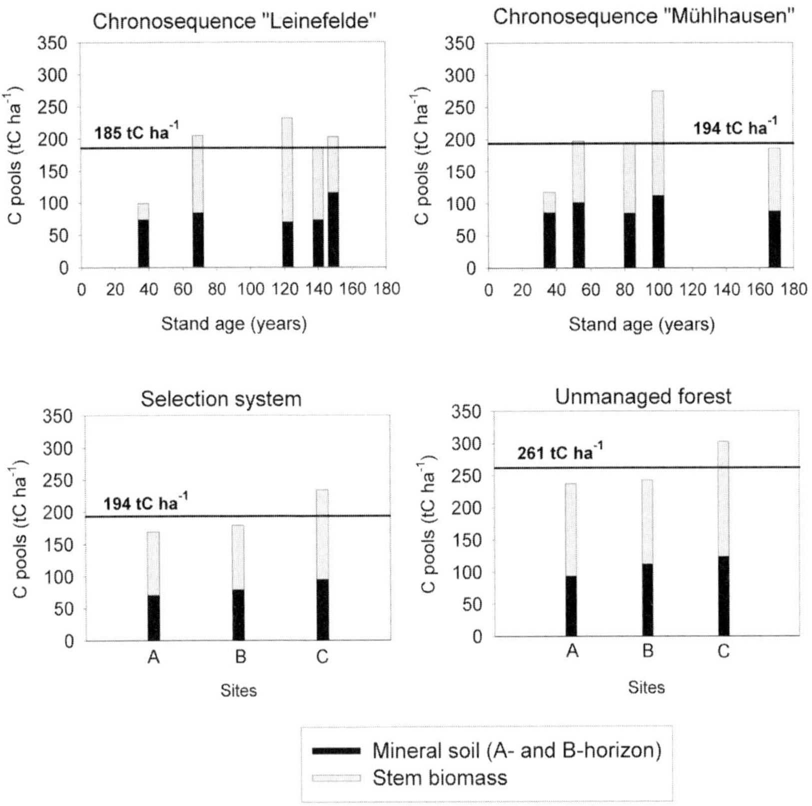

Fig. 10.7. Organic carbon pools (SOC and stem biomass) of different silvicultural systems and depending on stand age. (Mund 2004)

(SOC and stem biomass) of the unmanaged forests were on average about 26 % higher than those of the managed forests, mainly because of their high stem biomass (Fig. 10.7).

Stand age, which is a direct and most obvious effect of the shelterwood system, had no effect on SOC pools. Other case studies, similar to the investigation by Mund (2004) that represent chronosequences following moderate disturbances of forests, confirm this lack of an "age effect" (e.g., Davis et al. 2003: natural chronosequence after windthrow; Mund et al, in prep.: spruce chronosequence after strip clear-cutting; Fig. 10.6). In contrast, forest chronosequences that follow intensive disturbances such as afforestation of grasslands or croplands (Thuille et al. 2000; Turner and Lambert 2000; Paul et al. 2002; Vesterdal et al. 2002) or after stand replacing fires (Bhatti et al. 2002; Wirth et al. 2002), showed at least a moderate effect of stand age on SOC pools.

10.6.3 Interactions of Soil Properties, Climate and Land-Use History

A comparison with other chronosequence studies, representing typical silvicultural systems in Germany, revealed how complex the interactions of forest management, vegetation type, soil characteristics, climate, and probably also land-use history are (Fig. 10.6; Mund 2004). Climatic effects associated with the higher elevation of the "Fichtelgebirge" (720–800 m a.s.l.) and the vegetation type could explain the higher SOC pools of the spruce chronosequence at that site, in comparison to the beech chronosequences in central Thuringia (about 400 m a.s.l.). The single beech stand at the Fichtelgebirge and the beech stands at Solling (both characterized by similar elevation and montane climate) have similar or even higher SOC pools than the spruce stands at the Fichtelgebirge. Differences in soil texture could cause the lower SOC pools of the "Eichsfeld" chronosequence compared to the "Leinefelde" and "Mühlhausen" chronosequence and the unmanaged forest, all growing at the same climate (Fig. 10.8).

However, Mund (2004) pointed out that historical use of the study sites may have reduced SOC pools in the past and that, because of their extent, these reductions and subsequent recoveries interact with climate and soil properties. Historical forestry reports, old forestry maps, and historical sources on settlements, churches, monasteries, mines, glassworks, etc., indicate a ranking of the probability, intensity, and duration of forest degradation of the study sites due to very intensive and destructive historical use (e.g., forest pasture, coppice systems, litter raking): Eichsfeld >Leinefelde >Mühlhausen >Fichtelgebirge >Solling. Also, agricultural use before the 16th century cannot be excluded at the chronosequences Eichsfeld and Leinefelde, because their silty soils and the moderate climate offered quite suitable conditions for cropping. Considering

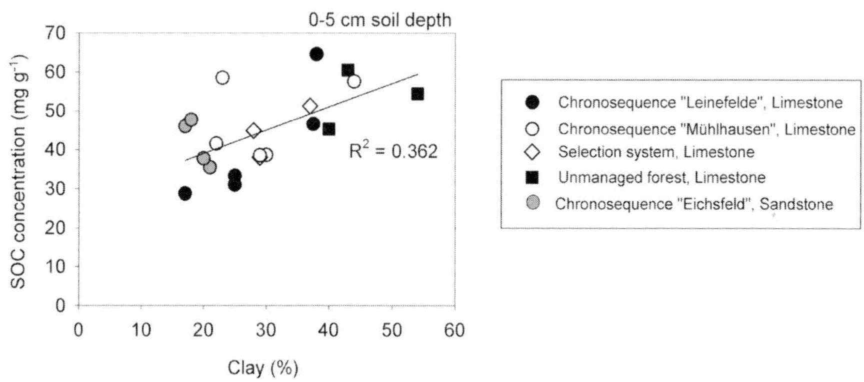

Fig. 10.8. Soil organic carbon concentrations at 0–5 cm soil depth depending on the clay content of the soil. All soils are forested with beech or mixed beech stands. (Mund 2004)

the interactions of soil texture with the stabilization process of SOC (Bonde et al. 1992; Christensen 1992; Sollins et al. 1996; Balesdent et al. 1998; Christensen 2001) as well as the activity and diversity of soil fauna (Gleixner, Chap. 9, this Vol.), SOC pools of clay soils may be less sensitive to and may recover faster from historical disturbances than silty or sandy soils. Thus, the SOC pools presented in Fig. 10.6 are likely the result of synergistic effects of climate, soil texture, species identity, historical use, and recent management. An increasing number of studies consider the probability of a very long-lasting (but difficult to quantify) historical dimension of major disturbances (e.g., Ulrich and Puhe 1994; Koerner et al. 1997; Caspersen et al. 2000; Thuille et al. 2000; Goodale and Aber 2001; Janssens et al. 2001; Berger et al. 2002; Dupouey et al. 2002; Rothe et al. 2002b). Nevertheless, the most critical point of this hypothesis is that all factors discussed influence SOC pools in the same direction ("multiple co-linearity"), and it will be a big challenge to separate site effects from biodiversity and management effects.

10.7 Conclusions

- Silvicultural activities affect the biogeochemistry of forest ecosystems. Living and dead biomass, the forest floor, and nitrogen and cation fluxes especially are susceptible to disturbances. However, the effects according to intensity or duration of disturbances, and the effects on element pools in the mineral soil, in particular SOC pools, still need to be quantified.
- Climate, soil conditions, and probably historical use can superimpose or at least modify impacts of present forest management or biodiversity on the biogeochemistry of forest ecosystems. Thus, site-specific factors and land-use history are likely responsible for the high variability of SOC pools.
- Changes in biodiversity are always associated with management and site conditions, and the available datasets are too limited to separate these effects. Therefore, it remains an open question if biodiversity could influence silviculture or biogeochemical cycles significantly.
- Furthermore, we assume that the effects of biodiversity, management, and site-specific factors cannot be separated by observational case studies. We strongly recommend establishing well-controlled experimental studies that exclude site-specific effects.
- In order to increase our mechanistic understanding of SOC changes, it will be necessary to fractionate the organic carbon of the bulk soil into: (1) a pool susceptible to increased decomposition by moderate disturbances or changes in biodiversity, and (2) a recalcitrant pool affected only by very intensive soil disturbances. This separation should be based on chemical properties of the organic substances.

Acknowledgments: We thank Andrew Manning for his helpful comments on a previous draft of the manuscript.

References

Apps MJ, Price DT (eds) (1996) Forest ecosystems, forest management and the global carbon cycle. Proceedings of the NATO Advanced Research Workshop "The role of global forest ecosystems and forest resource management in the global cycle", held in Banff, Canada, 12–16 Sept 1994. Springer, Berlin Heidelberg New York

Baldock JA, Oades JM, Waters AG, Peng X, Vassallo AM, Wilson MA (1992) Aspects of the chemical structure of soil organic materials as revealed by solid-state C13 NMR spectroscopy. Biogeochemistry 16:1–42

Balesdent J, Besnard E, Arrouays D, Chenu C (1998) The dynamics of carbon in particle-size fractions of soil in a forest-cultivation sequence. Plant Soil 201:49–57

Barg AK, Edmonds RL (1999) Influence of partial cutting on site microclimate, soil nitrogen dynamics, and microbial biomass in Douglas-fir stands in western Washington. Can J For Res 29:705–713

Bauhus J, Bartsch N (1995) Mechanisms of carbon and nutrient release and retention in beech forest gaps. I. Microclimate, water balance and seepage water chemistry. Plant Soil 169:579–584

Berger TW, Neubauer C, Glatzel G (2002) Factors controlling soil carbon and nitrogen stores in pure stands of Norway spruce (*Picea abies*) and mixed species stands in Austria. For Ecol Manage 159:3–14

Bhatti JS, Apps MJ, Jiang H (2002) Influence of nutrients, and site conditions on carbon stocks along a boreal forest transect in central Canada. Plant Soil 242:1–14

Binkley D (1992) Mixtures of $nitrogen_2$-fixing and $non\text{-}nitrogen_2$-fixing tree species. In: Cannell MGR, Malcolm DC, Robertson PA (eds) The ecology of mixed-species stands of trees. Blackwell, Oxford, pp 99–123

Binkley D, Giardina C (1998) Why do tree species affect soils? The warp and woof of tree – soil interactions. Biogeochemistry 42:89–106

Black TA, Harden JW (1995) Effect of timber harvest on soil carbon storage at Blodgett experimental forest, California. Can J For Res 25:1385–1396

Block R, van Rees KCJ, Pennock DJ (2002) Quantifying harvesting impacts using soil compaction and disturbance regimes at a landscape scale. Soil Sci Soc Am J 66:1669–1676

Bonde TA, Christensen, BT, Cerri CC (1992) Dynamics of soil organic matter as reflected by natural ^{13}C abundance in particle size fractions of forested and cultivated oxisols. Soil Biol Biochem 24:275–277

Bormann FH, Likens GE (1979) Pattern and process in a forested ecosystem. Springer, Berlin Heidelberg New York

Bradley RL, Titus BD, Hogg K (2001) Does shelterwood harvesting have less impact on forest floor nutrient availability and microbial properties than clearcutting? Biol Fertil Soil 34:162–169

Brumme R (1995) Mechanisms of carbon and nutrient release and retention in beech forest gaps. III. Environmental regulation of soil respiration and nitrous oxide emissions along a microclimatic gradient. Plant Soil 169:593–600

Burschel P, Huss J (1987) Grundriß des Waldbaus. Parey, Hamburg

Burschel P, Kürsten E, Larson BC (1993) Die Rolle von Wald und Forstwirtschaft im

Kohlenstoffhaushalt – eine Betrachtung für die Bundesrepublik Deutschland. Forstwissenschaftliche Fakultät der Universität München und Bayerische Forstliche Versuchs- und Forschungsanstalt, München

Cannell MGR, Dewar RC, Thornley JHM (1992) Carbon flux and storage in European forests. In: Teller A, Mathy P, Jeffers JNR (eds) Responses of forest ecosystems to environmental changes. Elsevier, London, pp 256–271

Caspersen JP, Pacala SW, Jenkins JC, Hurtt GC, Moorcroft PR, Birdsey RA (2000) Contributions of land-use history to carbon accumulation in U.S. forests. Science 290:1148–1151

Chen JQ, Franklin JF, Spies TA (1995) Growing-season microclimatic gradients from clear-cut edges into old-growth Douglas-fir forests. Ecol Appl 5:74–86

Christensen BT (1992) Physical fractionation of soil and organic matter in primary particle size and density separates. Adv Soil Sci 20:1–90

Christensen BT (2001) Physical fractionation of soil and structural and functional complexity in organic matter turnover. Eur J Soil Sci 52:345–353

Clayton JL, Kennedy DA (1985) Nutrient losses from timber harvest in the Idaho Batholith. Soil Sci Soc Am J 49:1041–1049

Cohen WB, Harmon ME, Wallin DO, Fiorella M (1996) Two decades of carbon flux from forests of the Pacific Northwest. Bioscience 46:836–844

Covington WW (1981) Changes in forest floor organic matter and nutrient content following clear cutting in northern hardwood. Ecology 62:41–48

Crow TR, Buckley DS, Nauertz EA, Zasada JC (2002) Effects of management on the composition and structure of northern hardwood forests in Upper Michigan. For Sci 48:129–145

Czimczik CI, Schmidt MWI, Schulze E-D Dynamics of black carbon, organic carbon, nitrogen, and sulfur stocks in podzolic soils of Siberian Scots pine forests following fire. Eur J Soil Sci (submitted)

Dai KH, Johnson CE, Driscoll CT (2001) Organic matter chemistry and dynamics in clear-cut and unmanaged hardwood forest ecosystems. Biogeochemistry 54:51–83

Davis MR, Allen RB, Clinton PW (2003) Carbon storage along a stand development sequence in a New Zealand *Nothofagus* forest. For Ecol Manage 177:313–321

Desjardins T, Andreux F, Volkoff B, Cerri CC (1994) Organic carbon and ^{13}C contents in soils and soil size-fractions, and their changes due to deforestation and pasture installation in eastern Amazonia. Geoderma 61:103–118

Dupouey JL, Dambrine E, Laffite JD, Moares C (2002) Irreversible impact of past land use on forest soils and biodiversity. Ecology 83:2978–2984

Duvall MD, Grigal DF (1999) Effects of timber harvesting on coarse woody debris in red pine forests across the Great Lakes states, USA. Can J For Res 29:1926–1934

Edwards NT, Ross-Todd BM (1983) Soil carbon dynamics in a mixed deciduous forest following clear-cutting with and without residue removal. Soil Sci Soc Am J 47:1014–1021

Ellenberg H, Mayer R, Schauermann J (eds) (1986) Ökosystemforschung – Ergebnisse des Sollingprojekts 1966–1986. Eugen Ulmer, Stuttgart

Ellenberg H (1996) Vegetation Mitteleuropas mit den Alpen. Ulmer, Stuttgart

Ellert BH, Gregorich EG (1995) Management-induced changes in the actively cycling fractions of soil organic matter. In: McFee WW, Kelly JM (eds) Carbon forms and functions in forest soils. Soil Science Society of America, Madison, Wisconsin, pp 119–138

Entry JA, Emmingham WH (1998) Influence of forest age on forms of carbon in Douglas-fir soils in the Oregon Coast Range. Can J For Res 28:390–395

Fleming RL, Black TA, Adams RS, Stathers RJ (1998) Silvicultural treatments, microclimatic conditions and seedling response in Southern interior clearcuts. Can J Soil Sci 78:115–126

Fleming TL, Freedman B (1998) Conversion of natural, mixed-species forests to conifer plantations: implications for dead organic matter and carbon storage. Ecoscience 5:213–221

Franklin JF, Spies TA, van Pelt R, Carey AB, Thornburgh DA, Berg DR, Lindenmayer DB, Harmon ME, Keeton WS, Shaw DC, Bible K, Chen J (2002) Disturbances and structural development of natural forest ecosystems with silvicultural implications, using Douglas-fir forests as an example. For Ecol Manage 155:399–423

Garten CT, Post IWM, Hanson PJ, Cooper LW (1999) Forest soil carbon inventories and dynamics along an elevation gradient in the southern Appalachian Mountains. Biogeochemistry 45:115–145

Gebauer G, Zeller B, Schmidt G, May C, Buchmann N, Colin-Belgrand M, Dambrine E, Martin F, Schulze E-D, Bottner P (2000) The fate of ^{15}N-labelled nitrogen inputs to coniferous and broadleaf forests. In: Schulze E-D (ed) Carbon and nitrogen cycling in European forest ecosystems. Ecological Studies, vol 142. Springer, Berlin Heidelberg New York, pp 144–170

Gerighausen U (2002) Dynamik der Kohlenstoffvorräte bewirtschafteter Buchenwälder auf Buntsandstein. Diplomarbeit am Max-Planck-Institut für Biogeochemie und an der Friedrich-Schiller-Universität Jena, Jena

Goodale CL, Aber JD (2001) The long-term effects of land-use history on nitrogen cycling in northern hardwood forests. Ecol Appl 11:253–267

Gray AN, Spies TA, Easter MJ (2002) Microclimatic and soil moisture responses to gap formation in coastal Douglas-fir forests. Can J For Res 32:332–343

Grier CC, Logan RS (1977) Old-growth *Pseudotsuga menziesii* communities of a western Oregon watershed: Biomass distribution and production budgets. Ecol Monogr 47:373–400

Grigal DF (2000) Effects of extensive forest management on soil productivity. For Ecol Manage 138:167–185

Grigal DF, Ohmann LF (1992) Carbon storage in upland forests of the Lake States. Soil Sci Soc Am J 56:935–943

Guggenberger G, Christensen BT, Zech W (1994) Land-use effects on the composition of organic matter in particle-size separates of soil. I. Lignin and carbohydrate signature. Eur J Soil Sci 45:449–458

Guggenberger G, Zech W, Haumaier L, Christensen BT (1995) Land use effects on the composition of organic matter in particle size separates of soils: II. CPMAS and solution 13C NMR analysis. Eur J Soil Sci 46:147–158

Guo LB, Gifford RM (2002) Soil carbon stocks and land use change: a meta analysis. Global Change Biol 8:345–360

Harmon ME, Franklin JF, Swanson FJ, Sollins P, Gregory SV, Lattin JD, Anderson NH, Cline SP, Aumen NG, Sedell JR, Lienkaemper GW, Cromack KJ, Cummins KW (1986) Ecology of coarse woody debris in temperate ecosystems. Adv Ecol Res 15:133–302

Harmon ME, Ferrell WK, Franklin JF (1990) Effects on carbon storage of conversion of old-growth forests to young forests. Science 247:699–702

Harvey AE, Larsen M, Jurgensen M (1981) Rate of woody residue incorporation into northern Rocky Mountain forest soils. USDA Forest Service Research Paper INT-282:0–5

Heinsdorf D (1986) Entwicklung der C- und N-Vorräte nach Kahlschlag auf Sandböden unter Kiefer. Ber Wiss Tagung Tharandt vom 8.-10. Oktober 1986, pp 98–109

Heinsdorf D, Krauß HH (1974) Ergebnisse eines Meliorationsversuches zu Kiefer und Roteiche auf einem humusarmen Sandboden im Tieflandsgebiet der DDR. Beitr Forstwirtsch 1:25–37

Houghton RA, Hobbie JE, Melillo JM, Moore B, Peterson BJ, Shaver GR, Woodwell GM (1983) Changes in the carbon content of terrestrial biota and soils between 1860 and 1980 – a net release of CO_2 to the atmosphere. Ecol Monogr 53:235–262

Huntington TG, Ryan DF (1990) Whole-tree-harvesting effects on soil nitrogen and carbon. For Ecol Manage 31:193–204

Janssens IA, Lankreijer H, Matteucci G, Kowalski AS, Buchmann N, Epron D, Pilegaard K, Kutsch W, Longdoz B, Grunwald T, Montagnani L, Dore S, Rebmann C, Moors EJ, Grelle A, Rannik U, Morgenstern K, Oltchev S, Clement R, Gudmundsson J, Minerbi S, Berbigier P, Ibrom A, Moncrieff J, Aubinet M, Bernhofer C, Jensen NO, Vesala T, Granier A, Schulze E-D, Lindroth A, Dolman AJ, Jarvis PG, Ceulemans R, Valentini R (2001) Productivity overshadows temperature in determining soil and ecosystem respiration across European forests. Global Change Biol 7:269–278

Johnson CE (1995) Soil nitrogen status 8 years after whole-tree clear-cutting. Can J For Res 25:1346–1355

Johnson DW, Curtis PS (2001) Effects of forest management on soil C and N storage: meta analysis. For Ecol Manage 140:227–238

Johnson DW, Henderson P (1995) Effects of forest management and elevated carbon dioxide on soil carbon storage. In: Lal R, Kimble J, Livine E, Stewart BA (eds) Soil management and greenhouse effect. CRC Lewis, Boca Raton, pp 137–145

Johnson CE, Johnson AH, Huntington TG, Siccama TG (1991) Whole-tree clear-cutting effects on soil horizons and organic matter pools. Soil Sci Soc Am J 55:497–502

Johnson CE, Driscoll CT, Fahey TJ, Siccama TG, Hughes JW (1995) Carbon dynamics following clear-cutting of a northern hardwood forest. In: McFee WW, Kelly JM (eds) Carbon forms and functions in forest soils. Soil Science Society of America, Madison Wisconsin, pp 463–488

Jurgensen MF, Harvey AE, Graham RT, Page-Dumroese DS, Tonn JR, Larsen MJ, Jain TB (1997) Impacts of timber harvesting on soil organic matter, nitrogen, productivity, and health of Inland Northwest forests. For Sci 43:234–251

Karjalainen T (1996) Dynamics and potential of carbon sequestration in managed stands and wood products in Finland under changing climatic conditions. For Ecol Manage 80:113–132

Khanna PK, Ludwig B, Bauhus J, O'Hara C (2001) Assessment and significance of labile organic C pools in forest soils. In: Lal R, Kimble JM, Follett RF, Stewart BA (eds) Assessment methods for soil carbon. Lewis, Boca Raton, pp 167–182

Kirby KJ, Reid CM, Thomas RC, Goldsmith FB (1998) Preliminary estimates of fallen dead wood and standing dead trees in managed and unmanaged forests in Britain. J Appl Ecol 35:148–155

Koerner W, Dupouey JL, Dambrine E, Benoît M (1997) Influence of past land use on the vegetation and soils of present day forest in the Vosges mountain, France. J Ecol 85:351–358

Krankina ON, Harmon ME, Kukuev YA, Treyfeld RF, Kashpor NN, Kresnov VG, Skudin VM, Protasov NA, Yatskov M, Spycher G, Povarov ED (2002) Coarse woody debris in forest regions of Russia. Can J For Res 32:768–778

Laiho R, Sanchez F, Tiarks A, Dougherty PM, Trettin CC (2003) Impacts of intensive forestry on early rotation trends in site carbon pools in the southeastern US. For Ecol Manage 174:177–189

Landsberg JJ, Gower ST (1997) Applications of physiological ecology to forest management. Academic Press, San Diego

Laporte MF, Duchesne LC, Morrison IK (2003) Effect of clearcutting, selection cutting, shelterwood cutting and microsites on soil surface CO_2 efflux in a tolerant hardwood ecosystem of northern Ontario. For Ecol Manage 174:565–575

Liechty HO, Holmes MJ, Reed DD, Mroz GD (1992) Changes in microclimate after stand conversion in two northern hardwood stands. For Ecol Manage 50:253–264

Lindenmayer D, McCarthy MA (2002) Congruence between natural and human forest disturbance: a case study from Australian montane ash forests. For Ecol Manage 155:319–335

Lundström US, van Breemen N, Bain D (2000) The podzolization process. A review. Geoderma 94:91–107
Marshall VG (2000) Impacts of forest harvesting on biological processes in northern forest soils. For Ecol Manage 133:43–60
Matson PA, Parton WJ, Power WJ, Swift MJ (1997) Agricultural intensification and ecosystem properties. Science 277:504–508
Matthews JD (1989) Silvicultural systems. Oxford Univ Press, Oxford
Mattson KG, Smith HC (1993) Detrital organic matter and soil CO_2 efflux in forest regeneration from cutting in West Virginia. Soil Biol Biochem 25:1241–1248
Mattson KG, Swank WT, Waide JB (1987) Decomposition of woody debris in a regenerating, clear-cut forest in the Southern Appalachians. Can J For Res 17:712–721
Matzner E (1988) Der Stoffumsatz zweier Waldökosysteme im Solling. Berichte des Forschungszentrums Waldökosysteme/Waldsterben, Reihe A, Bd 40. Wissenschaft und Schrifttum eV Göttingen, Göttingen
McFee WW, Stone EL (1966) The persistence of decaying wood in the humus layer of northern forests. Soil Sci Soc Am Proc 30:513–516
Messina MG, Schoenholtz SH, Lowe MW, Wang ZY, Gunter DK, Londo AJ (1997) Initial responses of woody vegetation, water quality, and soils to harvesting intensity in a Texas bottomland hardwood ecosystem. For Ecol Manage 90:201–215
Mitscherlich G (1981) Wald, Wachstum und Umwelt. 2. Waldklima und Wasserhaushalt. Sauerländer, Frankfurt/Main
Morgan JL, Campbell JM, Malcolm DC (1992) Nitrogen relations of mixed-species stands on oligotrophic soils. In: Cannell MGR, Malcolm DC, Robertson PA (eds) The ecology of mixed-species stands of trees. Blackwell, Oxford, pp 65–85
Mund M (2004) Carbon pools of European beech forests (*Fagus sylvatica*) under different silvicultural management. Dissertation, Universität Göttingen
Mund M, Buchmann N, Schulze E-D Impacts of stand age and tree harvesting on soil organic carbon stocks of temperate spruce forests (*Picea abies*) (in prep.)
Nihlgård B (1971) Pedological influence of spruce planted on former beech forest soils in Scania, South Sweden. Oikos 22:302–314
Nyland RD (1996) Silviculture. Concepts and applications. McGraw-Hill, Boston
Olsson BA, Staaf H, Lundkvist H, Jan B, Rosén K (1996) Carbon and nitrogen in coniferous forest soil after clear-felling and harvest of different intensity. For Ecol Manage 82:19–32
Paul KI, Polglase PJ, Nyakuengama J, Khanna PK (2002) Change in soil carbon following afforestation. For Ecol Manage 168:241–257
Pedlar JH, Pearce JL, Venier LA, McKenney DW (2002) Coarse woody debris in relation to disturbance and forest type in boreal Canada. For Ecol Manage 158:189–194
Persson T, van Oene H, Harrison AF, Karlsson PS, Bauer GA, Cerny J, Coûteaux M-M, Dambrine E, Högberg P, Kjøller A, Matteucci G, Rudebeck A, Schulze E-D, Paces T (2000) Experimental sites in the NIPHYS/CANIF Project. In: Schulze E-D (ed) Carbon and nitrogen cycling in European forest ecosystems. Ecological Studies, vol 142. Springer, Berlin Heidelberg New York, pp 14–44
Prescott CE (2002) The influence of forest canopy on nutrient cycling. Tree Physiol 22:1193–1200
Prescott CE, Maynard DG, Laiho R (2000a) Humus in northern forests: friend or foe? For Ecol Manage 133:23–36
Prescott CE, Vesterdal L, Pratt J, Venner KH, de Montigny LM, Trofymow JA (2000b) Nutrient concentrations and nitrogen mineralization in forest floors of single species conifer plantations in coastal British Columbia. Can J For Res 30:1341–1352
Prescott CE, Zabek LM, Staley CL, Kabzems R (2000 c) Decomposition of broadleaf and needle litter in forests of British Columbia: influences of litter type, forest type, and litter mixtures. Can J For Res 30:1742–1750

Pypker TG, Fredeen AL (2002) Ecosystem CO_2 flux over two growing seasons for a sub-boreal clearcut 5 and 6 years after harvest. Agric For Meteorol 114:15–30

Quesnel HJ, Curran MP (2000) Shelterwood harvesting in root-disease infected stands – post- harvest soil disturbance and compaction. For Ecol Manage 133:89–113

Rannik Ü, Altimir N, Raittila J, Suni T, Gaman A, Hussein T, Hölttä T, Lassila H, Latokartano M, Lauri A, Natsheh A, Petäjä T, Sorjamaa R, Ylä-Mella H, Keronen P, Berninger F, Vesala T, Hari P, Kulmala M (2002) Fluxes of carbon dioxide and water vapour over Scots pine forest and clearing. Agr For Meteorol 111:187–202

Rehfuess KE (1990) Waldböden. Parey, Hamburg

Reynolds PE, Simpson JA, Lautenschlager RA, Bell FW, Gordon AM, Buckley DA, Gresch DA (1997) Alternative conifer release treatments affect below- and near- ground microclimate. For Chron 73:75–82

Röhrig E, Gussone HA (1990) Waldbau auf ökologischer Grundlage. Zweiter Band: Baumartenwahl, Bestandesbegründung und Bestandespflege. Parey, Hamburg

Rollinger JL, Strong TF, Grigal DF (1998) Forested soil carbon storage in landscapes of the northern Great Lakes region. In: Lal R, Kimble JM, Follett RF, Stewart BA (eds) Management of carbon sequestration in soil. CRC, Boca Raton, pp 335–350

Rothe A, Binkley D (2001) Nutritional interactions in mixed species forests: a synthesis. Can J For Res 31:1855–1870

Rothe A, Huber C, Kreutzer K, Weis W (2002a) Deposition and soil leaching in stands of Norway spruce and European beech: results from the Höglwald research in comparison with other European case studies. Plant Soil 240:33–45

Rothe A, Kreutzer K, Kuchenhoff H (2002b) Influence of tree species composition on soil and soil solution properties in two mixed spruce-beech stands with contrasting history in Southern Germany. Plant Soil 240:47–56

Schulze E-D (1982) Plant life forms and their carbon, water, and nutrient relations. In: Lange OL, Nobel PS, Osmond CB, Ziegler H (eds) Physiological plant ecology II. Springer, Berlin Heidelberg New York, pp 616–676

Schulze E-D, Bazzaz FA, Nadelhoffer KJ, Koike T, Takatsuki S (1996) Biodiversity and ecosystem function of temperate deciduous broad-leaved forests. In: Mooney HA, Cushman JH, Medina E, Sala OE, Schulze E-D (eds) Functional roles of biodiversity – a global perspective. Wiley, Chichester, pp 71–98

Schulze E-D, Lloyd J, Kelliher FM, Wirth C, Rebmann C, Lühker B, Mund M, Knohl A, Milykova IM, Schulze W, Ziegler W, Varlagin AB, Sogachev AF, Valentini R, Dore S, Grigoriev S, Kolle O, Panfyorov MI, Tchebakova N, Vygodskaya NN (1999) Productivity of forests in the Eurosiberian boreal region and their potential to act as a carbon sink – a synthesis. Global Change Biol 5:703–722

Schulze E-D, Högberg O, van Oene H, Persson T, Harrison AF, Read D, Kjøller A, Matteucci G (2000) Interactions between the carbon- and nitrogen cycle and the role of biodiversity: a synopsis of a study along a north-south transect through Europe. In: Schulze E-D (ed) Carbon and nitrogen cycling in European forest ecosystems. Ecological Studies, vol 142. Springer, Berlin Heidelberg New York, pp 468–491

Schwarze FWMR, Engels J, Mattheck C (1999) Holzzersetzende Pilze in Bäumen. Rombach, Freiburg im Breisgau

Seymour RS, White AS, deMaynadier PG (2002) Natural disturbance regimes in northeastern North America – evaluating silvicultural systems using natural scales and frequencies. For Ecol Manage 155:357–367

Sollins P, Homann P, Caldwell BA (1996) Stabilization and destabilization of soil organic matter: mechanisms and controls. Geoderma 74:65–105

Thuille A, Buchmann N, Schulze E-D (2000) Carbon stocks and soil respiration rates during deforestation, grassland use and subsequent Norway spruce afforestation in the Southern Alps, Italy. Tree Physiol 20:849–857

Trofymow JA, Blackwell BA (1998) Changes in ecosystem mass and carbon distributions in coastal forest chronosequences. Northwest Sci 72:40–42

Turner J, Lambert M (2000) Change in organic carbon in forest plantation soils in eastern Australia. For Ecol Manage 133:231–247

Ulrich B, Puhe J (1994) Studie B: Auswirkungen der zukünftigen Klimaveränderung auf mitteleuropäische Waldökosysteme und deren Rückkopplungen auf den Treibhauseffekt. In: Enquete-Kommission Schutz der Erdatmosphäre des Deutschen Bundestages (ed) Studienprogramm: Enquete-Kommission Schutz der Erdatmosphäre des Deutschen Bundestages, Bd 2. Wälder. Economica, Bonn

UN (United Nations) (1997) Kyoto protocol to the United Nations framework convention on climate change. United Nations

UN (United Nations) (2001) Report of the conference of the parties on the second part of its sixth session, held at Bonn from 16–27 July 2001. FCCC/CP/2001/5

Vesterdal L, Dalsgaard M, Felby C, Raulund-Rasmussen K, Jørgensen BB (1995) Effects of thinning and soil properties on accumulation of carbon, nitrogen and phosphorus in the forest floor of Norway spruce stands. For Ecol Manage 77:1–10

Vesterdal L, Raulund-Rasmussen K (1998) Forest floor chemistry under seven tree species along a soil fertility gradient. Can J For Res 28:1636–1647

Vesterdal L, Ritter E, Gundersen P (2002) Change in soil organic carbon following afforestation of former arable land. For Ecol Manage 169:137–147

Weber M (2001) Kohlenstoffspeicherung in Lenga- (*Nothofagus pumilio*) Primärwäldern Feuerlands und Auswirkungen ihrer Überführung in Wirtschaftswald auf den C-Haushalt. Kessel, Remagen-Oberwinter

Wirth C, Schulze E-D, Lühker B, Grigoriev S, Siry M, Hardes G, Ziegler W, Backor M, Bauer G, Vygodskaya NN (2002) Fire and site type effects on the long-term carbon and nitrogen balance in pristine Siberian Scots pine forests. Plant Soil 242:41–63

Wirth C, Schulze E-D, Schwalbe G, Tomczyk S, Weber G, Weller E (2004) Dynamik der Kohlenstoffvorräte in den Wäldern Thüringens. Abschlussbericht zur 1. Phase des BMBF-Projektes "Modelluntersuchungen zur Umsetzung des Kyoto-Protokolls", Mitteilungen 23/2004. Thüringer Landesanstalt für Wald, Jagd und Fischerei, Gotha

Worrell R, Hampson A (1997) The influence of some forest operations on the sustainable management of forest soils – a review. Forestry 70:61-85

Part D Animals, Pests, and Disturbances

11 Linkages Between Tree Diversity, Soil Fauna and Ecosystem Processes

S. Scheu

11.1 Introduction

The soil is arguably the most interesting interface in ecosystems. Organic matter produced by plants ultimately enters the detrital system where it is recycled and feeds back to plant production. The processes involved in the recycling of plant residues are driven by a vast diversity of organisms, coupled by extraordinarily complex interactions. It is understandable, therefore, that the mechanisms involved in the recycling process of detritus are often regarded as very difficult to find. Commonly, the detrital system is assumed to function solely as a mineralization mechanism, the actual actors are ignored and processes explained as fluxes of carbon and nutrients driven by abiotic factors, such as temperature and moisture. Ignorance of the structure of belowground animal communities and their interactions involved in element cycling appeared to be justified as abiotic notions gave way to considerations in the second half of the last century that carbon and energy flow, as documented by ecosystem studies, is almost exclusively due to microbial activity. However, more recently, it has been stressed by soil ecologists, particularly those with a soil animal background, that the contribution of decomposer invertebrates to element cycling considerably exceeds their direct effect via their own metabolism. Rather, soil invertebrates indirectly modify decomposition processes by changing the structure and activity of the microbial community (Coleman et al. 1983; Anderson 1987; Wolters 1991). This view has altered the perspective to some extent and has greatly stimulated the investigation of faunal microbial interactions. Still, the feedback to plants continued to be based on carbon and nutrient fluxes. It was just in the last 10–20 years that plant roots were found to be integrated as components of the decomposer system, driving the structure and the activity of the belowground food web that finds its way back to plant growth. It is now becoming increasingly clear that the above- and belowground communities are much more closely linked than previously assumed, and that neither can be understood without consid-

Ecological Studies, Vol. 176
M. Scherer-Lorenzen, Ch. Körner, and E.-D. Schulze (Eds.)
Forest Diversity and Function: Temperate and Boreal Systems

ering the complex interactions of the organisms involved (Scheu 2001; van der Putten et al. 2001; Scheu and Setälä 2002; Wardle 2002).

This review explores how the structure and functioning of the belowground food web are affected by the composition of the aboveground plant community, and how this feeds back to the structure of the plant community, and thus to that of the whole ecosystem.

A comprehensive review of this very wide topic is beyond the scope of a single paper. Therefore, I will focus on certain aspects of the interrelationship between the community structure of plants and that of the belowground system. Three topics will be explored: (1) the relationship between plant (tree) diversity and the diversity and community structure of the decomposer system, (2) the dependency of ecosystem processes on the community structure of the belowground system, and (3) the feedback mechanisms of the belowground community as they influence the plant community structure. I will ignore feedbacks due to direct soil fauna – root interactions, e.g., root herbivory. This is not to imply that these interactions are less important, but they have been reviewed in detail recently (Clay and van der Putten 1999; Mortimer et al. 1999; Strong 1999; van der Putten et al. 2001). I must stress that despite the increasing attention paid to root herbivore–plant interactions in nonarable systems such as grasslands, knowledge of their role for plant growth and plant community composition in natural systems such as forests is virtually nonexistent.

11.2 Plant (Tree) Diversity as Determinant of the Belowground Food Web

Soils harbor an exceptionally high number of organisms interconnected by complex trophic and non-trophic interactions. Uncovering these interactions and understanding their importance for food web structure has been hampered by the lack of appropriate methodology, e.g., only a few of the bacterial species in soil can be cultured on artificial media in the laboratory. However, with the development and use in soil food web studies of new methodological techniques such as molecular markers, phospholipids, and stable isotopes, a new era has begun (Mogge et al. 2000; Bridge and Spooner 2001; Ruess et al. 2002; Scheu 2002). These new concepts, combined with rigorous experimental manipulations of soil communities, promise much progress in understanding the structure and functioning of soil food webs in the near future (Scheu et al. 1999; Wardle 2002). However, as baseline data on which hypotheses for explaining the diversity of soil organisms can be built, detailed descriptive studies of soil animal communities of different habitats are necessary.

11.2.1 Descriptive Studies

The basal energy source and nutrient resource of soil communities are provided by dead organic matter, and plants and decomposers are not directly linked via feeding interactions. Specific interactions between plant species and decomposer organisms presumably contribute little to soil animal species diversity. Most decomposer soil animal species are food generalists rather than specialists. Commonly, species are aggregated to very general feeding groups according to taxonomic similarity, such as primary decomposers, bacterial feeders, fungal feeders, and predators. Explaining why each of these groups consists of a great diversity of species is a challenge for soil ecological research (cf. Scheu and Setälä 2002). Schaefer (1999) discussed 18 hypotheses for explaining soil animal species diversity. Only some of these hypotheses are related to the structure of the plant community. Certainly, the relationship between plant diversity and soil animal species diversity is not as close as that between plants and herbivores.

In one of the very few forests in the world in which the soil fauna community has been studied comprehensively, the Göttinger Wald, a total of 1,918 animal species have been found, which is about 5% of the regional species pool of Germany (Schaefer 1991a). Of these 1,918 species, 254 are phytophagous (including rhizophagous species), and the vast majority of the species live below the ground in the litter layer and the upper mineral soil horizons feeding on detritus, microorganisms, and animal prey. Considering that the plant diversity in this forest is low (ca. 30 species), with the tree canopy built almost exclusively by a single species, the high diversity of soil animal species below the ground is surprising. The basal resources, comprised almost exclusively of beech leaf litter, are very uniform, and it therefore is evident that the relationship between the diversity below ground and that of plants is weak.

Of the 1,918 species of the Göttinger Wald, 380 are microbivores or detritivores (excluding Diptera) and therefore closely connected to litter decomposition. Most interestingly, a majority of species are either predators (440 species) or parasitoids (682 species), which have little connection to the community structure of plants (Schaefer 1999). Noteworthy is that the microbivore and detritivore species of the Göttinger Wald represent 17.4% of the regional pool of species of these trophic groups, and with 11.5% predators also represent a considerable proportion of the regional pool of this predator/parasitoid group. In contrast, herbivores comprise of only 2.4% of the regional herbivore species number. This underlines the exceptionally high diversity of microbivores, detritivores, and predators in soil, but it also indicates that compared to herbivores the β-diversity is low. Obviously, very different plant communities harbor similar microbivore, detritivore, and also predator communities below ground. This again highlights the weak relationship between plant diversity and soil animal species diversity.

The diet of many soil invertebrates is notoriously variable; decomposers often feed on a complex mixture of resources and most of what is ingested is egested, only little assimilated. The integration of the morphology of mouthparts with food resources may be weak (Scheu 2002). More recently, natural variations in stable isotope signatures, most importantly those of ^{15}N, have been used to study the trophic structure of soil animal communities (Ponsard and Arditi 2000; Scheu and Falca 2000). A major result of these studies was the finding that higher taxonomic units are of very limited use in depicting trophic groups, except for very general categories such as predators and detritivores. Microbivorous or detritivorous groups such as earthworms, millipedes, isopods, collembolans, and oribatid mites comprise of species that predominantly feed on litter (primary decomposers), while others mostly feed on microorganisms and microbial residues (secondary decomposers). Similarly, predator taxa such as centipedes, spiders, gamasid mites, and staphylinid beetles comprise of species that form a gradient from those predominantly feeding on primary decomposers to those feeding on secondary decomposers. One of the most diverse group of soil animals with respect to its feeding relationships are dipterans. Stable isotope analysis indicate that they form a "gradient" from herbivores and primary decomposers to predators predominantly feeding on other predators (intra-guild predators; Fig. 11.1). Trophic interactions below ground therefore can only be understood if the animal community is analyzed with high taxonomic resolution. Due to the great

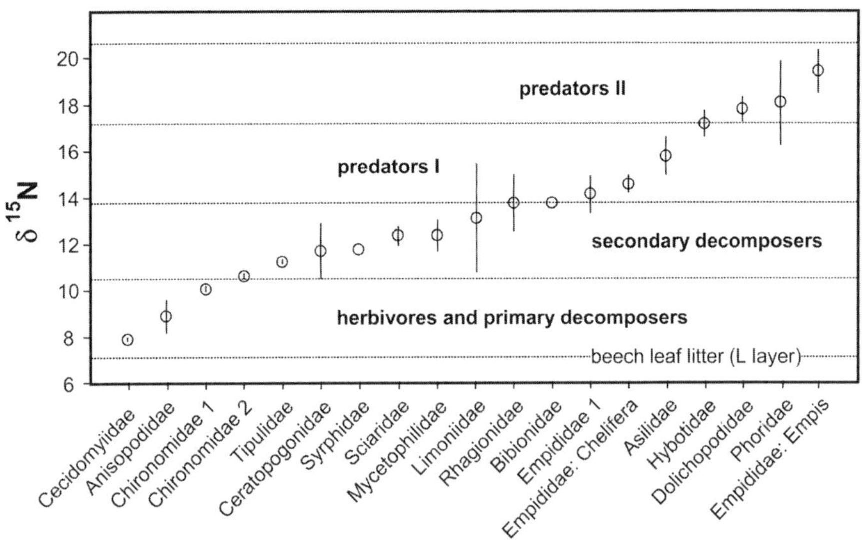

Fig. 11.1. Trophic structure of the community of dipteran larvae of an oak hornbeam forest (near Darmstadt, Germany) as indicated by natural variations in ^{15}N adult signatures of individuals soon after hatching. (S. Scheuermann, and S. Scheu, unpubl. data)

diversity of soil animal species and the difficult determination procedure in many taxa, this has rarely be done. If it has been done, only single stands or species-poor mixtures have been considered (cf. Schaefer and Schauermann 1990; Schaefer 1991a). Very little is known about the structure of soil animal communities in forests with a great diversity of tree species such as deciduous forest in Asia or North America. Knowledge on hyperdiverse systems such as tropical rain forests is extraordinary poor.

11.2.2 Manipulative Experiments

Compared to the herbivore community (Symstad et al. 2000; Haddad et al. 2001) few experiments have in a rigorous way explored the relationship between plant and soil animal diversity; the relationship between tree diversity and soil animal diversity has not been studied at all. However, from existing studies on grasslands and differently managed forest ecosystems, characteristics of the relationship between tree diversity and soil animal community structure may be inferred.

Collembolans are a diverse soil animal group consisting mainly of fungal feeders, although some species also feed on plants and others on algae and detritus (Hopkin 1997; Petersen 2002). Grasslands typically harbor 15–30 species. At the Swiss site of the BIODEPTH experiment in which plant species diversity had been experimentally manipulated, including 1, 2, 4, 8, 16, and 32 species' treatments (cf. Diemer et al. 1997; Spehn et al. 2000), collembolan species diversity correlated little with plant species diversity (Fig. 11.2). On average, single-plant species plots harbored 7.2 species, and 32 plant species' plots 8.0 species of collembolans (Salamon et al. 2003). Similarly, the number of plant functional groups, which varied between one and three and included legumes, grasses and non-legume herbs, did not significantly affect collembolan species' numbers.

Similar to collembolans, the diversity of macrofauna was not significantly affected by plant species diversity in the Swiss BIODEPTH experiment (D. Felzmann, pers. comm.). However, the number of microbivore and detritivore taxa significantly increased with plant functional groups, indicating that the diversity of microbivores and detritivores is related to the structure of the plant community. The number of macrofauna predators was even less firmly related to the structure of the plant community; it was significantly affected neither by plant species diversity nor by plant functional group diversity (D. Felzmann, pers. comm.). However, again, it was not independent of plant community structure, as indicated by a significant increase in predator species numbers in plant monocultures compared to polycultures. Overall, the results of this experiment underline the weak relationship between plant species diversity and belowground animal diversity, although the latter diversity is not independent of plant community structure. The results actually indicate

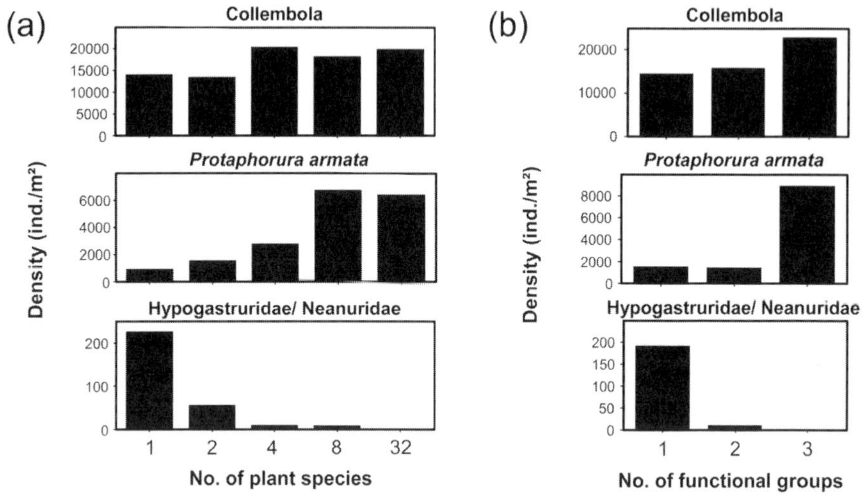

Fig. 11.2. Changes in density of different collembolan taxa with **a** plant species and **b** plant functional group diversity in the Swiss BIODEPTH experiment. (Modified from Salamon et al. 2003)

that the structure of the soil animal community responds more sensitively to plant species and functional group diversity than does its species composition. As outlined above, the diversity of collembolans was not significantly affected by plant species diversity. However, both plant species and functional group diversity strongly affected the community structure of collembolans (Salamon et al. 2003). For example, the density of *Protaphorura armata* increased strongly with plant species and functional group diversity whereas that of Hypogastruridae/Neanuridae declined (Fig. 11.2).

The few experimental studies that have investigated relationships between plant (litter) and soil animal species diversity in forests also indicate that this relationship is weak. Hansen and Coleman (1998), Hansen (1999) and Kaneko and Salamanca (1999) reported a significantly lower species number of oribatid mites in litterbags containing monotypic litters compared to litter mixtures of different tree species. Similarly, Migge et al. (1998) reported oribatid species numbers to be slightly higher in mixed compared to pure stands of beech and spruce. Blair et al. (1995) also reported that soil invertebrate community structure in litter mixtures differs from that of monotypic litter. Especially fungivorous nematodes were stimulated in litter mixtures, while other groups including microarthropods were little affected. In each of these studies, however, the authors conclude that the observed patterns may not primarily be due to the more diverse litter materials but to specific traits of certain litter types (see below). In contrast to findings in these studies focused on decomposer invertebrates, it appears that predators may benefit significantly from planting of tree species mixtures. The density and diversity of soil-surface-hunting predators such as carabid beetles have been documented to be increased in conifer plantations by the presence of broadleaf species (Butterfield and Benitez Malvido 1992). Similarly, birds have been shown to benefit from the admixture of broadleaf species in conifer plantations (Peck 1989).

11.2.3 Plant Traits Affecting Soil Animal Community Structure

In forests in particular the soil animal community has been documented to vary strongly with soil acidity, and soil acidity is known to be affected by tree species (Huhta et al. 1986; Ponge et al. 1997; Augusto et al. 1998). As documented frequently, coniferous tree species such as spruce may strongly acidify the soil (Ellenberg et al. 1986; Ulrich and Sumner 1991; Ulrich 1994) and the structure of the soil animal community between (e.g., beech and spruce) forests is known to differ strongly (Ellenberg et al. 1986; Scheu et al. 2003). This, however, may not only be due to differences in soil acidity but also to other environmental factors such as habitat structure, soil moisture, and chemical characteristics of litter materials.

Of the number of traits of plant litter materials that affect litter decomposition and the structure of the decomposer community, plant nutrient and plant polyphenol content are considered to be of particular importance (Swift et al. 1979; Blair et al. 1990; Enriquez et al. 1993; Harborne 1997; Berg 2000; Hättenschwiler and Vitousek 2000). Results from the Swiss BIODEPTH experimental site indicate that changes in the decomposer activity and community structure vary with plant diversity and plant functional types, and this has been attributed in large part to the presence of legumes, which compared to other herbs and grasses are rich in nitrogen (Spehn et al. 2000; Salamon et al. 2003; Felzmann, unpubl. data). Experiments manipulating plant species diversity at the Cedar Creek Natural History area, Minnesota, USA, also indicate that effects of plant species diversity on litter decomposition are caused by variations in litter nitrogen concentration (Knops et al. 2001). Similar results have been reported from the English BIODEPTH experiment at Silwood Park (Hector et al. 2000). Wardle et al. (1993, 1999) documented that weeds may beneficially affect the decomposer community due to higher nitrogen concentrations compared to agricultural crop plants. Also, due to selective grazing on nutrient-rich understory vegetation, vertebrate browsers have been shown to detrimentally affect the decomposer community (Wardle et al. 2001). Wardle et al. (1994) concluded that compared to competitors (sensu Grime 1979) ruderal plants support higher numbers of decomposers; they ascribed this to the higher nutrient concentration in ruderals compared to competitors. However, results of the study of Bardgett et al. (1999), in which both plant species diversity and nitrogen availability had been manipulated, suggest that plant species traits (other than nitrogen concentration) are more important than nitrogen availability in regulating microbial activity and abundance. Hooper and Vitousek (1997) stressed that functional characteristics of plant species, such as litter nitrogen concentration, may be at least as important as the number of species in maintaining critical processes and services in any ecosystem.

The polyphenol concentration of litter materials detrimentally affects its digestibility for both herbivores and detritivores (Swift et al. 1979; Kuiters 1990). In soil, plant polyphenol concentration significantly affects litter decomposition and decomposer activity and community structure (Harborne 1997; Northup et al. 1998; Lorenz et al. 2000). In boreal forests, plant polyphenol concentration likely is one of the most important factors determining decomposition rates, accumulation of organic matter, and decomposer community structure (Wardle et al. 1997, 1998). Plant polyphenol concentration presumably is a key factor for understanding the linkage between traits of plant litter and nitrogen transformations in soil. Plant polyphenols probably reduce nitrogen losses by forming complexes with soluble organic nitrogen compounds (Northup et al. 1995, 1998). Presumably, this fosters mycorrhizal plant associations and affects the activity and community structure of saprophytic microorganisms. Therefore, polyphenols may not only regulate litter decomposition but also the structure of the belowground community, which is likely to feed back to plant growth and plant community structure (Hättenschwiler and Vitousek 2000).

The dependency of the decomposer community on plant litter traits suggests that mixing of litter significantly affects the structure of decomposer communities, despite perhaps having little influence on soil animal species diversity. Indeed, although soil animal species diversity varies little with plant litter diversity (see above), soil animal community structure has been shown to significantly differ in pure and mixed stands (Saetre et al. 1999; Scheu et al. 2003). While the response of the soil fauna community to litter mixtures may not be directly linked to litter traits such as nitrogen and polyphenol concentration, other factors such as changes in the community structure of microorganisms may be important. Indeed, it has been documented that, e.g., the structure of the mycorrhizal community differs significantly in pure and mixed litters of oak and pine (Conn and Dighton 2000). It has also been shown frequently that decomposition processes differ in pure and mixed litters (Briones and Ineson 1996; Wardle et al. 1997; Salamanca et al. 1998; Anderson and Hetherington 1999; Hansen 1999; Prescott et al. 2000; Hättenschwiler, Chap. 8, this Vol.). These differences are likely associated with variations in the structure of the saprophytic microbial community (cf. Blair et al. 1990; Elliott et al. 1993; Saetre and Baath 2000). Generally, mixing of tree species appears to beneficially affect the decomposer community (Chapman et al. 1988; Blair et al. 1990, 1994; Salamanca et al. 1998; Prescott et al. 2000). In the Solling, a mountain range in northern Germany, effects of mixing of beech and spruce have been studied in detail. As documented by Scheu et al. (2003), the structure of the soil animal community differs strongly between spruce and beech stands although the species composition differed only slightly (Alphei and Klages 1997; Platner et al. 1997; Salamon et al. 1997; Migge et al. 1998; Salamon and Alphei 2001).

11.2.4 Case Study: The Decomposer Community in Pure and Mixed Stands of Beech and Spruce

In central Europe, much of the native beech forest has been replaced by spruce for economical reasons (Kazda and Pichler 1998). Spruce needles differ from beech leaves by a number of traits including high polyphenol content. Both beech leaves and spruce needles are known to decompose slowly, but very different reasons may be responsible for the slow decay rates. Beech leaves are high in lignin, whereas spruce needles are characterized by high polyphenol concentration (Swift et al. 1979). Replacement of beech by spruce has been documented to result in an accumulation of litter and the formation of thick organic layers (Nihlgard 1971; Ellenberg et al. 1986; Mardulyn et al. 1993; Berger and Hager 2000). Obviously, litter decomposition in spruce stands is retarded compared to beech stands. However, this retardation is not due to the fact that spruce needles are more recalcitrant than beech leaves. When exposed in a beech stand, spruce needles decompose as fast as beech leaves (Albers et al. 2003). Rather, since both spruce needles and beech leaves exposed in spruce stands decompose much slower than when exposed in beech stands, specific conditions of spruce stands retard litter decay (Albers et al. 2003). High concentrations of polyphenols in spruce litter layers among other factors contribute to this pattern.

Litter accumulation and the formation of mor soils in spruce forests result in reduced concentrations of microbial biomass and microbial energy efficiency (Scheu et al. 2003). These changes in the microbial community propagate into the decomposer animal community and these changes propagate further into predators (Scheu et al. 2003). Biomass of most of the microbivorous and detritivorous soil animal groups studied (but not that of total microbial biomass) was significantly higher in beech compared to spruce forests (Fig. 11.3). This indicates that in the forests studied, neither the amount of organic matter in L/F and H/Ah layers, nor the amount of microbial biomass, controlled microbivores and detritivores. Rather, the quality of litter materials and the concentration of microbial biomass therein appeared to be most important.

Herbivores and predators also were favored by beech: the biomass of one of the herbivorous groups studied (rhizophagous nematodes) was significantly increased in beech stands; none of the groups was increased in spruce stands. Parallel to the response of microbivores and detritivores, the biomass of most predatory groups studied was increased in beech stands (Fig. 11.4). Interest-

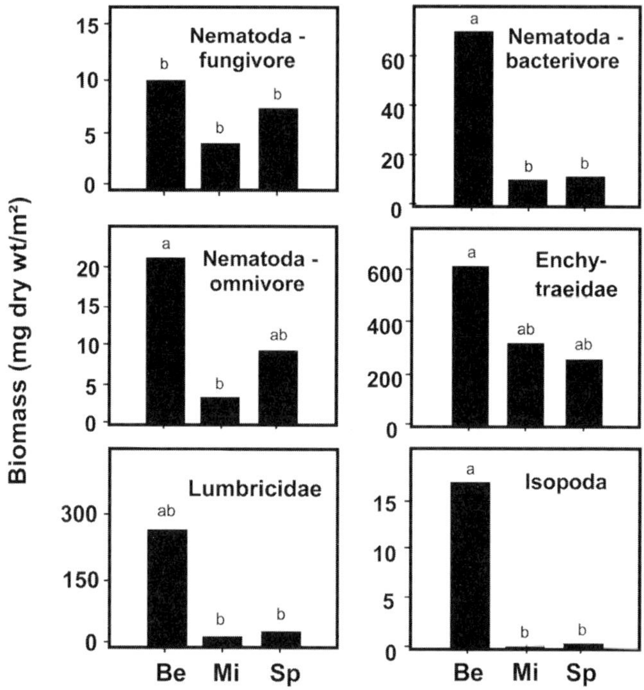

Fig. 11.3. Biomass of selected microbivorous and detritivorous (incl. omnivorous) soil invertebrates in pure and mixed stands of beech and spruce 120 years old. *Bars* sharing the *same letter* do not differ significantly ($P>0.05$, Tukey's minimum significant difference test; graph based on data given in Scheu et al. 2003); *Be* beech, *Mi* mixed, *Sp* spruce

ingly, the biomass ratio between prey and predators was at a minimum in mature beech and mixed stands, indicating more intense top-down control in these forests. Mixed stands were more similar to spruce stands with respect to the biomass of soil animal groups, but predator–prey interactions appeared to be more alike in mature beech and mixed stands. Obviously, replacement of beech by spruce strongly alters the structure and internal regulating forces of the soil food web. Weaker top-down forces in spruce forests are likely to be related to the thicker organic layer, which functions as a refuge for prey species. Decomposing spruce needles form an interwoven network of debris that is hard to penetrate for larger soil animal species. In fact, predators that were found to differ most between beech and spruce consist of some of the larger species of Chilopoda, Cantharidae, Araneida, and Carabidae. The biomass of small microarthropod predators, such as staphylinid beetles and gamasid mites, differed little between beech and spruce. Overall, this study shows that the structure of the below ground food web is a more sensitive indicator of changes in plant community structure than of soil animal species

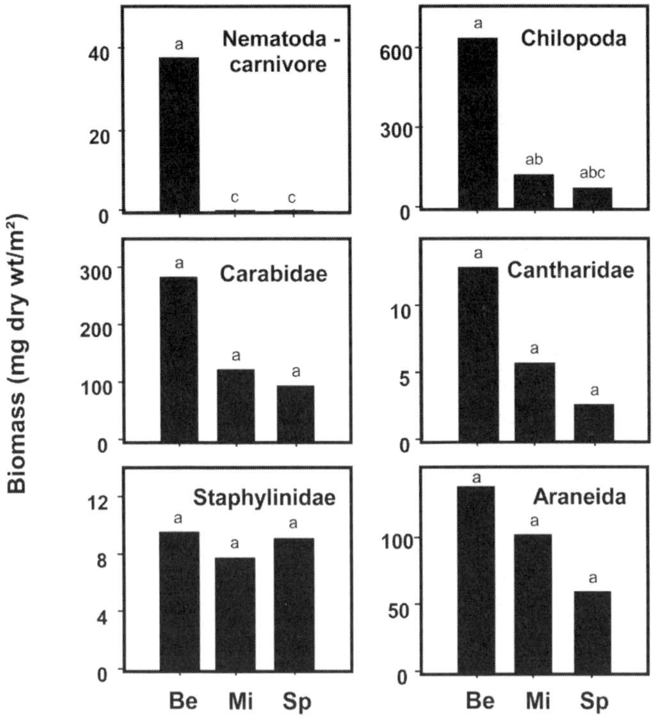

Fig. 11.4. Biomass of selected predator soil invertebrates in pure and mixed stands of beech and spruce 120 years old. *Bars* sharing the *same letter* do not differ significantly ($P>0.05$, Tukey's minimum significant difference test; graph based on data given in Scheu et al. 2003); *Be* beech, *Mi* mixed, *Sp* spruce

richness. Obviously, litter quality and the changes in physical topsoil properties, such as bulk density and soil pore volume, that are associated with the plantation of mixed stands (Berger and Hager 2000) greatly alters the dominance structure within trophic groups, but it also alters the interactions between trophic levels. The more pronounced top-down control in beech and mixed stands of beech and spruce may significantly contribute to the stability of the decomposer community.

11.3 Effects of Soil Fauna on Ecosystem Processes

In natural ecosystems, and less so in agricultural systems also, the soil can be viewed as a habitat built through the action of animals; particularly primary and secondary decomposers form the habitat of the whole soil community (Anderson 1995; Lavelle et al. 1997; Waid 1999). It is scarcely surprising, therefore, that the decomposer soil fauna has been viewed as a major driving factor for microbial activity in soils (Coleman et al. 1983; Anderson 1987; Wolters 1991). It has been estimated that the direct contribution of soil animals to energy flow in terrestrial systems is low, usually <10% (Reichle et al. 1975; Schaefer 1991b). For mineralization of nutrients their direct contribution might be higher, but presumably is still below 30% (Verhoef and Brussaard 1990; de Ruiter et al. 1993). Generally, effects of soil organisms on soil processes are intimately linked to their size. Small organisms such as bacteria, fungi, and protozoa are the key drivers of energy and nutrient transformations. In contrast, the large decomposer organisms such as earthworms, millipedes, and isopods are the dominant habitat transformers (Lavelle 1997; Anderson 2000; Scheu and Setälä 2002).

It has been documented that the composition and diversity of plant species affects processes below ground that feed back to ecosystem properties such as resilience and resistance (Wardle et al. 2000). Nilsson et al. (1999) reported that the competition between plant species differs when plants are grown in humus formed by monotypic vs. mixed litters, although, the effects were small and tended to be idiosyncratic. Nevertheless, the study documents that plant litter has important afterlife effects. From a functional point of view, plants and decomposer organisms are mutualists. Two different flows of matter connect plants with the belowground system, litter input and root exudates. Litter materials (leaves, roots, woody debris) are discrete physical structures built of complex polymers, whereas root exudates constitute of soluble compounds that are easily decomposed. For processing plant litter both physical litter transformers and mineralizers are necessary. In contrast, root exudates are used directly by microorganisms that may be controlled by predators (see below). Hence, the soil community may be viewed as consisting of two different mutualistic systems that feed back to the aboveground system (Wall and Moore 1999; Bonkowski et al. 2001). It is not the diversity of soil organisms that intimately links the below- and aboveground systems, but rather keystone species that keep these mutualistic subsystems functioning (Wardle et al. 1997; Huhta et al. 1998; Bardgett and Shine 1999; Laakso and Setälä 1999; Griffiths et al. 2000, 2001). However, for certain functional groups such as mycorrhiza, it has been shown that species diversity matters (van der Heijden et al. 1998; Jonsson et al. 2001). Also, theoretical considerations suggest that microbial diversity in soil positively affects nutrient cycling and ecosystem processes through either greater intensity of microbial exploitation of organic compounds or functional-niche complementarity, much as in plants (Loreau 2001). Obviously, the insufficient congruence between theoretical and experimental results needs further attention.

11.4 Feedbacks from Belowground Community Composition to Plant Community Structure

Evaluating the role of soil invertebrates for ecosystem functioning by focusing on changes in carbon and nitrogen transfer is important, but other processes also need to be considered. Interactions between soil invertebrates and plants may significantly contribute to ecosystem properties. Plants are intimately associated with soil organisms in the rhizosphere, and roots form an essential part of the belowground food web. Interactions between soil invertebrates and plants, mediated by soil microorganisms, are particularly numerous and important, since they modify plant growth and vegetation structure and therefore the whole aboveground community (Fig. 11.5). They include grazing on mycorrhizal fungi (Klironomos and Kendrick 1995; Setälä 1995), grazing on plant pathogens (Curl et al. 1988; Lartey et al. 1994; Pussard et al. 1994), dispersing of plant growth-stimulating microorganisms such as rhizobia, mycorrhiza, and mutualistic rhizosphere bacteria (Gange 1993; Harinikumar and Bagyaraj 1994; Stephens et al. 1993, 1995; Lussenhop 1996), and, by microorganisms, antagonizing root pathogens (Stephens and Davoren 1997). Furthermore, detritivores indirectly modify plant growth by fashioning changes

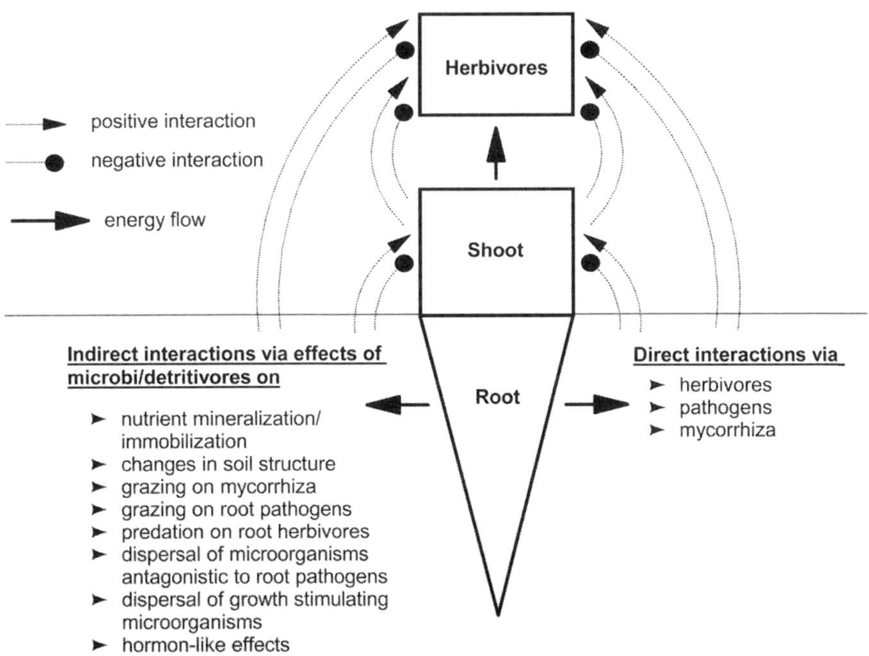

Fig. 11.5. Ways in which soil animals affect plant growth and thereby herbivore performance. (Modified from Scheu and Setälä 2002)

in soil structure (Hoogerkamp et al. 1983; Boyle et al. 1997) and by hormone-like effects (Jentschke et al. 1995; Muscolo et al. 1996, 1999). Other activities such as plant seed dispersal (Thompson et al. 1993; Willems and Huijsmans 1994) and plant damage by root herbivores and pathogens (Masters et al. 1993; Sarathchandra et al. 1996; Strong et al. 1996; Zunke and Perry 1997) add to the spectrum of links between soil animals and plants. As emphasized by Strong (1999), root herbivores are embedded in a complex predator/parasite community that includes organisms as different as parasitic nematodes (as antagonists of insect root herbivores) and nematode-trapping fungi (as antagonists of root-feeding nematodes), each and all of which may exert top-down effects on the plant root systems and therefore on plant growth. Similar arguments hold for plant pest species, since most of them colonize the soil at certain life stages, integrating themselves into the decomposer food web (see below). Rhizosphere interactions intimately link the below- and aboveground communities. However, despite their great importance, information on how these interactions affect plant growth, vegetation structure, and the aboveground food web is very limited (van der Putten et al. 2001; Scheu and Setälä 2002; Wardle 2002).

In addition to bottom-up controls of plant community structure and the aboveground food web via modifications in plant growth, belowground organisms may affect the plant–herbivore system by modifying top-down forces (Fig. 11.6). As indicated above, many herbivore species do colonize the soil at certain life stages and therefore, at least at some period of time, are integrated into the belowground food web. There, they may be consumed by predators that generally live on prey from the decomposer community. Furthermore, predators such as spiders, carabid and staphylinid beetles in their early life stages live on a diet of decomposer animals and, when becoming adult, may leave the soil and consume herbivores in the aboveground vegetation, thereby fostering top-down control of herbivores in their own habitat. Both processes may significantly contribute to fostering top-down forces exerted on plant herbivores and therefore help in preventing outbreaks of pest species. A basic requirement for this interconnection of the below- and aboveground food webs is that soil predators are generalist feeders which in fact is the case. Generalist feeding including polyphagy, omnivory, and intraguild predation appears to be a characteristic feature of soil predators (Scheu and Setälä 2002). This allows switching of prey from the decomposer food web to components of the aboveground system, as exemplified by the studies of Settle et al. (1996), Symondson et al. (2000), and Halaj and Wise (2002). Switching from decomposer prey organisms to pest species has been proposed to be a very general and important mechanism linking below- and aboveground communities in arable and forest ecosystems (Scheu 2001).

In forest ecosystems, tree species diversity and composition may significantly affect the susceptibility to insect outbreaks (Watt 1992), and this is likely to be caused by increased densities of natural enemies (cf. Watt et al.

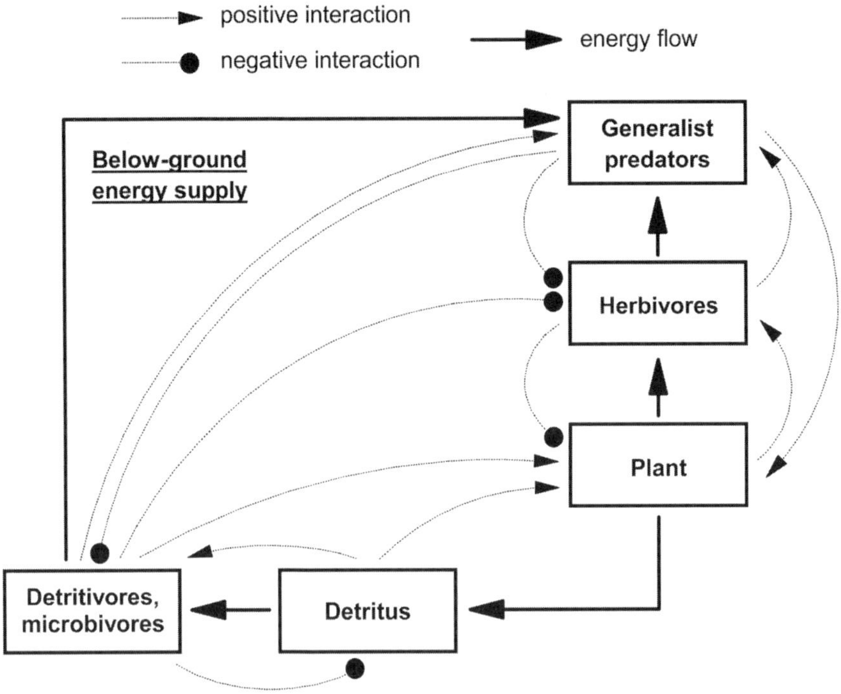

Fig. 11.6. Strengthening of the trophic cascade between generalist predators and plants via belowground energy supply. Note that increased input of detritus associated with increased belowground production may help to reduce plant damage by herbivores via supporting generalist predator populations. (Modified from Scheu 2001)

1997; Rieske and Buss 2001; Raymond et al. 2002; Jactel, Chap. 12, this Vol.). As documented above, predators may reach considerably larger populations in pure and mixed stands of beech compared to spruce monocultures, which likely increase the internal control mechanisms of prey populations. Forest management practices that foster internal predator–prey interactions therefore are likely to contribute to the stability of forest ecosystems and, as indicated by Scheu et al. (2003), the admixture of beech to spruce monocultures likely functions in this way.

11.5 Conclusions

The decomposer food web essentially relies on resources built up by primary producers above the ground. While the diversity of the decomposer organisms is not intimately linked to plant diversity, plant traits such as nitrogen and polyphenol concentration are key drivers for belowground community structure. The recycling of plant residues is driven by soil organisms; the above- and belowground food webs therefore complement each other and the interrelationship may be viewed as mutualistic partnership. Traits of plant litter may have evolved to support or suppress certain decomposer organisms and their associated functions. Also, plants may be able to direct resources to foster certain mutualists in the rhizosphere, such as mycorrhiza or bacteria and their associated grazers. Activity and structure of the decomposer food web feed back to plant growth, vegetation development, and the aboveground food web in a variety of ways. Two feedback scenarios have been distinguished: (1) the detritivore–plant–herbivore pathway, i.e., soil organism-mediated changes in plant growth viewed as bottom-up control of the aboveground food web; (2) the detritivore-generalist predator–herbivore pathway, i.e., the fostering of top-down effects exerted on plant herbivores by generalist predators, which also live on prey from the decomposer community. These complex feedback mechanisms suggest that changes in the decomposer community caused by tree diversity and species mixtures feed back to trees themselves and the whole aboveground food web. The decomposer food web therefore not only contributes to ecosystem properties associated with belowground matter transfer, such as nutrient losses and humus formation, but also to aboveground processes such as the control of pest species. Management practices intending to conserve the biodiversity of forests and the functioning of forest ecosystems therefore need to more closely consider the structure and driving forces of the decomposer food web.

Acknowledgments: The study in part was supported by the German Science Foundation (DFG; FOR 456) and the European Commission (CONSIDER).

References

Albers D, Migge S, Schaefer M, Scheu S (2003) Decomposition of beech *(Fagus sylvatica)* and spruce *(Picea abies)* litter in pure and mixed stands of beech and spruce. Soil Biol Biochem 36:155–164

Alphei J, Klages U (1997) Die Gemeinschaftsstruktur freilebender Bodennematoden in Rein- und Mischbeständen von Buche und Fichte auf Buntsandstein. Mitt Dtsch Bodenkundl Ges 85 II:461–464

Anderson JM (1987) Interactions between invertebrates and microorganisms: noise or necessity for soil processes? In: Fletcher M, Grag TRG, Jones JG (eds) Ecology of microbial communities. Cambridge Univ Press, Cambridge, pp 125–145

Anderson JM (1995) Soil organisms as engineers: microsite modulation of macroscale processes. In: Jones CG, Lawton JH (eds) Linking species to ecosystems. Chapman and Hall, New York, pp 94–106

Anderson JM (2000) Foodweb functioning and ecosystem processes: problems and perceptions of scaling. In: Coleman DC, Hendrix PF (eds) Invertebrates as webmasters in ecosystems. CABI, Wallingford, pp 3–24

Anderson JM, Hetherington SL (1999) Temperature, nitrogen availability and mixture effects on the decomposition of heather [*Calluna vulgaris* (L.) Hull] and bracken [*Pteridium aquilinum* (L.) Kuhn] litters. Funct Ecol 13 [Suppl 1]:116–124

Augusto L, Bonnaud P, Ranger J (1998) Impact of tree species on forest soil acidification. For Ecol Manage 105:67–78

Bardgett RD, Shine A (1999) Linkages between plant litter diversity, soil microbial biomass and ecosystem function in temperate grasslands. Soil Biol Biochem 31:317–321

Bardgett RD, Mawdsley JL, Edwards S, Hobbs PJ, Rodwell JS, Davies WJ (1999) Plant species and nitrogen effects on soil biological properties of temperate upland grasslands. Funct Ecol 13:650–660

Berg B (2000) Litter decomposition and organic matter turnover in northern forest soils. For Ecol Manage 133:13–22

Berger TW, Hager H (2000) Physical top soil properties in pure stands of Norway spruce (*Picea abies*) and mixed species stands in Austria. For Ecol Manage 136:159–172

Blair JM, Parmelee RW, Beare MH (1990) Decay rates, nitrogen fluxes, and decomposer communities of single- and mixed-species foliar litter. Ecology 71:1976–1985

Blair JM, Parmelee RW, Wyman RL (1994) A comparison of the forest floor invertebrate communities of four forest types in the northeastern U.S. Pedobiologia 38:146–160

Blair JM, Parmelee RW, Lavelle P (1995) Influence of earthworms on biogeochemistry. In: Hendrix PF (ed) Earthworm ecology and biogeography in North America. Lewis, Boca Raton, pp 127–158

Bonkowski M, Jentschke G, Scheu S (2001) Contrasting effects of microbial partners in the rhizosphere: interactions between Norway spruce seedlings (*Picea abies* Karst.), mycorrhiza (*Paxillus involutus* (Batsch) Fr.) and naked amoebae (protozoa). Appl Soil Ecol 18:193–204

Boyle KE, Curry JP, Farrell EP (1997) Influence of earthworms on soil properties and grass production in reclaimed cutover peat. Biol Fertil Soils 25:20–26

Bridge P, Spooner B (2001) Soil fungi: diversity and detection. Plant Soil 232:147–154

Briones MJI, Ineson P (1996) Decomposition of eucalyptus leaves in litter mixtures. Soil Biol Biochem 28:1381–1388

Butterfield J, Benitez Malvido J (1992) Effect of mixed-species tree planting and the distribution of soil invertebrates. In: Cannell MGR, Malcolm DC, Robertson PA (eds) The ecology of mixed-species stands of trees. Blackwell, Oxford, pp 255–265

Chapman K, Whittaker JB, Heal OW (1988) Metabolic and faunal activity in litters of tree mixtures compared with pure stands . Agric Ecosyst Environ 24:33–40

Clay K, van der Putten WH (1999) Pathogens and plant life histories. In: Buorisalo TO, Mutikainen PK (eds) Life history evolution in plants. Kluwer, Dordrecht, pp 275–301

Coleman DC, Reid CPP, Cole CV (1983) Biological strategies of nutrient cycling in soil systems. In: MacFadyen A, Ford ED (eds) Advances in ecological research. Academic Press, New York, pp 1–55

Conn C, Dighton J (2000) Litter quality influences on decomposition, ectomycorrhizal community structure and mycorrhizal root surface acid phosphatase activity. Soil Biol Biochem 32:489–496

Curl EA, Lartey R, Peterson CM (1988) Interactions between root pathogens and soil microarthropods. Agric Ecosyst Environ 24:249–261

De Ruiter PC, Moore JC, Zwart KB, Bouwman LA, Hassink J, Bloem J, De Vos JA, Marinissen JCY, Midden WAM (1993) Simulation of nitrogen mineralization in the belowground food webs of two winter wheat fields. J Appl Ecol 30:95–106

Diemer M, Joshi J, Körner C, Schmid B, Spehn E (1997) An experimental protocol to assess the effects of plant diversity on ecosystem functioning utilized in a European research network. Bull Geobot Inst ETH 63:95–107

Ellenberg H, Mayer R, Schauermann J (1986) Ökosystemforschung – Ergebnisse des Sollingprojektes. Ulmer, Stuttgart

Elliott WM, Elliott NB, Wyman RL (1993) Relative effect of litter and forest type on rate of decomposition. Am Midl Nat 129:85–97

Enriquez S, Duarte CM, Sand-Jensen K (1993) Patterns in decomposition rates among photosynthetic organisms: the importance of detritus C:N:P content. Oecologia 94:457–465

Gange AC (1993) Translocation of mycorrhizal fungi by earthworms during early succession. Soil Biol Biochem 25:1021–1026

Griffiths BS, Ritz K, Bardgett RD, Cook R, Christensen S, Ekelund F, Sorensen SJ, Baath E, Bloem J, De Ruiter PC, Dolfing J, Nicolardot B (2000) Ecosystem response of pasture soil communities to fumigation-induced microbial diversity reductions: an examination of the biodiversity-ecosystem function relationship. Oikos 90:279–294

Griffiths BS, Ritz K, Wheatley R, Kuan HL, Boag B, Christensen S, Ekelund F, Sorensen SJ, Muller S, Bloem J (2001) An examination of the biodiversity-ecosystem function relationship in arable soil microbial communities. Soil Biol Biochem 33:1713–1722

Grime JP (1979) Plant strategies and vegetation processes. Wiley, Chichester

Haddad NM, Tilman D, Haarstad J, Ritchie M, Knops JMH (2001) Contrasting effects of plant richness and composition on insect communities: a field experiment. Am Nat 158:17–35

Halaj J, Wise DH (2002) Impact of a detrital subsidy on trophic cascades in a terrestrial grazing food web. Ecology 83: 3141–3151

Hansen RA (1999) Red oak litter promotes a microarthropod functional group that accelerates its decomposition. Plant Soil 209:37–45

Hansen RA, Coleman DC (1998) Litter complexity and composition are determinants of the diversity and species composition of oribatid mites (Acari: Oribatida) in litterbags. Appl Soil Ecol 9:17–23

Harborne JB (1997) Role of phenolic secondary metabolites in plants and their degradation in nature. In: Cadisch G, Giller KE (eds) Driven by nature: plant litter quality and decomposition. CAB International, Wallingford, pp 67–74

Harinikumar KM, Bagyaraj DJ (1994) Potential of earthworms, ants, millipedes, and termites for dissemination of vesicular-arbuscular mycorrhizal fungi in soil. Biol Fertil Soils 18:115–118

Hättenschwiler S, Vitousek PM (2000) The role of polyphenols in terrestrial ecosystem nutrient cycling. TREE 15:238–243

Hector A, Beale AJ, Minns A, Otway SJ, Lawton JH (2000) Consequences of the reduction of plant diversity for litter decomposition: effects through litter quality and microenvironment. Oikos 90:357–371

Hoogerkamp M, Rogaar H, Eijsackers HJP (1983) Effect of earthworms on grassland on recently reclaimed polder soils in the Netherlands. In: Satchell JE (ed) Earthworm ecology form Darwin to vermiculture. Chapman and Hall, London, pp 85–105

Hooper DU, Vitousek PM (1997) The effects of plant composition and diversity on ecosystem processes. Science 277:1302–1305

Hopkin SP (1997) Biology of the springtails – Insecta: Collembola. Oxford Univ Press, Oxford

Huhta V, Hyvönen R, Koskenniemi A, Vilkamaa PKP, Sulander M (1986) Response of soil fauna to fertilization and manipulation of pH in coniferous forests. Acta For Fenn 195:1–30

Huhta V, Persson T, Setälä H (1998) Functional implications of soil fauna diversity in boreal forests. Appl Soil Ecol 10:277–288

Jentschke G, Bonkowski M, Godbold DL, Scheu S (1995) Soil protozoa and forest tree growth: non-nutritional effects and interaction with mycorrhizae. Biol Fertil Soils 19:263–269

Jonsson LM, Nilsson MC, Wardle DA, Zackrisson O (2001) Context dependent effects of ectomycorrhizal species richness on tree seedling productivity. Oikos 93:353–364

Kaneko N, Salamanca EF (1999) Mixed leaf litter effects on decomposition rates and soil microarthropod communities in an oak-pine stand in Japan. Ecol Res 14:131–138

Kazda M, Pichler M (1998) Priority assessment for conversion of Norway spruce forests through introduction of broadleaf species. For Ecol Manage 102:245–258

Klironomos JN, Kendrick WB (1995) Stimulative effects of arthropods on endomycorrhizas of sugar maple in the presence of decaying litter. Funct Ecol 9:528–536

Knops JMH, Wedin D, Tilman D (2001) Biodiversity and decomposition in experimental grassland ecosystems. Oecologia 126:429–433

Kuiters AT (1990) Role of phenolic substances from decomposing forest litter in plant-soil interactions. Acta Bot Neerl 39:329–348

Laakso J, Setälä H (1999) Sensitivity of primary production to changes in the architecture of belowground food webs. Oikos 87:57–64

Lartey RT, Curl EA, Peterson CM (1994) Interactions of mycophagous collembola and biological control fungi in the suppression of *Rhizoctonia solani*. Soil Biol Biochem 26:81–88

Lavelle P (1997) Faunal activities and soil processes: adaptive strategies that determine ecosystem function. Adv Ecol Res 27:93–122

Lavelle P, Bignell D, Lepage M, Wolters V, Roger P, Ineson P, Heal OW, Dhillion S (1997) Soil function in a changing world: the role of invertebrate ecosystem engineers. Eur J Soil Biol 33:159–193

Loreau M (2001) Microbial diversity, producer-decomposer interactions and ecosystem processes: a theoretical model. Proc R Soc Lond B 268:303–309

Lorenz K, Preston CM, Raspe S, Morrison IK, Feger KH (2000) Litter decomposition and humus characteristics in Canadian and German spruce ecosystems: information from tannin analysis and ^{13}C CPMAS NMR. Soil Biol Biochem 32:779–792

Lussenhop J (1996) Collembola as mediators of microbial symbiont effects upon soybean. Soil Biol Biochem 28:363–369

Mardulyn P, Godden B, Amiano Echezarreta P, Penninckx M, Gruber W, Herbauts J (1993) Changes in humus microbiological activity induced by the substitution of the natural beech forest by Norway spruce in the Belgian Ardennes. For Ecol Manage 59:15–27

Masters GJ, Brown VK, Gange AC (1993) Plant mediated interactions between above- and below-ground insect herbivores. Oikos 66:148–151

Migge S, Maraun M, Scheu S, Schaefer M (1998) The oribatid mite community (Acarina) of pure and mixed stands of beech (*Fagus sylvatica*) and spruce (*Picea abies*) of different age. Appl Soil Ecol 9:115–121

Mogge B, Loferer C, Agerer R, Hutzler P, Hartmann A (2000) Bacterial community structure and colonization patterns of *Fagus sylvatica* L. - ectomycorrhizospheres as determined by fluorescence in situ hybridization and confocal laser scanning microscopy. Mycorrhiza 9:271–278

Mortimer SR, Van der Putten WH, Brown VK (1999) Insect and nematode herbivory below ground: interactions and role in vegetation succession. In: Olff H, Brown VK, Drent RH (eds) Herbivores: between plants and predators. Br Ecol Soc Symp 38. Blackwell, Oxford, pp 205–238

Muscolo A, Panuccio MR, Abenavoli MR, Concheri G, Nardi S (1996) Effect of molecular complexity and acidity of earthworm faeces humic fractions on glutamate dehydrogenase, glutamine synthetase, and phosphoenolpyruvate carboxylase in *Daucus carota* alpha II cells. Biol Fertil Soils 22:83–88

Muscolo A, Bovalo F, Gionfriddo F, Nardi S (1999) Earthworm humic matter produces auxin-like effects on *Daucus carota* cell growth and nitrate metabolism. Soil Biol Biochem 31:1303–1311

Nihlgard B (1971) Pedological influence of spruce planted on former beech forest soils in Scania, South Sweden. Oikos 22:302–314

Nilsson MC, Wardle DA, Dahlberg A (1999) Effects of plant litter species composition and diversity on the boreal forest plant-soil system. Oikos 86:16–26

Northup RR, Yu Z, Dahlgren RA, Vogt KA (1995) Polyphenol control of nitrogen release from pine litter. Nature 377:227–229

Northup RR, Dahlgren RA, McColl JG (1998) Polyphenols as regulators of plant-litter-soil interactions in northern California's pygmy forest: a positive feedback? Biogeochemistry 42:189–220

Peck KM (1989) Tree species preferences shown by foraging birds in forest plantations in northern England. Biol Cons 48:41–57

Petersen H (2002) General aspects of collembolan ecology at the turn of the millennium. Pedobiologia 46:246–260

Platner C, Scheu S, Schaefer M, Schauermann J (1997) Untersuchungen zur Bodenmakrofauna von Rein- und Mischbeständen von Buche und Fichte unterschiedlichen Alters auf Buntsandstein. Mitt Dtsch Bodenkundl Ges 85(II):575–578

Ponge JF, Arpin P, Sondag F, Delecour F (1997) Soil fauna and site assessment in beech stands of the Belgian Ardennes. Can J For Res 27:2053–2064

Ponsard S, Arditi R (2000) What can stable isotopes (δ^{15}N and δ^{13}C) tell about the food web of soil macro-invertebrates? Ecology 81:852–864

Prescott CE, Zabek LM, Staley CL, Kabzems R (2000) Decomposition of broadleaf and needle litter in forests of British Columbia: influences of litter type, forest type, and litter mixtures. Can J For Res 30:1742–1750

Pussard M, Alabouvette C, Levrat P (1994) Protozoa interactions with soil microflora and possibilities for biocontrol of plant pathogens. In: Darbyshire JF (ed) Soil protozoa. CAB International, Wallingford, pp 123–146

Raymond B, Vanbergen A, Watt A, Hartley SE, Cory JS, Hails RS (2002) Escape from pupal predation as a potential cause of outbreaks of the winter moth, *Operophtera brumata*. Oikos 98:219–228

Reichle DE, McBrager JF, Ausums S (1975) Ecological energetics of decomposer invertebrates in a deciduous forest and total respiration budget. In: Vanek J (ed) Progress in soil zoology. Proceedings of the 5th International Soil zoology Colloquium. Academia Prague, Prague, pp 283–291

Rieske LJ, Buss LJ (2001) Effects of gypsy moth suppression tactics on litter- and ground-dwelling arthropods in the central hardwood forests of the Cumberland Plateau. For Ecol Manage 149:181–195

Ruess L, Häggblom MM, Zapata EJG, Dighton J (2002) Fatty acids of fungi and nematodes – possible biomarkers in the soil food chain? Soil Biol Biochem 34:745–756

Saetre P, Baath E (2000) Spatial variation and patterns of soil microbial community structure in a mixed spruce-birch stand. Soil Biol Biochem 32:909–917

Saetre P, Brandtberg PO, Lundkvist H, Bengtsson J (1999) Soil organisms and carbon, nitrogen and phosphorus mineralisation in Norway spruce and mixed Norway spruce – birch stands. Biol Fertil Soils 28:382–388

Salamanca EF, Kaneko N, Katagiri S (1998) Effects of leaf litter mixtures on the decomposition of *Quercus serrata* and *Pinus densiflora* using field and laboratory microcosm methods. Ecol Eng 10:53–73

Salamon J, Alphei J (2001) Diversität und Struktur der Collembolengemeinschaften in Buche-Fichte-Mischbeständen versus monospezifischen Beständen. Mitt Dtsch Bodenkundl Ges 95:92–95

Salamon J, Albers D, Schauermann J (1997) Vergleich der Collembolengemeinschaften im Boden und in Litterbags in einem Gradienten von Buche (*Fagus sylvatica*) zu Fichte (*Picea abies*) auf Buntsandstein. Mitt Dtsch Bodenkundl Ges 85(II):595–598

Salamon J, Schaefer M, Alphei J, Schmid B, Scheu S (2003) Effects of plant diversity on Collembola in an experimental grassland ecosystem. Oikos 106:51–60

Sarathchandra SU, Watson RN, Cox NR, Dimenna ME, Brown JA, Burch G, Neville FJ (1996) Effects of chitin amendment of soil on microorganisms, nematodes, and growth of white clover (*Trifolium repens* L) and perennial ryegrass (*Lolium perenne* L). Biol Fertil Soils 22:221–226

Schaefer M (1991a) The animal community: diversity and resources. In: Röhrig E, Ulrich B (eds) Temperate deciduous forests. Ecosystems of the world 7. Elsevier, Amsterdam, pp 51–120

Schaefer M (1991b) Ecosystem processes: secondary production and decomposition. In: Röhrig E, Ulrich B (eds) Temperate deciduous forests. Ecosystems of the world 7. Elsevier, Amsterdam, pp 175–218

Schaefer M (1999) The diversity of the fauna of two beech forests: some thoughts about possible mechanisms causing the observed patterns. In: Kratochwil A (ed) Biodiversity in ecosystems. Kluwer, Dordrecht, pp 39–57

Schaefer M, Schauermann J (1990) The soil fauna of beech forests: comparison between a mull and a moder soil. Pedobiologia 34:299–314

Scheu S (2001) Plants and generalist predators as links between the below-ground and above-ground system. Basic Appl Ecol 2:3–13

Scheu S (2002) The soil food web: structure and perspectives. Eur J Soil Biol 38:11–20

Scheu S, Falca M (2000) The soil food web of two beech forests (*Fagus sylvatica*) of contrasting humus type: stable isotope analysis of a macro- and a mesofauna-dominated community. Oecologia 123:285–296

Scheu S, Setälä H (2002) Multitrophic interactions in decomposer communities. In: Tscharntke T, Hawkins BA (eds) Multitrophic level interactions. Cambridge Univ Press, Cambridge, pp 223–264

Scheu S, Alphei J, Bonkowski M, Jentschke G (1999) Soil food web interactions and ecosystem functioning: experimental approaches with systems of increasing complexity. In: Rastin N, Bauhus J (eds) Going underground – ecological studies in forest soils. Research Signpost, Trivandrum, India, pp 1–32

Scheu S, Albers D, Alphei J, Buryn R, Klages U, Migge S, Platner C, Salamon JA (2003) The soil fauna community in pure and mixed stands of beech and spruce of different age: trophic structure and structuring forces. Oikos 101:223–265

Setälä H (1995) Growth of birch and pine seedlings in relation to grazing by soil fauna on ectomycorrhizal fungi. Ecology 76:1844–1851

Settle WH, Ariawan H, Astuti ET, Cahyana W, Hakim AL, Hindayana D, Sri Lestari A, Sartanto P (1996) Managing tropical rice pests through conservation of generalist natural enemies and alternative prey. Ecology 77:1975–1988

Spehn EM, Joshi J, Schmid B, Diemer M, Körner C (2000) Above-ground resource use increases with plant species richness in experimental grassland ecosystems. Funct Ecol 14:326–337

Stephens PM, Davoren CW (1997) Influence of the earthworms *Aporrectodea trapezoides* and *A. rosea* on the disease severity of *Rhizoctonia solani* on subterranean clover and ryegrass. Soil Biol Biochem 29:511–516

Stephens PM, Davoren CW, Ryder MH, Doube BM (1993) Influence of the lumbricid earthworm *Aporrectodea trapezoides* on the colonization of wheat roots by *Pseudomonas corrugata* strain 2140R in soil. Soil Biol Biochem 25:1719–1724

Stephens PM, Davoren CW, Hawke BG (1995) Influence of barley straw and the lumbricid earthworm *Aporrectodea trapezoides* on *Rhizobium meliloti* L5-30R, *Pseudomonas corrugata* 2140R, microbial biomass and microbial activity in a red-brown earth soil. Soil Biol Biochem 27:1489–1497

Strong DR (1999) Predator control in terrestrial ecosystems: the underground food chain of bush lupine. In: Olff H, Brown VK, Drent RH (eds) Herbivores: between plants and predators. Blackwell, Oxford, pp 577–602

Strong DR, Kaya HK, Whipple AV, Child AL, Kraig S, Bondonno M, Dyer K, Maron JL (1996) Entomopathogenic nematodes: natural enemies of root-feeding caterpillars on bush lupine. Oecologia 108:167–173

Swift MJ, Heal OW, Anderson JM (1979) Decomposition in terrestrial ecosystems. Blackwell, Oxford

Symondson WOC, Glen DM, Erickson ML, Liddell JE, Langdon CJ (2000) Do earthworms help to sustain the slug predator *Pterostichus melanarius* (Coleoptera: Carabidae) within crops? Investigations using monoclonal antibodies. Mol Ecol 9:1279–1292

Symstad AJ, Siemann E, Haarstad J (2000) An experimental test of the effect of plant functional group diversity on arthropod diversity. Oikos 89:243–253

Thompson L, Thomas CD, Radley JMA, Williamson S, Lawton JH (1993) The effect of earthworms and snails in a simple plant community. Oecologia 95:171–178

Ulrich B (1994) Nutrient and acid-base budget of Central European forest ecosystems. In: Godbold DL, Hüttermann A (eds) Effects of acid rain on forest processes. Wiley, New York, pp 1–50

Ulrich B, Sumner ME (1991) Soil activity. Springer, Berlin Heidelberg New York

Van der Heijden MGA, Klironomos JN, Ursic M, Moutoglis P, Streitwolf-Engel R, Boller T, Wiemken A, Sanders IR (1998) Mycorrhizal fungal diversity, ecosystem variability and productivity. Nature 396:69–72

Van der Putten WH, Vet LEM, Harvey JA, Wäckers FL (2001) Linking above- and below-ground multitrophic interactions of plants, herbivores, pathogens, and their antagonists. TREE 16:547–554

Verhoef HA, Brussaard L (1990) Decomposition and nitrogen mineralization in natural and agroecosystems: the contribution of soil animals . Biogeochemistry 11:175–212

Waid JS (1999) Does soil biodiversity depend upon metabiotic activity and influences? Appl Soil Ecol 13:151–158

Wall DH, Moore JC (1999) Interactions underground – soil biodiversity, mutualism, and ecosystem processes. Bioscience 49:109–117

Wardle DA (2002) Linking the aboveground and belowground components. Princeton Univ Press, Princeton

Wardle DA, Nicholson KS, Yeates GW (1993) Effect of weed management strategies on some soil-associated arthropods in maize and asparagus ecosystems. Pedobiologia 37:257–269

Wardle DA, Nicholson KS, Rahman A (1994) Influence of herbicide applications on the decomposition, microbial biomass, and microbial activity of pasture shoot and root litter. NZ J Agric Res 37:29–39

Wardle DA, Zackrisson O, Hornberg G, Gallet C (1997) The influence of island area on ecosystem properties. Science 277:1296–1299

Wardle DA, Barker GM, Bonner KI, Nicholson KS (1998) Can comparative approaches based on plant ecophysiological traits predict the nature of biotic interactions and individual plant species effects in ecosystems? J Ecol 86:405–420

Wardle DA, Nicholson KS, Bonner KI, Yeates GW (1999) Effects of agricultural intensification on soil-associated arthropod population dynamics, community structure, diversity and temporal variability over a seven-year period. Soil Biol Biochem 31:1691–1706

Wardle DA, Bonner KI, Barker GM (2000) Stability of ecosystem properties in response to above-ground functional group richness and composition. Oikos 89:11–23

Wardle DA, Barker GM, Yeates GW, Bonner KI, Ghani A (2001) Introduced browsing mammals in New Zealand natural forests: aboveground and belowground consequences. Ecol Monogr 71:587–614

Watt AD (1992) Insect pest population dynamics: effects of tree species diversity. In: Cannell MGR, Malcolm DC, Robertson PA (eds) The ecology of mixed-species stands of trees. Blackwell, Oxford, pp 267–275

Watt AD, Stork NE, McBeath C, Lawson GL (1997) Impact of forest management on insect abundance and damage in a lowland tropical forest in southern Cameroon. J Appl Ecol 34:985–998

Willems JH, Huijsmans KGA (1994) Vertical seed dispersal by earthworms: a quantitative approach. Ecography 17:124–130

Wolters V (1991) Soil invertebrates – effects on nutrient turnover and soil structure – a review. Z Pflanzenern Bodenkd 154:389–402

Zunke U, Perry RN (1997) Nematodes: harmful and beneficial organisms. In: Benckiser G (ed) Fauna in soil ecosystems. Recycling processes, nutrient fluxes, and agricultural production. Dekker, New York, pp 85–133

12 A Test of the Biodiversity–Stability Theory: Meta-analysis of Tree Species Diversity Effects on Insect Pest Infestations, and Re-examination of Responsible Factors

H. Jactel, E. Brockerhoff, and P. Duelli

12.1 Introduction

The notion that biodiversity leads to stability of ecosystems is the subject of an ongoing debate in ecology (McCann 2000). An early proponent of this theory (Elton 1958) was partly inspired by observations on the stability of insect populations in different forest ecosystems, such as the apparently lower frequency of outbreaks of phytophagous insects in complex tropical forests than in simplified communities such as boreal forests. Numerous studies focusing on ecosystem resistance as a measure of stability have since addressed this theory, mainly in agricultural ecosystems, where the cultivation of single crop species in large monocultures is thought to aggravate pest problems as compared with polycultures. Although the validity of the principle has been criticized on the basis of theoretical models (May 1973), or because it was thought that a lack of diversity per se was not necessarily related to pest abundance (van Emden and Williams 1974), a review of 150 studies showed that diverse agroecosystems had lower pest populations than monocultures in 62 % of the cases (Risch et al. 1983). Similarly, using a meta-analysis of 21 independent studies, Tonhasca and Byrne (1994) found that pest insect densities were significantly lower in diversified crop than in monocultures in almost 60 % of the comparisons. Moreover, targeted diversification of agricultural ecosystems is increasingly being used, with success, to control pests by reducing the concentration of host plants or by providing habitat characteristics that improve the action of natural enemies (Russell 1989; Tshernyshev 1995; Landis et al. 2000).

The evidence from agricultural monocultures is often extended to account for the apparently greater frequency of pest outbreaks in forest monocultures, an explanation widely noted in ecology textbooks (e.g., Krebs 1985; Speight and Wainhouse 1989; Begon et al. 1996; Speight and Wylie 2001). However,

Ecological Studies, Vol. 176
M. Scherer-Lorenzen, Ch. Körner, and E.-D. Schulze (Eds.)
Forest Diversity and Function: Temperate and Boreal Systems

contrary to agroecosystems, previous reviews of the effects of biodiversity on the stability of forest pest populations have not provided consistent evidence of the applicability of the theory. Although the reviews of Gibson and Jones (1977), Schwerdtfeger (1981) and Barthod (1994) report examples of pest outbreaks in single-species plantation forests that appear to support the idea, the authors acknowledge that there are few direct comparisons that could ascertain a causal relationship with tree species diversity. Watt (1992) also argues that there are few direct comparisons between the relative abundance of insect pests in pure and mixed-species stands, and that pest problems in forest monocultures are not necessarily related to the stand composition, but to other causes such as the even-aged structure or the use of intensive silvicultural practices. Speight and Wylie (2001) give several reasons for the expectation that forest pest problems should be more abundant in monoculture forests, such as the greater availability of resources, but they also state that clear evidence is hard to find. Occasionally, the discussion has also suffered from a lack of appreciation of several confounding factors, such as the origin of the tree species planted. For example, plantation forests of exotic species can be remarkably pest free (Bain 1981; Gagdil and Bain 1999), particularly when they are planted far from the region of their origin and if the local flora does not contain any closely related species that could act as a source of pests, such as radiata pine in the southern hemisphere. On the other hand, once colonized by an introduced pest, plantation of exotic trees may be more vulnerable than monoculture of native trees, due to the lack of local natural enemies; e.g., *Cinara cedri*, an exotic aphid introduced in *Cedrus* plantations in France (Fabre 1988) and *Rhyacionia buoliana* in radiata pine plantations in Chile (Lanfranco et al. 2000).

It is regrettable that the debate on the effects of biodiversity on resistance of forest ecosystems to pest insects has so far suffered from a lack of objective, quantitative assessments. Hence, we aim at providing a quantitative review of all available studies that compared measures of pest abundance or damage in single-species stands and in mixed-species stands. We use meta-analysis (Gurevitch and Hedges 1993; Koricheva et al. 1998a,b) to examine a dataset derived from 54 independent studies that are likely to represent most of the published experiments on this subject. Then we discuss the potential ecological mechanisms that can explain the relationship between tree species diversity and stand resistance to pest insects. Here, we focus mainly on the composition of canopy tree species at the stand level, but also use experiments that examine the effects of differences in understory species diversity. Finally, we discuss spatial scales (e.g., stand vs. landscape level) that are relevant for future studies in this area.

12.2 Comparing Insect Pest Damage in Pure vs. Mixed Stands of Trees: A Meta-analysis

12.2.1 The Meta-analysis Method

We compiled a list of suitable experimental or observational studies published from 1966 to 2000 by making keyword searches in CABI Bioscience and Tree CD bibliographic databases and by searching for suitable references cited in the literature. Studies were included in the analysis if they met the following three criteria: (1) the resistance of a particular tree species to insect pests was compared in single-tree species and mixed stands, in the same area and time period. Mixed stands consisted of a mixture of canopy species or a mixture of overstory tree species and understory species of shrubs and herbs or grasses; (2) either pest abundance, pest damage, or tree mortality was recorded as response variable; (3) the mean of the response variable, a measure of variance, and the sample size were reported in the text or graphically. We chose the single tree species stands as the experimental group and the mixed stands as the control. When results for a response variable were reported in the same paper for several pest species or different mixtures, the data for each insect and treatment type were included as for an individual study. Results for several sampling dates were not used as replicates unless data were obtained in independent tree samples.

The analyses were carried out using *MetaWin* 2.0 statistical software (Rosenberg et al. 2000). For each individual study an estimate of the magnitude of the treatment effect, Hedges's d effect size, was calculated as the difference between the mean of the experimental group (pure stand) and the control group (mixed stand) divided by the pooled standard deviation and multiplied by a correction factor that accounts for small sample size. Positive d therefore indicates higher densities or damage in pure stands as compared to mixed stands. Effect sizes across all studies were combined using the random effects model to provide the grand mean effect size (d_{++}), an estimate of the overall effect of mixed vs. single-species stand composition on the susceptibility to pest insects. The effect was considered as statistically significant if the bootstrap confidence interval, calculated with 999 iterations, did not bracket zero. Pest insects were classified according to order, feeding guild, and host specificity. Six feeding guilds were used: chewing, boring, sucking, galling, mining, and cone feeding. Insects were categorized as oligophagous when they were host specific to trees within a genus or a family and as polyphagous when they feed on trees from more than one family. Using *MetaWin*, the mean effect size (d_{+}) was then calculated for each class of insect order, feeding guild, or host specificity, and the between-classes heterogeneity was tested against a Chi-square distribution to evaluate the significance of the class effect. Finally, the "'file-drawer problem'" (publication bias resulting from the greater likelihood of studies with statistically significant results to be published) was addressed by calculating a fail-safe sample size which represents an estimate of the number of non-significant, unpublished, or missing studies that would need to be added to the analysis in order to make the overall test of an effect statistically non-significant.

12.2.2 Results of the Meta-analysis

A total of 54 individual studies were included in the database, originating from 29 publications published in 22 journal articles or books. They covered a wide range of tree species (30) and forest insect species (28) from all continents and biomes. However some tree species are more often represented, such as European oaks, while some important tree species like Norway spruce are missing, due to the apparent lack of relevant published studies. Among the 54 studies, 15 showed a negative effect size and 39 a positive effect size of monoculture (Fig. 12.1, Appendix). Overall, the grand mean effect size equaled 0.65 and was significantly different from zero, with a bootstrap confi-

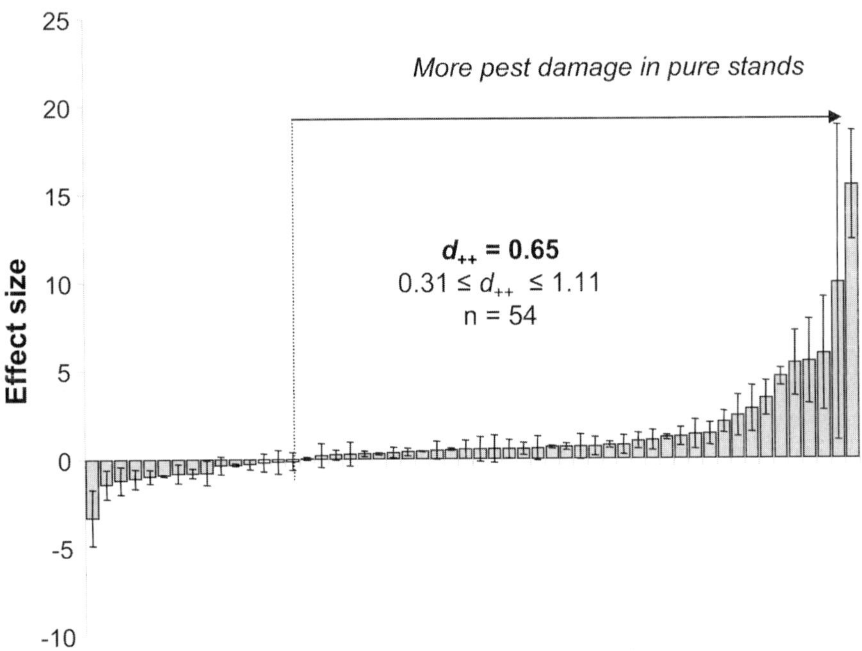

Fig. 12.1. Hedges's effect size and variance of individual studies on the response of pest damage or abundance to pure vs. mixed stand management. In most cases, the effect is positive, indicating that lower tree species diversity results in greater insect pest abundance, density, or damage. Value of d_{++}, the grand mean effect size, is given with the bootstrap confidence interval

dence interval of 0.31–1.11. According to Cohen (1969), an effect size of 0.5–0.8 shows a medium effect and a large effect if it is over 0.8. The meta-analysis therefore indicates that planting or managing a tree species as a pure stand on average significantly increased the risk of insect pest damage as compared to a mixed stand. Tonhasca and Byrne (1994) made a similar meta-analysis to test the effect of crop diversity on densities of agricultural pest insect. They also found a significant reduction of insect densities in diversified crops as compared to monocultures, although the effect size was smaller (d_{++}=0.35). Rosenthal's method was used for our dataset to calculate the fail-safe number, and the latter was compared to the conservative critical value of $5n+10$ (Rosenthal 1979), where n is the sample size. Hence, for our database, this value is 5×54+10=280. According to the *Meta-Win* analysis, the fail-safe sample size was 426, and thus much greater than the conservative value of 280 and eight times the number of studies included in the meta-analysis. Thus, this result appears robust with respect to the "file-drawer problem".

We first compared the response of forest pest damage to pure vs. mixed stand management in different biomes (Table 12.1). The mean effect size was always positive and significant, though more important for boreal forests than for temperate or tropical forests (df=2, Q_M=8.22=, P=0.02; where Q_M is the variation in effect sizes explained by the categorical model; Rosenberg et al. 2000). One can assume that tree species diversity being lower in boreal forests, the addition of associated tree species to pure stands may have more dramatic effect on pest dynamics. However, the meta-analysis only included five studies and three different pest insects from boreal forests, thus limiting the scope of this particular result. Then we tested the influence of the pest taxon, guild, and host specificity on the response of forest pest damage or abundance to pure vs. mixed stand management (Table 12.1). The effects of pure or mixed stand composition on pest infestation did not vary significantly among insect orders (df=4, Q_B=2.00, P=0.74). The mean effect size was positive and significant for all taxa except for Homoptera. Apart from Hymenoptera and Diptera, where the sample sizes were small, large positive mean effect sizes of monoculture were obtained for Lepidoptera and Coleoptera (Table 12.1). This indicates that damage caused by these pests was significantly greater in pure stands. This result is of particular interest because defoliators and wood borers belonging to these orders cause most forest damage worldwide. The effect of feeding guild was not significant either (df = 5, Q_B = 9.02, P = 0.11). Gallers, chewers, and especially borers demonstrated a high positive response to the stand composition (Table 12.1). For the other three guilds, effect sizes were low or non-significant, indicating a variable response of the species to the stand composition and/or a statistical bias due to the small number of comparisons. However, the effect of host specificity was highly significant (df=1, Q_B=12.33, P<0.001). The mean effect size was positive and high for oligophagous pest insects (d_+=1.28), demonstrating that insects with a host range limited to a particular genus or family were very likely to increase their

Table 12.1. Results of the meta-analysis on the influence of forest biome, pest insect taxon, feeding guild and host specificity on pest damage or abundance response to pure vs. mixed forest stand management.

Variable	Class	Comparison sample size	Mean effect size (d_+)	Bootstrap confidence interval
Forest biome	Boreal	5	2.34[a]	1.27–3.93
	Temperate	32	0.51[a]	0.00–1.13
	Tropical	17	0.52[a]	0.07–1.08
All pests insects				
Pest taxon	*Lepidoptera*	30	0.70[a]	0.21–1.30
	Coleoptera	10	1.25[a]	0.03–3.35
	Homoptera	6	0.54	–0.41–1.89
	Hymenoptera	2	0.56[a]	0.54–0.58
	Diptera	2	0.16[a]	0.03–0.29
Feeding guild	Chewer	26	0.63[a]	0.09–1.29
	Borer	10	1.78[a]	0.76–3.80
	Sucker	7	0.29	–0.31–0.77
	Galler	5	0.37[a]	0.19–0.57
	Miner	4	–0.07	–2.15–1.66
	Seed feeder[b]	2	0.04	–0.28–0.36
Host specificity	Oligophagous[c]	26	1.28[a]	0.77–2.11
	Polyphagous[d]	23	0.04	-0.39–0.51
POLYPHAGOUS PESTS				
Presence of other host tree in the mixture	Non- or less-susceptible trees	9	0.80[a]	0.37–1.56
	More susceptible trees	8	–0.18	–0.72–0.38

[a] denotes a significant effect based on a bootstrap test
[b] Cone and seed insects
[c] Insects with a host range that includes at most one family of trees
[d] Insects with a host range that includes trees from more than one family

damage on a tree species grown in pure stands. The mean effect size for polyphagous insects, attacking tree species from more than one family, was almost nil (d_+=0.03) and non-significant. Among the 24 studies concerning such polyphagous pests, 12 showed a negative effect size, including all 4 studies on species that can feed on both conifers and broadleaf trees. Therefore, our analysis indicates that, contrary to the general trend, about half of these polyphagous forest pest insects can actually cause more damage in mixed than in pure stands. Studying agroecosystems, Andow (1991) also found that 59 % of the monophagous species had lower densities in polycultures than in monocultures but only 28 % of the polyphagous species showed the same pattern.

The meta-analysis thus substantiated the widespread belief that, overall, forest monocultures are more prone to pest insect infestation than more diverse forests. Based on a quantitative comparison, it clearly shows that, irrespective of forest biome or pest guild, the risk of pest damage on a particular tree species is on average about 65 % higher when trees are growing in pure stands than in mixed stands. Polyphagous insect herbivores represent the main exception, as they caused more damage in tree mixtures in almost 50 % of the published studies. To interpret these new findings, we subsequently analyzed how tree species richness may interfere with host selection and predation processes in herbivorous insects attacking forest trees.

12.3 Effect of Tree Species Diversity on Stand Resistance to Pest Insects: The Main Ecological Hypotheses

Numerous theories have been advanced to support the hypothesis that herbivory should be higher in pure stands. We propose that these theories largely fit into three categories of processes relating to (1) the accessibility of hosts, (2) the impact of natural enemies, and (3) pest shift among host tree species. The following review of these theories is based on studies used in our meta-analysis, as well as findings from other relevant publications.

12.3.1 Host Accessibility

12.3.1.1 The Quantitative Barrier to Resource Exploitation

The first main theory to explain why pure stands would experience higher pest infestation is related to the higher concentration of food resources in such stands which, in turn, would provide insects with unlimited accessibility. As emphasized by Kareiva (1983), herbivore abundance consistently increases with host plant patch size. The absolute amount of suitable host trees in the forest ecosystem can benefit pest species like bark beetles whose population dynamics are primarily resource dependent. *Ips typographus* and *Ips sexdentatus* develop a severe outbreak in European forest monocultures in the year following a storm. These species cannot normally breed in vigorous, healthy trees, where defense reactions prevent the development of the brood, but they can overcome this protection when mass attacks occur. After a storm, both *Ips* species build up their populations in windthrown trees and then can extend their damage to numerous healthy trees in pine or spruce plantations (Speight and Wainhouse 1989).

It is generally thought that insects locate their host plants more efficiently in stands where the resource is more concentrated (Root 1973; Russell 1989). Baliddawa (1985), reviewing 36 studies that reported lower herbivore abundance in agricultural polycultures, found 24 cases where the explanation appeared to be that the host was more difficult to locate in mixed systems. A more successful colonization of trees in large patches is likely for species that are poor dispersers. For example, the psyllid *Phytolima lata* exhibits a very limited radius of dispersal. It developed outbreaks in pure stands of *Chlorophora excelsa* because abundant host trees occurred in close proximity, enhancing the spread of the pest (Roberts 1969). For some forest pests, host location is thought to be a trial-and-error process of landing on trees until a suitable host is found. The more diverse a stand, the longer it will take to find a host species and the higher the risk of mortality during the host-location process (Jones 2001). The risk is even higher for insect species with wind-dispersed larvae. The spruce budworm *Choristoneura fumiferana*, a major defoliator of fir and spruce in Canada, causes more damage in pure stands of balsam fir (Cappucino et al. 1998). The dispersal success of young larvae, which spin silken threads and balloon to new trees, is important for populations of this species. Kemp and Simmons (1979) demonstrated that survival of dispersing larvae was lower in mixed stands of host and non-host species than in pure balsam fir stands.

Three types of obstacles can disrupt the location of host trees in mixtures: physical barriers, chemical disruption, and, less commonly, temporal misfits.

12.3.1.2 The Physical Barrier to Host Colonization

Forest pests can be prevented from finding the host tree in mixed stands because they are physically hidden (Watt 1992). Spruce budworm larval dispersal is disturbed when host trees grow beneath a hardwood overstory (Batzer et al. 1987) or when they are insulated by non-host ponderosa pines (Fauss and Pierce 1969). Bergeron et al. (1995) suggested that the presence of a deciduous canopy could prevent budworm females from finding and colonizing understory conifers. The presence of broadleaf trees is also thought to break the continuity between pines, reducing host tree discovery by *Dendroctonus frontalis* (Belanger and Malac 1980; Schowalter and Turchin 1993). Young *Pinus taeda* can be protected from *Rhyacionia frustrana* attacks by a shelter of herbs and shrubs (Warren 1963, in Berisford and Kulman 1967). Moore et al. (1991) could significantly relate the abundance of leafrollers on oak leaves to reduced leaf apparency in mixtures of oaks/alders and oaks/spruce. Host plants may also be difficult to distinguish against a background of non-host vegetation (Cromartie 1981). Female pine processionary moths, *Thaumetopoea pityocampa*, use the tree silhouette standing out against a clear background to locate their host tree. This behavior is thought to explain why hardwood borders can reduce processionary moth infestations in pine plantations (Demolin 1962). Floater and Zalucki (2000) also observed that increased ground cover around acacia trees reduces egg-laying by the processionary moth *Ochrogaster lunifer*, and they suggested that females use visual clues to colonize the host trees.

12.3.1.3 The Chemical Barrier to Host Location

Chemical stimuli from host and non-host trees can also affect the rate of habitat colonization. This process is well illustrated in bark beetles, which are attracted by volatiles emitted from host trees. Several non-host volatiles have also been identified for conifer-infesting bark beetles, including green leaf volatiles from angiosperm trees, bark alcohols, and trans-conophthorin. Such compounds have shown repellent effects to *Dendroctonus frontalis*, *Ips grandicollis* and *Ips avulsus* (Dickens et al. 1992), *Pityogenes chalcographus* (Byers et al. 1998), *Ips typographus* (Zhang et al. 1999), *Tomicus piniperda* and *Tomicus minor* (Schlyter et al. 2000), *Dendroctonus pseudotsugae* (Huber and Borden 2001a), *Dendroctonus ponderosae* (Huber and Borden 2001b), and *Ips duplicatus* (Zhang et al. 2001). A significant reduction in pine log infestation by the pine stenographer beetle *Ips sexdentatus* was observed in the field when stems of the non-host species *Betula pendula* surrounded pine bolts (Jactel et al. 2001). Zhang (2001) suggested that because mixed conifer and broad-leaved forests have greater semiochemical diversity than pure host stands, they disturb olfactory host selection, thereby reducing the probability of outbreaks of conifer-infesting bark beetles. More generally, chemical disruption would theoretically reduce residency time in the habitat and disturb the reproductive behavior, leading to an overall reduction of infestation rates in mixed stands (Risch 1981).

12.3.1.4 The Temporal Barrier to Host Utilization

A third type of obstacle may also disrupt the colonization process of forest pests in mixed stands: the temporal barrier. In several tree defoliators, the young larvae can only feed on young leaves and thus require that the phenology of the defoliator matches that of the required host stage. They require that bud burst coincide with egg hatch to build up their population (Satchell 1962; Witter and Waisnanen 1978; DuMerle and Mazet 1983; Wint 1983). In a mixed stand of several host species, the egg hatch is then likely to misfit with the bud burst period of at least some of the tree species, resulting in an overall decrease in pest damage. By contrast, host-specific adaptation could occur in a pure stand, improving the temporal synchronicity of egg hatching and consequently increasing the rate of defoliation (Mitter et al. 1979).

It is therefore not only the resource concentration that is reduced in mixed forests but also the ability of pests to locate host trees, both in space and time. For all these reasons, not exclusive but potentially additive, the probability of a pest insect to reach and attack host trees is lower in tree species mixtures, reducing the survival rate of pest populations and the risk of pest damage.

12.3.2 Impact of Natural Enemies

12.3.2.1 Alternative Hosts for Generalist Parasitoids and Predators

The second main reason for predicting that pure forests would be more prone to insect damage than tree mixtures has been proposed by Root (1973) as the "natural enemies hypothesis". This hypothesis, tested by Risch (1981) and reviewed by Russell (1989) in agroecosystems, rests on the assumption that "predators and parasitoids are more effective in complex environments than in simple ones" (Root 1973). It is thought that many natural enemies are better able to control pest insects when they can maintain or expand their populations on alternative prey or hosts during the time period when the suitable stage of the pest species is not present in the habitat. Because diverse plant communities usually provide habitat for more herbivorous species (Lawton and Strong 1981), mixed tree stands are likely to provide such alternative prey or hosts, and consequently could improve predation and parasitism rates of pests. Several studies showed that conifer stands mixed with hardwoods suffered less damage from spruce budworm, *Choristoneura fumiferana*, than pure conifer stands, and this could be related to increased parasitism rates. For example, parasitism from *Trichogramma minutum* increased with the density of non-host tree species due to a greater number of alternative hosts (Kemp and Simmons 1978). Another spruce budworm parasitoid, *Meteorus trachynotus*, can overwinter on other Lepidoptera larvae that feed on deciduous trees before it is able to attack *Choristoneura fumiferana* in mixed stands (Maltais et al. 1989). The spruce budworm pupal parasitoid *Itoplectis conquisitor* is also a natural enemy of the birch tube maker *Acrobasis betulella* (Cappucino and Martin 1997). Likewise, Cappucino et al. (1998) noticed that the tachinid *Actia interrupta* was bivoltine and would so probably depend on alternative hosts for overwintering. In a study on *Choristoneura jezoensis*, Higashiura (1991) attributed the greater abundance of the parasitoid *Lissonota saturator* in natural mixed forest than in single-species plantation forest to the presence of alternative hosts on broad-leaf trees. This mechanism does not necessarily depend on a mixture of canopy species because a rich understory can also provide alternative prey. For example, coccinellid predators of the walnut aphid *Chromaphis juglandicola* can prey on other aphids that feed on ground cover plants in walnut stands (Sluss 1967).

12.3.2.2 Complementary Food for Specialist Parasitoids

The availability of alternative hosts and prey is likely to benefit mostly generalist predators and parasitoids. However, numerous studies have shown that diverse plant communities can also provide a better supply of food resources such as pollen, nectar, and honeydew that can increase the effectiveness of specialized parasitoids (Russell 1989). Significantly more *Malacosoma americanum* and *Cydia pomonella* are parasitized in orchards with a rich understory of wild flowers than in orchards with poor floral abundance (Leius 1967). Mixed stands with a greater diversity of flowering herbs that potentially provided important nectar resources to parasitoids were found to have less damage from *Choristoneura fumiferana* (Simmons et al. 1975). Hassan (1967) observed that adult parasitoids of numerous forest insects fed not only on nectar of certain understory herbs but also on pollen of grasses and even on extrafloral nectaries of bracken fern (*Pteridium aquilinum*). Syme's (1975) classic study demonstrated that access to flowers could increase both fecundity and longevity of two native parasitoids of *Rhyacionia buoliana*. In a laboratory experiment, it was also observed that the provision of honey to water could dramatically increase the life span of two bark beetle parasitoids compared with those that had access only to pure water (Mendel 1988). Honeydew, which is mostly produced by tree aphids, can sometimes be an important food source for adult parasitoids. Contrary to sources of nectar, the presence of such food for parasitoids does not necessarily depend on the occurrence of additional tree or understory species. However, it has been shown that mixtures of tree species can potentially provide a more consistent supply of honeydew, because different tree species have different aphids that produce at different times and so complement each other (Zoebelein 1957). Thus, in summary, there is evidence that the complementary food sources in more diverse habitats can result in enhanced fecundity and longevity of parasitoids which can (1) extend their foraging period, (2) potentially reduce problems of temporal mismatch between parasitoid and host, and (3) increase total parasitism per parasitoid. Collectively, these can lead to a multifactorial increase in the effectiveness of specialized parasitoids.

12.3.2.3 Diversity of Shelter for Natural Enemies

Theoretically, diverse plant communities also provide better habitat for natural enemies because they offer more variation in microhabitats and microclimate, and thus better shelter for escaping adverse conditions. In a comprehensive review of studies in agroecosystems, Landis et al. (2000) conclude that (1) overwintering habitats are critical to ensure biological control the following year, (2) adequate shelters are required during summer to escape detrimental high temperature and low humidity, and (3) non-crop vegetation may be used by predators as oviposition sites. There is less evidence from experiments in forest ecosystems to support the occurrence of these processes, but the same principles are likely to apply. For instance, Zach and Falls (1975) and Dickson (1979) argued that the presence of hardwood species in conifer stands would provide more favorable microhabitats for insectivorous birds preying on *Choristoneura fumiferana*. Studies in agroecosystems indicate that a rich understory or ground cover are beneficial for ground predators such as carabids (Cromartie 1981; Brust et al. 1986), which are known to feed on the pupal stage of defoliators overwintering in the forest litter. Recently, Finke and Denno (2002) also demonstrated that complex plant associations can provide a greater range of refuges for predators, reducing intra-guild predation (super-predator impact) and thereby enhancing prey control.

12.3.2.4 Potential Negative Effects of Tree Species Mixtures

It is important to note that a diverse vegetation can also have negative effects on natural enemies. The main drawback appears to be the potential reduction of foraging effectiveness in complex-structured plants associations (Sheehan 1986) including a disruptive effect of chemical stimuli. Some predators and parasitoids locate their prey or host by the odor of the plant on which the host feeds (Cromartie 1981). The bark beetle parasitoid *Roptocerus xylophagorum* is not attracted by the larval or pupal host odor alone but actually needs both tree and bark beetle odor together to respond (Sullivan et al. 2000). The observations suggest that non-host repellents would direct foraging parasitoids away from trees attacked by bark beetles. Theoretically, more diverse habitats could also reduce the effectiveness of generalist natural enemies to control a particular pest insect if they are preferentially preying on other herbivores, i.e., alternative hosts, although this disruption hypothesis appears not to have been documented to date in forest ecosystems.

12.3.3 Pest Shift Among Host Tree Species

A third theory can explain why herbivory can be higher in pure stands but, interestingly, also give an explanation for the few cases where more forest pest damage is observed in tree mixtures. This concerns the heteroecious insect species that need two alternate, i.e., obligatory, hosts to complete their life cycle and the polyphagous insect pests that can shift from one host species to an alternative, i.e., facultative, host.

12.3.3.1 Mixtures of Two Host Tree Species Can Benefit Heteroecious Pests: The Succession Process

Several forest pests require two hosts to complete their life cycle. This succession process mainly applies to the adelgids which often have sexual and asexual stages on different host species. The most commonly cited example is that of *Adelges cooleyi,* which makes galls on spruce and lays eggs on Douglas fir (Barbosa and Wagner 1988). Likewise, *Pachypappa tremulae* move between aspen and Norway spruce (Stroyan 1975), *Prociphilus fraxini* between ash and white fir (Stary 1982), and *Pemphigus bursarius* between poplar and herbaceous plants (Phillips et al. 1999). Recently, an invasive species of cynipid in the United Kingdom, *Andricus quercuscalicis,* has been observed alternating on *Quercus cerris* and *Quercus robur* (Baksha 2000; Schonrogge et al. 2000). It is obvious that damage by these heteroecious pests is likely to occur in tree mixtures that include the two obligatory hosts, and unlikely in pure stands. On the other hand, some species that were originally heteroecious have changed to an anholocyclic life history, i.e., they reproduce entirely parthenogenetically on the secondary host as well. This has been documented for *Dreyfusia nordmannianae* (Kitzberger and Führer 1993), which therefore could also create problems in monocultures.

12.3.3.2 Mixture of Host Tree Species Reducing Polyphagous Pest Damage: The Diversion Process

Polyphagous pest damage on a particular tree species can also be reduced by the association with a more palatable secondary tree species that would be infested first. Such a diversion process in tree mixture has been observed for *Amblypelta cocophaga.* Damage in *Eucalyptus deglupta* plantations was significantly reduced by maintaining in the inter-row more suitable alternative host shrubs of this insect (Bigger 1985). The presence of Scots pine as trap trees is also recommended to decrease *Pissodes strobus* infestation in white pine plantations (Belyea 1923, in Barbosa and Wagner 1988). A mixed planting of other eucalyptus species more suitable to *Chrysophtharta bimaculata* showed that these trees could attract the beetles away from *Eucalyptus nitens* (Elek 1997).

12.3.3.3 Mixture of Host Tree Species Increasing Polyphagous Pest Damage: The Contagion Process

The association of several hosts species can, however, also lead to the opposite effect; that is, an increase of forest pest damage in tree mixtures. Such a process has sometimes been called "associational susceptibility" (Brown and Ewel 1987; White and Whitham 2000). This seems to occur when a polyphagous pest builds up its population on a more palatable host, exploits the main part of that resource and then "spills over" to the associated host. For instance, Brown and Ewel (1987) explained the higher rate of herbivory in *Cordia alliodora* grown in mixtures than when it is grown in monocultures by the presence of heavily consumed plant species surrounding the trees.

This contagion process is well documented for the gypsy moth, *Lymantria dispar*. Although its preferred hosts are broad-leaved species such as oak, poplar, and birch (Lechowitz and Mauffette 1986), it can feed on pines and other conifers once its preferred hosts have been defoliated (Montgomery et al. 1989). As a result of these feeding preferences, white pine growing in mixed stands with oaks are more heavily defoliated by the gypsy moth than in pure stands (Brown et al. 1988). Gottschalk and Twery (1989) suggested the following mechanism: early gypsy moth instars cannot develop as well on pine needles as on oak foliage, so the presence of oaks increases the survival of larvae to the older stages which can later move to pines and feed on needles. The timing of defoliation in the mixed oak/pine stands demonstrates the contagion process, as the oaks are defoliated first, and then the pines.

In pure stands, *Curculio elephas* caused more damage in cork oak, *Quercus suber*, than in holm oak, *Quercus rotundifolia* (Soria et al. 1995), indicating that the former are more susceptible. When mixed with *Quercus suber*, *Quercus rotundifolia* showed a higher rate of acorn attacks than in pure stands, probably due to contagion of weevils from *Quercus suber*.

In a comprehensive study, Moore et al. (1991) compared the damage caused by several pest insects on oak trees grown in pure stands, and in oak/alder and oak/spruce mixtures. The weevil *Phyllobius argentatus*, a highly polyphagous pest that can feed on broad-leaved trees and on spruce (Parry 1981), caused more damage on oaks in both mixtures than in pure oak stands (Moore et al. 1991). Interestingly, the moth *Stigmella* sp. and the leafhopper *Eurhadina* sp., which are also polyphagous but can only feed on hardwood species, did more damage in mixed oak/alder than in pure oak stands, but less damage in oak/spruce mixtures, i.e., where no other suitable hosts were present to allow contagion.

However, the best demonstration of the contagion process in a polyphagous forest pest is provided by the elegant study of White and Whitham (2000). Using both observational and experimental data, they showed that cottonwoods (*Populus angustifolia* × *Populus fremontii*) located under box elder (*Acer negundo*) suffered about twice as much defoliation from the fall cankerworm *Alsophila pometaria* than cottonwoods under other cottonwoods. Furthermore, they clearly demonstrated that this associational susceptibility was the result of three factors: (1) the presence of a generalist herbivore, i.e., a polyphagous pest that could feed on both cottonwood and box elder; (2) a difference in host preferences with the fall cankerworm feeding preferably on box elders, and (3) a high density of the polyphagous insect, leading the pest population to deplete the preferred resource (box elder), and then to spill over onto cottonwood, the less-preferred host.

12.3.3.4 Polyphagous Pests, an Exception to the Diversity–Resistance Paradigm

Because of the singularity of polyphagous pest behavior in mixtures of host tree species, we tried to compare their average damage in pure vs. mixed stands. Using the studies collected for the meta-analysis (Table 12.1), we found that the mean effect size is high and positive for oligophagous forest pests but low and insignificant for polyphagous pests, which confirms a high variability of the response of polyphagous pests to stand composition. To test the effect of the presence of another host species in the mixture we classified the individual studies concerning polyphagous pests according to absence of host or presence of only less susceptible hosts vs. presence of more susceptible hosts (when information of the relative susceptibility was available). For the nine individual studies where no other or only less sensitive host species were present in the mixture, the overall effect size was positive and significant (d_+=0.80), indicating that mixed forests had a lower risk of damage from polyphagous pest when no other more palatable host species is present (Table 12.1). In contrast, the overall effect size for the other eight studies was negative, although non-significant. This suggests that there is a risk of increasing damage from polyphagous pests when a more sensitive host tree is associated with the main tree species. Because the contagion process would only start when most of the primary host resources have been exploited, one could expect that the effect of mixing a crop species with a more sensitive species would depend on pest abundance: a reduction of pest damage in the main tree species at low pest abundance (diversion), an increase in pest damage in the main tree species at high pest abundance (contagion).

12.4 Tree Species Diversity and Pest Damage at the Landscape Level

All the studies that contributed to our meta-analysis concerned a comparison between pure and mixed-species stands of trees. However, recent evidence from agroecosystem experiments shows that vegetation diversity at the landscape scale could be even more important than stand composition in reducing pest damage (Mensah 1999; Thies and Tscharntke 1999; Landis et al. 2000). Adjacent vegetation seems to be particularly beneficial (Duelli et al. 1990; Dennis and Fry 1992; Thomas et al. 1992; Boatman 1994; Altieri 1999). An analogue effect is likely to apply to the forest mosaic.

12.4.1 Interference of Forest Mosaic Heterogeneity with Meta-population Dynamics

Most forest pest dynamics may be interpreted according to the meta-population theory (Hanski and Gilpin 1997; Coulson et al. 1999). Populations are distributed among patches of suitable habitat separated by unfavorable areas within the forest mosaic, and their survival would depend on their dispersal ability, i.e., the probability of exchanging individuals with other sub-populations or the probability of finding new resources. Accordingly, the more isolated stands of host trees are within a forest matrix of non-host species, the lower the risk of outbreak. More generally, in a forest mosaic consisting of different forest types and/or a heterogeneous distribution of patches, the paradigm of the "balance of nature" (Pimm 1991) predicts that no single species will become absolutely dominant over a larger area, or over a longer period of time, because there is always an antagonist at hand to reduce the dominance of a pest organism at the start of an outbreak. At a conceptual level, a landscape mosaic, consisting of different habitat types and/or a heterogeneous habitat patch distribution, contains higher species richness than a uniform landscape (Duelli 1997). The underlying mechanisms are meta-community dynamics (Wilson 1992), where the complex interplay of all the meta-population dynamics of the various species present in a landscape mosaic is being considered. Species diversity increases with habitat diversity because more niches are available. However, biodiversity also increases with habitat heterogeneity, even with the same number of habitat types, because an increase in habitat patchiness results in more habitat borders with ecotonal structures (Duelli 1997). Mutual exchanges between patches by dispersal of individuals and species can also increase species richness. Consequently, the dominance distribution of herbivores and their antagonists may change from patch to patch due to stochastic events of immigration and local extinction. Thus, habitat diversity and heterogeneity can be seen as an insurance against pest outbreaks: it may not always work, but in general the damage will be lower.

12.4.2 Reduction of Pest Damage by Habitat Diversity in Forest Landscapes

A study on the population dynamics of the acacia processionary moth (*Ochrogaster lunifer*) in Australia suggests that more diverse habitat with mixed plant species would increase pest population stability and consequently lower the occurrence of outbreaks (Floater and Zalucki 2000). Also, in Australia, a shelter belt of mixed species was established along a blue gum *Eucalyptus globulus* plantation to maximize habitat for beneficial insects, lizards, frogs, and birds. Blue gum trees surrounded by a 600-m monoculture of the same species were significantly more damaged by leaf-chewing moths (*Mnesapela privata* and *Roeselia lugens*) and beetles (*Anoplagnathus* spp.) than trees within 150 m of the mixed-species shelterbelt (Dufty et al. 2000). A study on spruce budworm demonstrated that balsam fir experienced lower mortality in stands surrounded by non-host deciduous forest than in stands within large conifer-dominated forest, which shows that tree species diversity at the landscape scale can influence the level of pest infestation (Cappucino et al. 1998). Similarly, Jactel et al. (2002) found that pure stands of maritime pine bordered by a mixed woodland of broad-leaved species were less attacked by the stem borer *Dioryctria sylvestrella* than pure stands among a monoculture of pine trees. By contrast, damage by the banana-spotting bug *Amblypelta lutescens* on papaw trees significantly decreased with increasing distance to the forest remnants where the pest can feed on native alternative hosts (Ryan 1994). The rainforest vegetation provides refuge for the pest insects and subsequently acts as source of infestation for nearby crop trees.

The studies on *Choristoneura fumiferana* (Cappucino et al. 1998) and *Dioryctria sylvestrella* (Jactel et al. 2002) furthermore demonstrated a similar effect of habitat diversity on parasitism rates whereby it was lower in the conifer monoculture. Floater and Zalucki (2000) suggested a different mechanism – lower host-tree apparency – to explain the reduction of *Ochrogaster lunifer* defoliation in mixed forest. The banana-spotting bug is a polyphagous species that can spill over from native bush species to papaw tree plantations. These examples clearly show that the ecological hypotheses such as the host tree accessibility, the impact of natural enemies and the contagious potential of polyphagous pests can also apply to the landscape scale.

12.5 Conclusions

12.5.1 New Support for the Diversity–Stability Theory

Based on the evidence from the over 50 comparative studies, our quantitative review lends good support to earlier proposals that the biodiversity–stability theory applies to trophic interactions in forest ecosystems. This refutes the findings of earlier studies concluding that there was no evidence for such an effect, based on mathematical models (May 1973; Pimm and Lawton 1978; Michalski and Arditi 1999) or on bibliographic reviews that suggested a lack of experimental evidence (Watt 1992; Barthod 1994; Landmann 1998).

Our meta-analysis and review of ecological mechanisms were able to demonstrate that:

- Tree species growing in mixed stands overall suffer less pest damage, or have lower pest populations, than pure stands;
- Three main ecological mechanisms can account for the lower damage in tree mixtures: reduced accessibility to pests of their host trees, greater impact of natural enemies, and diversion from a less susceptible to a more susceptible tree species;
- The main exception to the diversity–resistance paradigm is the case of polyphagous pest insects. Such insects can first build up their populations on a preferred host tree species and then spill over onto an associated host tree species, according to the contagion process.

12.5.2 Implications for Forest Pest Management

The evidence we present on the relationship between tree diversity and forest resistance also has potentially important implications for the management of forests, whereby increasing forest biodiversity may not only have conservation benefits but also result in reduced pest impact. Following the successful use of ecological crop management to control pests, such as intercropping (Landis et al. 2000) or development of beetle banks (Wratten et al. 1998), mixed-species tree plantations as well as the conservation of plant diversity (including understory) in secondary forests could be advocated. Of course, benefits from mixtures in terms of protection against pest damage will have to be weighed against the potentially increased costs of silvicultural operations. However, increasing forest diversity could also be achieved by promoting or restoring mixed-species woodlands in parts of the forest mosaic where site conditions or stand accessibility make intensive forest management for production less profitable.

12.5.3 Needs for Further Research

12.5.3.1 Effect of Species Composition

Our review shows that while the quantitative component of forest biodiversity is important, the choice of tree species in the mixture is critical. The "quality" of tree diversity needs to meet the requirements on which the relevant ecological mechanisms rest, such as the provision of alternative insect prey for key natural enemies, or the lack of an alternative host tree for polyphagous herbivores. These intricacies are highly relevant for the broader debate about the relative significance of species complementarity (Tilman 2000) and species composition (Hooper and Vitousek 1997) in ecosystem functioning. To better understand the relationship between biodiversity and forest stability, future studies should make use of proper experimental designs, such as coupling gradients of tree species diversity with manipulative herbivores release. As modeling is becoming more sophisticated and realistic, approaches such as the modeling of food webs (e.g., Wilby and Thomas 2002) are likely to provide additional insights into the relevant interspecific interactions.

12.5.3.2 Biodiversity and the Population Stability of Forest Herbivores

One limitation of many of the studies we reviewed was that pest populations or damage were not assessed over successive years. Such studies may not be able to adequately take account of the temporal component of forest insect population dynamics, which is particularly important to assess ecosystem stability on the basis of resilience. Recent experimental and theoretical studies indicated that diversity gives rise to ecosystem stability but not necessarily to population-level stability, a process which could depend on the strength of consumer–resources interactions (McCann 2000). Long term entomological studies examining, for instance, the duration of outbreaks in forests of different species diversity would definitively contribute to the debate.

Acknowledgments: We are grateful for constructive comments on earlier versions of the manuscripts from L. Berndt, J. Koricheva, and B. Wermelinger, Ch. Körner, M. Scherer-Lorenzen, and two anonymous reviewers.

Appendix: Summary of the data included in the meta-analysis and corresponding individual effect size (*d*)

Tree species in the monoculture	Associated tree species[a]	Forest biome[b]	Herbivorous insect species[c,d]	*d*	Study
Abies balsamea	MI	BOR	*Choristoneura fumiferana*	5.38	Su et al. (1996)
Abies balsamea	MI	BOR	*Choristoneura fumiferana*	1.25	Bergeron et al. (1995)
Abies balsamea	MI	BOR	*Choristoneura fumiferana*	1.00	McLean (1980)
Abies balsamea	BR	TEM	*Choristoneura fumiferana*	4.60	Batzer et al. (1987)
Albizia guachapele	BR	TRO	Lepidoptera sp.	–0.13	Montagnini et al. (1995)
Callophylum brasiliense	BR	TRO	Lepidoptera sp.	–0.36	Montagnini et al. (1995)
Cordia alliodora	BR	TRO	?	–0.93	Brown et al. (1988)
Corylus avellana	BR	TEM	*Myzocallis coryli*	0.52	Gantner (2000)
Dipteryx panamensis	BR	TRO	Lepidoptera sp.	0.56	Montagnini et al. (1995)
Eucalyptus botryoides	BR	TEM	*Cardiaspina fiscella*	9.96	Smith et al. (1989)
Eucalyptus cloeziana	BR	TEM	Lepidoptera sp.	2.06	Zanuncio et al. (1998)
Eucalyptus deglupta	BR	TEM	*Amblypelta cocophaga*	0.28	Bigger (1985)
Eucalyptus regnans	BR	TEM	*Chrysoptharta bimaculata*	0.56	Greaves (1966)
Genipa americana	BR	TRO	Lepidoptera sp.	1.05	Montagnini et al. (1995)
Milicia excelsa	BR	TRO	*Phytolyma lata*	0.57	Nichols et al. (1999)
Picea glauca	BR	BOR	*Pissodes strobi*	2.79	Taylor et al. (1996)
Picea sitchensis	BR	TEM	*Pissodes strobi*	0.68	McLean (1989)
Pinus densiflora	BR	TEM	*Thecodiplosis japonensis*	0.29	Bae et al. (1997)
Pinus densiflora	BR	TEM	Acari sp.	0.78	Bae et al. (1997)
Pinus densiflora	BR	TEM	Homoptera sp.	–0.19	Bae et al. (1997)
Pinus densiflora	BR	TEM	Homoptera sp.	0.53	Bae et al. (1997)
Pinus nigra laricio	BR	TEM	*Thaumetopoea pytiocampa*	1.19	Géri (1980)
Pinus silvestris	SH	BOR	*Hylobius abietis*	3.40	Stadniskii (1978)
Pinus strobus	BR	TEM	*Lymantria dispar*	–1.11	Gottschalk and Twery (1989)
Pinus strobus	BR	TEM	*Lymantria dispar*	–0.98	Brown et al. (1988)
Pinus strobus	BR	TEM	*Pissodes strobi*	15.49	Katovich (1992)
Pinus taeda	MI	TEM	*Dendroctonus frontalis*	0.69	Schowalter and Turchin (1993)

Pinus taeda	SH	TEM	*Rhyacionia frustrana*	0.24	Miller et al. (1983)
Pinus taeda	BR	TEM	*Rhyacionia frustrana*	0.39	Berisford and Kulman (1967)
Populus angustifolia	BR	TEM	*Alsophila pometaria*	−0.80	White and Whitham (2000)
Pseudotsuga menziesii	CO	TEM	*Choristoneura occidentalis*	0.77	Fauss and Pierce (1969)
Quercus petraea	BR	TEM	Homoptera sp.	−1.42	Moore et al. (1991)
Quercus petraea	CO	TEM	Homoptera sp.	1.35	Moore et al. (1991)
Quercus petraea	BR	TEM	Lepidoptera sp.	0.47	Moore et al. (1991)
Quercus petraea	CO	TEM	Lepidoptera sp.	0.27	Moore et al. (1991)
Quercus petraea	BR	TEM	*Neuroterus* sp.	0.54	Moore et al. (1991)
Quercus petraea	CO	TEM	*Neuroterus* sp.	0.58	Moore et al. (1991)
Quercus petraea	BR	TEM	*Phyllobius argentatus*	−0.78	Moore et al. (1991)
Quercus petraea	CO	TEM	*Phyllobius argentatus*	−1.21	Moore et al. (1991)
Quercus petraea	BR	TEM	*Phyllonorycter* sp.	0.21	Moore et al. (1991)
Quercus petraea	CO	TEM	*Phyllonorycter* sp.	2.41	Moore et al. (1991)
Quercus petraea	BR	TEM	*Stigmella* sp.	−3.30	Moore et al. (1991)
Quercus petraea	CO	TEM	*Stigmella* sp.	−0.15	Moore et al. (1991)
Quercus rotundifolia	BR	TEM	*Curculio elephas*	−0.28	Soria et al. (1995)
Quercus suber	BR	TEM	*Curculio elephas*	0.36	Soria et al. (1995)
Sonneratia apetala	BR	TRO	*Zeuzera conferta*	0.41	Wazihullah et al. (1996)
Sonneratia apetala	BR	TRO	*Zeuzera conferta*	0.67	Wazihullah et al. (1996)
Stryphnodendron microstachyum	BR	TRO	Cecidomyiidae sp.	0.03	Folgarait et al. (1995)
Stryphnodendron microstachyum	BR	TRO	*Euclystis* sp.	−0.31	Folgarait et al. (1995)
Stryphnodendron microstachyum	BR	TRO	Lepidoptera sp.	5.49	Montagnini et al. (1995)
Toona ciliata	BR	TRO	*Hypsipyla robusta*	5.94	Keenan et al. (1995)
Virola koschnyi	BR	TRO	Lepidoptera sp.	1.39	Montagnini et al. (1995)
Vochysia ferruginea	BR	TRO	Lepidoptera sp.	0.49	Montagnini et al. (1995)
Vochysia guatemalensis	BR	TRO	Lepidoptera sp.	−0.81	Montagnini et al. (1995)

[a] Associated tree species: *BR* broad-leaved trees, *CO* conifers, *MI* mixture of conifers and broad-leaved trees, *SH* shrubs
[b] Forest biome: *BOR* boreal, *TEM* temperate, *TRO* tropical
[c] Different results for the same insect – tree interaction correspond to different mixture compositions
[d] ? indicates that the insect species responsible was not identified in the study

References

Altieri MA (1999) The ecological role of biodiversity in agro-ecosystems. Agric Ecosyst Environ 74:19–31
Andow DA (1991) Vegetational diversity and arthropod population response. Annu Rev Entomol 36:561–586
Bae WI, Shin SC, Kim ZS (1997) Difference in occurrence of pine needle gall midge and sucking insects in pure-pine and mixed-pine stands. FRI J For Sci 56:37–43
Bain J (1981) Forest monocultures – how safe are they? An entomologists view. NZ J For Sci 26:37–42
Baksha MW (2000) Sexual generation of the gall wasp, *Andricus quercuscalicis*, its parasitoids and their abnormal sex ratio. Bangladesh J For Sci 27:82–89
Baliddawa CW (1985) Plant species diversity and crop pest control. An analytical review. Insect Sci Appl 6:479–487
Barbosa P, Wagner MR (1988) Introduction to forest and shade tree insects. Academic Press, San Diego
Barthod C (1994) Sylviculture et risques sanitaires dans les forêts tempérées - 1ère partie. Rev For Fr 46:609–628
Batzer HO, Benzie JW, Popp MP (1987) Spruce budworm damage in aspen/balsam fir stands affected by cutting methods. North J Appl For 4:73–75
Begon M, Harper JL, Townsend CR (1996) Ecology: individuals, populations and communities. Blackwell Science, Oxford
Belanger RP, Malac BF (1980) Silviculture can reduce losses from the southern pine beetle. USDA Agric Handb 576. USDA, Washington, DC
Bergeron Y, Leduc A, Morin H, Joyal C (1995) Balsam fir mortality following the last spruce budworm outbreak in northwestern Quebec. Can J For Res 25:1375–1384
Berisford CW, Kulman HM (1967) Infestation rate and damage by the Nantucket pine tip moth in six loblolly pine stand categories. For Sci 13:429–438
Bigger M (1985) The effect of attack by *Amblypelta cocophaga* (Hemiptera: Coreidae) on growth of Eucalyptus deglupta in the Solomon Islands. Bull Entomol Res 75:595–608
Boatman N (1994) Field margins: integrating agriculture and conservation. British Crop Protection Council, Surrey
Brown BJ, Ewel JJ (1987) Herbivory in complex and simple tropical successional ecosystems. Ecology 68:108–116
Brown JH, Cruickshank VB, Gould WP, Husband TP (1988) Impact of gypsy moth defoliation in stands containing white pine. North J Appl For 5:108–111
Brust GE, Stinner BR, McCartney DA (1986) Predation by soil inhabiting arthropods in intercropped and monoculture agroecosystems. Agric Ecosyst Environ 18:145–154
Byers JA, Zhang QH, Schlyter F, Birgersson G (1998) Volatiles from non-host birch trees inhibit pheromone response in spruce bark beetles. Naturwissenschaften 85:557–561
Cappucino N, Martin MA (1997) The birch tube-maker *Acrobasis betulella* in a fragmented habitat: the importance of patch isolation and edges. Oecologia 110:69–76
Cappucino N, Lavertu D, Bergeron Y, Regniere J (1998) Spruce budworm impact, abundance and parasitism rate in a patchy landscape. Oecologia 114:236–242
Cohen J (1969) Statistical power analysis for the behavioural sciences. Academic Press, New York
Coulson RN, McFadden BA, Puley PE, Lovelady CN, Fitzgerald JW, Jack SB (1999) Heterogeneity of forest landscapes and the distribution of abundance of the southern pine beetle. For Ecol Manage 114:471–485
Cromartie WJ (1981) The environmental control of insects using crop diversity. In: Pimentel D (ed) Handbook of pest management. Chemical Rubber Company in Agriculture, Boca Raton, pp 223–251

Demolin G (1962) Comportement des adultes de *Thaumetopoea pityocampa*. CR Acad Sci D Nat 255:2838–2839

Dennis P, Fry GLA (1992) Field margins: can they enhance natural enemy populations densities and general arthropod diversity on farms? Agric Ecosyst Environ 40:95–115

Dickens JC, Billings RF, Payne TL (1992) Green leaf volatiles interrupt aggregation pheromone response in bark beetles infecting pines. Experientia 48:523–524

Dickson JG (1979) Seasonal population of insectivorous birds in mature bottomland hardwood forest in south Louisiana. In: Dickson JG, Connor RN, Fleet RR, Jackson JA, Kroll JC (eds) The role of insectivorous birds in forest ecosystems. Academic Press, New York, pp 261–269

Duelli P (1997) Biodiversity evaluation in agricultural landscapes: An approach at two different scales. Agric Ecosyst Environ 62:81–91

Duelli P, Studer M, Marchand I, Jakob S (1990) Population movements of arthropods between natural and cultivated areas. Biol Conserv 54:193–207

Dufty AC, Senanayake FR, Jack JB (2000) Analogue forestry – a total ecosystem management approach that maximises biodiversity within plantation agriculture. Planter 76:671–696

DuMerle P, Mazet R (1983) Phénologies comparées du chêne pubescent, du chêne vert et de *Tortrix viridana* (Lepidoptera: Tortricidae). Mise en évidence chez l'insecte de deux populations sympatriques adaptées chacune à l'un des chênes. Acta Oecol Oec Appl 4:55–74

Elek JA (1997) Assessing the impact of leaf beetles in eucalyptus plantations and exploring options for their management. Tasforests 9:139–154

Elton CS (1958) The ecology of invasions by animals and plants. Methuen, London

Fabre JP (1988) Possibilités d'infestation par les pucerons *Cedrobium laportei* et *Cinara cedri* chez le genre *Cedrus*. Ann For Sci 45:125–140

Fauss DL, Pierce WR (1969) Stand conditions and spruce budworm damage in a western Montana forest. J For 67:322–325

Finke DL, Denno RF (2002) Intraguild predation diminished in complex-structured vegetation: implications for prey suppression. Ecology 83:643–652

Floater GJ, Zalucki MP (2000) Habitat structure and egg distributions in the processionary caterpillar *Ochrogaster lunifer*: lessons for conservation and pest management. J Appl Ecol 37:87–99

Folgarait PJ, Marquis RJ, Ingvarsson P, Bracker HE, Arguedas M (1995) Pattern of attack by insect herbivores and fungus on saplings in a tropical tree plantation. Environ Entomol 24:1487–1494

Gagdil PD, Bain J (1999) Vulnerability of planted forests to biotic and abiotic disturbances. New For 17:227–238

Gantner M (2000) Aphidofauna of hazel bushes (*Corylus* L.) on a protected plantation, an unprotected plantation and in a forest. Annales-Universitatis-Mariae-Curie-Sklodowska, Sectio-EEE, Horticultura, vol 8, pp 155–167

Géri C (1980) Application des méthodes d'études démécologiques aux insectes défoliateurs forestiers: cas de *Diprion pini* et dynamique des populations de la processionnaire du pin en Corse. Thesis, Université de Paris-Sud

Gibson IAS, Jones TM (1977) Monoculture as the origin of major pests and diseases. In: Cherrett JM, GR Sagar (eds) Origins of pest, parasite, disease and weed problems. 8th Symposium of the British Ecological Society, Bangor, 12–14 April 1976. Blackwell, Oxford, pp 139–161

Gottschalk KW, Twery MJ (1989) Gypsy moth in pine-hardwood mixtures. In: Waldrop TA (ed) Pine-hardwood mixtures: a symposium on management and ecology of the type. USDA Forest Service General Technical Report SE 58, pp 50–58

Greaves R (1966) Insect defoliation of eucalyptus regrowth in the Florentine Valley, Tasmania. Appita 19:119–126

Gurevitch J, Hedges LV (1993) Meta-analysis: Combining the results of independent experiments. In: Scheiner SM, Gurevitch J (eds) Design and analysis of ecological experiments. Chapman and Hall, New York, pp 378–425

Hanski I, Gilpin ME (1997) Metapopulation biology: ecology, genetics and evolution. Academic Press, London

Hassan E (1967) Untersuchungen über die Bedeutung der Kraut- und Strauchschicht als Nahrungsquelle für Imagines entomophager Hymenopteren. Z Angew Entomol 60:238–265

Higashiura Y (1991) Pest control in natural and man-made forests in northern Japan. For Ecol Manage 39:55–64

Hooper DU, Vitousek PM (1997) The effects of plant composition and diversity on ecosystem processes. Science 277:1302–1305

Huber DPW, Borden JH (2001a) Angiosperm bark volatiles disrupt response of Douglas-fir beetle, *Dendroctonus pseudotsugae*, to attractant-baited traps. J Chem Ecol 27:217–233

Huber DPW, Borden JH (2001b) Protection of lodgepole pines from mass attack by mountain pine beetle, *Dendroctonus ponderosae*, with nonhost angiosperm volatiles and verbenone. Entomol Exp Appl 99:131–141

Jactel H, Goulard M, Menassieu P, Goujon G (2002) Habitat diversity in forest plantations reduces infestations of the pine stem borer *Dioryctria sylvestrella*. J Appl Ecol 39:618–628

Jactel H, van Halder I, Menassieu P, Zhang QH, Schlyter F (2001) Non-host volatiles disrupt the response of the stenographer bark beetle, *Ips sexdentatus* (Coleoptera: Scolytidae) to pheromone baited traps and maritime pine logs. IPM Rev 16:197–207

Jones RE (2001) Mechanisms for locating resources in space and time: impacts on the abundance of insect herbivores. Aust Ecol 26:518–524

Kareiva P (1983) Influence of vegetation texture on herbivore populations: resource concentration and herbivore movement. In: Denno RF, McClure MS (eds) Variable plants and herbivores in natural and managed systems. Academic Press, New York, pp 259–289

Katovich SA (1992) White pine weevil response to oak overstorey girdling – Results from a 16-year-old study. North J Appl For 9:51–54

Keenan R, Lamb D, Sexton G (1995) Experience with mixed species rainforest plantations in North Queensland. Commonw For Rev 74:315–321

Kemp WP, Simmons GA (1978) The influence of stand factors on parasitism of spruce budworm eggs by *Trichogramma minutum*. Environ Entomol 7:685–688

Kemp WP, Simmons GA (1979) Influence of stand factors on survival of early instar spruce budworm. Environ Entomol 8:993–996

Kitzberger P, Führer E (1993) Decline of silver fir regeneration caused by the silver fir woolly aphid. Allg Forst Zeit 48:1369–1372

Koricheva J, Larsson S, Haukioja E (1998a) Insect performance on experimentally stressed woody plants: a meta-analysis. Annu Rev Entomol 43:195–216

Koricheva J, Larsson S, Haukioja E, Keinanen M (1998b) Regulation of woody plant secondary metabolism by resource availability: hypothesis testing by means of meta-analysis. Oikos 83:212–226

Krebs CJ (1985) Ecology: the experimental analysis of distribution and abundance. Harper and Row, New York

Landis DA, Wratten SD, Gurr GM (2000) Habitat management to conserve natural enemies of arthropod pests in agriculture. Annu Rev Entomol 45:175–201

Landmann G (1998) Forest health, silviculture and forest management. In: Montoya R (ed) Problemas sanitarios en los sistemas forestales: de los espacios protegidos a los cultivos de especies de crecimiento rapido Colleccion Tecnica. Publicaciones del Organismo Autonomo Parques Nacionales, Madrid, pp 155–183

Lanfranco D, Klasmer V, Botto E, Ide S (2000) Risk to South America radiata pine resources by the shoot moth, *Rhyacionia buoliana*. XXIth international congress of entomology, Foz de Iguazù, Brasil. 20–26 Aug 2000

Lawton JH, Strong DR (1981) Community patterns and competition in folivorous insects. Am Nat 118:317–338

Lechowitz MJ, Mauffette Y (1986) Host preferences of the gypsy moth in eastern North American versus European forests. Rev Entomol Québec 31:43–51

Leius K (1967) Influence of wild flowers on parasitism of tent caterpillar and codling moth. Can Entomol 99:444–446

Maltais J, Regnière J, Cloutier C, Hebert C, Perry DF (1989) Seasonal biology of *Meteorus trachynotus* (Hymenoptera: Braconidae) and of its overwintering host *Choristoneura rosaceana* (Lepidoptera: Tortricidae). Can Entomol 121:745–756

May RM (1973) Stability and complexity in model ecosystems. Princeton Univ Press, Princeton

McCann KS (2000) The diversity-stability debate. Nature 405:228–233

McLean DA (1980) Vulnerability of fir-spruce stands during uncontrolled spruce budworm outbreaks: a review and discussion. For Chron 56:213–221

McLean JA (1989) Effect of red alder overstorey on the occurrence of *Pissodes strobi* during the establishment of a Sitka spruce plot. In: Alfaro RI, Glover SG (eds) Insects affecting reforestation: biology and damage. Proceedings of a meeting of the IUFRO working group on insects affecting reforestation, Forestry Canada, Vancouver, 3–9 July 1988, Canada, pp 167–176

Mendel Z (1988) Effects of food, temperature and breeding conditions on the life span of adults of three co-habiting bark beetle (Scolytidae) parasitoids (Hymenoptera). Environ Entomol 17:293–298

Mensah RK (1999) Habitat diversity: implications for the conservation and use of predatory insects of *Helicoverpa* spp. in cotton systems in Australia. Int J Pest Manage 45:91–100

Michalski J, Arditi R (1999) The complexity-stability problem in food web theory. What can we learn from exploratory models? In: Blasco F, Weill A (eds) Advances in environmental and ecological modelling. Elsevier, Paris, pp 91–119

Miller FD, Stephen FM (1983) Effects of competing vegetation on Nantucket pine tip moth (Lepidoptera: Tortricidae) populations in loblolly pine plantations in Arkansas. Environ Entomol 12:101–105

Mitter C, Futuyama DJ, Schneider JC, Hare JD (1979) Genetic variation and host plant relations in a parthenogenetic moth. Evolution 33:777–790

Montagnini F, Gonzalez E, Porras C, Rheingans R (1995) Mixed and pure forest plantations in the humid neotropics: a comparison of early growth, pest damage and establishment costs. Commonw For Rev 74:306–314

Montgomery ME, McManus ML, Berisford CW (1989) The gypsy moth in pitch pine-oak mixtures: predictions for the South based on experiences in the North. In: Waldrop TA (ed) Pine-hardwood mixtures: a symposium on management and ecology of the type. USDA For Serv Gen Techn Rep SE 58:43–49

Moore R, Warrington S, Whittaker JB (1991) Herbivory by insects on oak trees in pure stands compared with paired mixtures. J Appl Ecol 28:290–304

Nichols DJ, Ofori D, Wagner MR, Bosu P, Cobbinah JR (1999) Survival, growth and gall formation by *Phytolyma lata* on *Milicia excelsa* established in mixed-species tropical plantations in Ghana. Agric For Entomol 1:137–141

Parry WH (1981) Weevils (Coleoptera: Curculionidae) on foliage of Sitka spruce in north east Scotland. Scot For 35:96–101

Phillips SW, Bale JS, Tatchell GM (1999) Escaping an ecological dead-end: asexual overwintering and morph determination in the lettuce root aphid *Pemphigus bursarius* L. Ecol Entomol 24:336–344

Pimm SL (1991) The balance of nature? Ecological issues in the conservation of species and communities. Univ Chicago Press, Chicago

Pimm SL, Lawton JH (1978) On feeding on more than one trophic model. Nature 275:542–544

Risch SJ (1981) Insect herbivore abundance in tropical monocultures and polycultures: an experimental test of two hypotheses. Ecology 62:1325–1340

Risch SJ, Andow D, Altieri MA (1983) Agroecosystem diversity and pest control: data, tentative conclusions, and new research directions. Environ Entomol 12:625–629

Roberts H (1969) Forest insects of Nigeria with notes on their biology and distribution. Commonwealth Forest Institute, Dept For, Oxford 44:1–206

Root RB (1973) Organisation of a plant-arthropod association in simple and diverse habitats: the fauna of collards (*Brassica oleracae*). Ecol Monogr 43:94–125

Rosenberg MS, Adams DC, Gurevitch J (2000) MetaWin: statistical software for meta-analysis. Version 2.0. Sinauer Associates, Sunderland, Massachusetts

Rosenthal R (1979) The "file drawer problem" and tolerance for null results. Psychol Rep 12:491–511

Russell EP (1989) Enemies hypothesis: a review of the effect of vegetational diversity on predatory insects and parasitoids. Environ Entomol 18:590–599

Ryan MA (1994) Damage to papaw trees by the banana-spotting bug, *Amblypelta lutescens* (Hemiptera: Coreidae), in North Queensland. Int J Pest Manage 40:280–282

Satchell JE (1962) Resistance in oaks to defoliation by *Tortrix viridana*. Ann Appl Biol 50:431–442

Sheehan W (1986) Response by specialist and generalist natural enemies to agroecosystem diversification: a selective review. Environ Entomol 15:456–461

Schlyter F, Zhang QH, Andersson P, Byers JA, Wadhams L, Löfqvist J, Birgersson G (2000) Electrophysiological and behavioral responses of pine shoot beetles, *Tomicus piniperda* and *T. minor* (Coleoptera: Scolytidae), to non-host leaf and bark volatiles. Can Entomol 132:965–981

Schonrogge K, Walker P, Crawley MJ (2000) Parasitoid and inquiline attack in the gall of four alien, cynipid gall wasps: host switches and the effect on parasitoid sex ratios. Ecol Entomol 25:208–219

Schowalter TD, Turchin P (1993) Southern pine beetle infestation development: interaction between pine and hardwood basal areas. For Sci 39:201–210

Schwerdtfeger F (1981) Waldkrankheiten. Parey, Hamburg

Simmons GA, Leonard DE, Chen CW (1975) Influence of tree species density and composition on parasitism of the spruce budworm, *Choristoneura fumiferana*. Environ Entomol 4:832–836

Sluss RR (1967) Population dynamics of the walnut aphid *Chromaphis juglandicola* in northern California. Ecology 48:41

Smith IW, Marks GC, Featherston GR, Geary PW (1989) Effects of inter-planted wattles on the establishment of eucalyptus planted on forest sites affected by *Phytophthora cinnamomi*. Aust For 52:74–81

Soria FJ, Villagran R, DelTio R, Ocete ME (1995) Incidencia de *Curculio elephas* (Coleoptera: Curculionidae) en alcornocales y encinares del parque natural Sierra Norte de Sevilla. Bol San Veg Plagas 21:195–201

Speight MR, Wainhouse D (1989) Ecology and management of forest insects. Oxford Science Publication, Clarendon Press, Oxford

Speight MR, Wylie FR (2001) Insect pests in tropical forestry. CABI, Oxon, UK

Stadnitskii GV (1978) Effect of the living ground vegetation on damage to Scots pine plantations by *Hylobius abietis*. Lesnoi Zhurnal 6:26–29

Stary P (1982) The role of ash (*Fraxinus*) as a reservoir of aphid parasitoids with description of a new species in Central Europe. Acta Entomol Bohemos 79:97–107

Stroyan HLG (1975) The life cycle and generic position of *Aphis tremulae* L. (Aphidoidea: Pemphiginae), with a description of the viviparous morphs and a discussion of spruce root aphids in the British Isles. Biol J Linn Soc 7:45–72

Su Q, MacLean DA, Needham TD (1996) The influence of hardwood content on balsam fir defoliation by spruce budworm. Can J For Res 26:1620–1628

Sullivan BT, Pettersson EM, Seltmann EM, Berisford CW (2000) Attraction of the bark beetle parasitoid *Roptrocerus xylophagorum* (Hymenoptera: Pteromalidae) to host-associated olfactory cues. Environ Entomol 29:1138–1151

Syme PD (1975) The effects of flowers on the longevity and fecundity of two native parasites of the European pine shoot moth in Ontario. Environ Entomol 4:337–346

Taylor SP, Alfaro RI, DeLong C, Rankin L (1996) The effect of overstory shading on white pine weevil damage to white spruce and its effects on spruce growth rates. Can J For Res 26:306–312

Thies C, Tscharntke T (1999) Landscape structure and biological control in agroecosystems. Science 285:893–895

Tilman D (2000) Causes, consequences and ethics of biodiversity. Nature 405:208–211

Thomas MB, Wratten SD, Sotherton NW (1992) Creation of "island" habitats in farmland to manipulate populations of beneficial arthropods: predator densities and species composition. J Appl Ecol 29:524–531

Tonhasca A, Byrne DN (1994) The effect of crop diversification on herbivorous insects: a meta-analysis approach. Ecol Entomol 19:239–244

Tshernyshev WB (1995) Ecological pest management (EPM): general approaches. J Appl Entomol 119:379–381

Van Emden HF, Williams G (1974) Insect stability and diversity in agro-ecosystems. Annu Rev Entomol 19:455–475

Warren LO (1963) Nantucket pine tip moth infestations. Severity of attacks as influenced by vegetative competition in pine stands. Arkansas Agriculture Experimental Station, Arkansas Farm Research, Nov–Dec 1963:4

Watt AD (1992) Insect pest population dynamics: effects of tree species diversity. In: Cannell MGR, Malcolm DC, Robertson PA (eds) The ecology of mixed-species stands of trees. Blackwell, Oxford, pp 267–275

Wazihullah AKM, Islam SS, Rahman F, Das S (1996) Reduced attack of keora (*Sonneratia apetala*) by stem borer in mixed-species plantations in coastal Bangladesh. J Trop For Sci 8:476–480

White JA, Whitham TG (2000) Associational susceptibility of cottonwood to a box elder herbivore. Ecology 81:1795–1803

Wilby A, Thomas MB (2002) Natural enemy diversity and pest control: patterns of pest emergence with agricultural intensification. Ecol Lett 5:353–360

Wilson DS (1992) Complex interactions in metacommunities, with implications for biodiversity and higher levels of selection. Ecology 73:1984–2000

Wint W (1983) The role of alternative host-plant species in the life of a polyphagous moth, *Operophtera brumata* (Lepidoptera: Geometridae). J Appl Ecol 52:439–450

Witter JA, Waisnanen LA (1978) The effect of differential flushing times among trembling aspen clones on tortricid caterpillar populations. Environ Entomol 7:139–143

Wratten SD, van Emden, HF, Thomas, MB (1998) Within-field and border refugia for the enhancement of natural enemies. In: Pickett CH, Bugg RL (eds) Enhancing the natural control of arthropod pests through habitat management. University of California Press, Berkeley, pp 375–404

Zach R, Falls JB (1975) Response of the ovenbird (Aves: Parulidae) to an outbreak of the spruce budworm. Can J Zool 53:1669–1672

Zanuncio JC, Mezzomo JA, Guedes RNC, Oliveira AC (1998) Influence of strips of native vegetation on Lepidoptera associated with *Eucalyptus cloeziana* in Brazil. For Ecol Manage 108:85–90

Zhang QH (2001) Olfactory recognition and behavioural avoidance of angiosperm non-host volatiles by conifer bark beetles. PhD thesis (Agraria 264). Swedish University of Agricultural Sciences, Alnarp, Sweden

Zhang QH, Schlyter F, Andersson P (1999) Green leaf volatiles interrupt pheromone response of spruce bark beetle, *Ips typographus*. J Chem Ecol 25:2847–2861

Zhang QH, Liu GT, Schlyter F, Birgersson G, Anderson P, Valeur P (2001) Olfactory responses of *Ips duplicatus* from Inner Mongolia, China to nonhost leaf and bark volatiles. J Chem Ecol 27:995–1009

Zoebelein G (1957) Die Rolle des Waldhonigtaus im Nahrungshaushalt forstlich nützlicher Insekten. Forstwiss Centralbl 76:24–34

13 Susceptibility to Fungal Pathogens of Forests Differing in Tree Diversity

M. Pautasso, O. Holdenrieder, and J. Stenlid

13.1 Terms of the Issue

Plant communities commonly have both host diversity and biotic diseases (Kimmins 1997a; Roy and Kirchner 2000), but to differing degrees. Disease and diversity are classically connected in the literature by the co-evolutionary explanation of sex (Hamilton 1980): recombinations of the host genome may produce disease-resistant genotypes, while preserving or enhancing diversity (Clay and Kover 1996; Kirchner and Roy 2000; Lively 2001). We argue here that this concept of species interaction has to be viewed in the context of both individuals and populations in order to make sense from the evolutionary (Tokeshi 1999) and forest economy viewpoints (Perry and Amaranthus 1997; Perry 1998). The general importance of disease and parasites in the dynamics of plant populations is well recognized (Harper 1977, 1990; Burdon 1987; Dobson and Crawley 1994; Garrett and Mundt 1999), but studies on disease dynamics in natural multi-species plant communities remain rare (Kranz 1990; Burdon 1993; Roy et al. 2000; Thrall and Burdon 2002).

Most forest ecosystems have been altered in their patterns and processes by human influences (Glatzel 1991; Edmonds et al. 2000; Wohlgemuth et al. 2002). Also biotic perturbations, including diseases, are increasingly perceived as an intrinsic part of the system (Ehrlich 1994; Schowalter et al. 1997; Spies and Turner 1999). Trees are an obvious structural component of ecosystems (Jones et al. 1997; Rao et al. 1997), but pathogenic and saprotrophic organisms also form an important part of biodiversity and contribute to the holistic functioning of forests (Ingram 1999; Courtecuisse 2001; Siitonen 2001). We refer here only to fungal pathogens of trees and thus do not focus on abiotic disturbances (Hansen and Rotella 1999; see Dhôte, Chap. 14, this vol.), or viruses, bacteria, and insects (Haack and Byler 1993; see Jactel et al., Chap. 12, this vol.), although, in many cases, these may be intimately interrelated with fungal pathogenesis (Hatcher 1995; Lundquist 1995; Maloney and Rizzo 2002).

Ecological Studies, Vol. 176
M. Scherer-Lorenzen, Ch. Körner, and E.-D. Schulze (Eds.)
Forest Diversity and Function: Temperate and Boreal Systems

Disease is "an impairment of the normal state of the living animal or plant body or of any of its components that interrupts or modifies the performance of the vital functions, being a response to environmental factors, to specific infective agents, to inherent defects of the organism, or the combination of these factors" (Webster's Dictionary, cited in Grogan 1987, p. 3). As normality is hard to define, disease is often a fuzzy concept with an arbitrary component (Holdenrieder 1991; Twery and Gottschalk 1996). The widely perceived stability of natural forest ecosystems (i.e., their resistance and resilience in response to disturbance, including disease) when compared with plantations, is frequently ascribed to their greater (tree species) diversity (Holling and Meffe 1996; McCann 2002). The term functional diversity has been suggested as collectively covering the mechanisms by which natural forests resist potential epidemics to some extent (Schmidt 1978). Functional diversity is not simply a surrogate of species richness (Dìaz and Cabido 2001; Petchey and Gaston 2002), but includes the inter- and intraspecific diversity of trees, pathogens, and antagonistic microorganisms. It is often correlated with species diversity and it could be considered also to include, or at least to be modified by, diversity in micro- and macroclimate and in edaphic factors.

The interrelationships between tree diseases and tree diversity have been reviewed from the perspectives of silviculture (Gibson and Jones 1977; Barthod 1994, 1995), plant pathology (van der Kamp 1991; Burdon 1994; Hansen 1999), and landscape ecology (Castello et al. 1995). Evolutionary ecology reviews have dealt with natural ecosystems, but have not specifically focused on forests (Dinoor and Eshed 1984; Burdon 1993; Jarosz and Davelos 1995; Gilbert 2002). Forest pathosystems provide a unique opportunity to improve our understanding of ecosystem function and to translate this knowledge into practice. We aim here to bring together ecology and forest pathology, with particular emphasis on fungal tree pathogens, in order to inspire further work on tree diversity and disease. A proper consideration of the functional role of pathogens is a necessity for forest, landscape, and carbon sequestration management.

13.2 Susceptibility as a Function of Tree Diversity

How may forests, their tree species diversity, and pathogens be brought together? Susceptibility is the key concept, because "susceptibility to disturbance, biotic or abiotic, must partially depend on taxon-specific traits that translate into differential metapopulation dynamics, speciation, and extinction" (Eble 1998, p. 139). The term "susceptibility" has been defined as the "inability of a plant to resist the effect of a pathogen or other damaging factor" (Shurtleff and Averre 1997 p. 321). Here, the concept will be extended to cover the overall susceptibility of trees to disease within a forest. In this con-

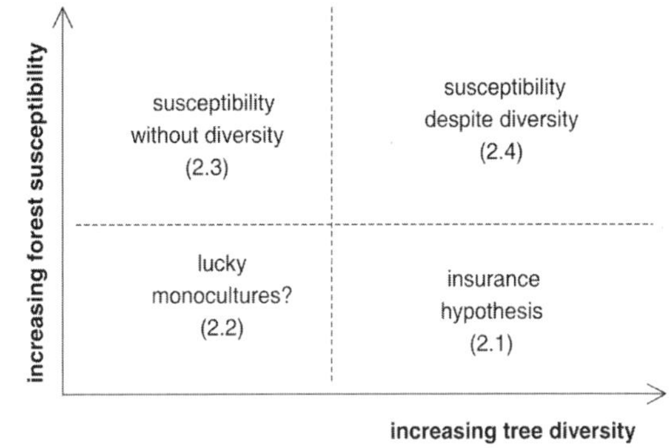

Fig. 13.1. Forest susceptibility as a function of tree diversity (see text for explanation)

text, the infection of an individual tree is largely a stochastic event and depends on the dispersal strategy of the pathogen.

The terms diversity and susceptibility can be connected in several ways, as van der Maarel (1993) has done for the terms disturbance and stability. First of all, diversity may guard against susceptibility (see Sect. 13.2.1). Thus, a lack of diversity appears to underlie the classic case of an ecosystem that is susceptible because it lacks diversity (see Sect. 13.2.3). On the other hand, there are cases in which forests with little tree diversity are not much affected by disease (see Sect, 13.2.2), although they may be in a transient state of disease escape, yet still representing a classic situation of high susceptibility due to lack of diversity. Finally, diversity and susceptibility may co-exist (see Sect. 13.2.4). These different configurations are shown in Fig. 13.1.

13.2.1 The Insurance Hypothesis

Most plants are susceptible to more than one pathogen, but not all plants are susceptible to all pathogens. The diversity of trees may therefore help in maintaining the physical structure of a forest following an outbreak of pathogens, or in restricting disease to a limited number of individuals. In mixed woodlands, if one species is struck by disease, even severely, others may fill the gaps and perpetuate the forest as a whole. Hence, a major argument for retaining diverse ecosystems is the insurance hypothesis; suggesting that high diversity maintains the overall integrity of an ecosystem while biotic and abiotic environmental conditions change over time (Yachi and Loreau 1999; Bengtsson et al. 2000; McCann 2002; Mitchell et al. 2002). Since species react differently to such challenges, a diverse ecosystem will respond in a more buffered manner than a less diverse one (Loreau et al. 2001). In the case of a forest ecosystem, the longevity of its component trees (Rajora and Mosseler 2001) makes it likely that numerous or long-term challenges will occur within the life spans of individuals, and will thus be compounded in their impact. These may include not only the introduction of new pathogens, perhaps introduced via worldwide anthropogenic traffic (Wingfield et al. 2001), but also climate change (Lonsdale and Gibbs 1995; Ayres and Lombardero 2000).

The disease-diversity hypothesis applies not only to tree species, but also to intraspecific variation; it is widely accepted that genetic diversity can be crucial to disease resistance in tree populations (Heybroek 1982; Han et al. 2000; Burdon 2001a). In this sense, genetic diversity also provides the best insurance against threats from introduced pathogens (see also Müller-Starck et al. Chap. 5, this Vol.). Age structure is another important aspect of diversity: Scots pine, for example, is only susceptible to *Phacidium infestans* snow blight at the seedling stage, when shoots and needles may still be covered by snow in winter (Roll-Hansen 1989). In forests with a mixed age structure, only a limited proportion of trees is at a susceptible age, so that spore dispersal is limited and the system is buffered against the effects of the disease (Burdon et al. 1992, 1994). This example shows that density-dependent effects may mediate the action of diversity on susceptibility.

Although there is a good theoretical basis for the role of diversity in protecting forests from pathogens, the evidence is often anecdotal, or extrapolated from studies of agricultural (Finckh and Wolfe 1998) and grassland (Knops et al. 1999; Mitchell et al. 2002) systems. In one such study, Zhu et al. (2000) assessed the susceptibility of rice mixtures in small-scale vs. large-scale experimental plots. They showed that crop diversification reduced the incidence of rice blast, but more so in the large-scale than in the small-scale plots. In various studies, mixtures of cultivars or species performed better than monocultures with regard to crop yields. In such cases, however, the mechanism involved could be a reduction not only in the number of pathogen propagules, but also in the intensity of intraspecific competition between the plants (Harper 1990; Wolfe 2000; Naeem 2002a).

In willow cropping systems, mixtures have been shown to moderate epidemics (Ramstedt 1999; Peacock et al. 2001). The main fungal pathogen studied in these systems is *Melampsora* rust, which consists of numerous species and pathotypes, each adapted to a specific range of willow species and hybrids (Pei et al. 1996; McCracken et al. 2000; Ramstedt et al. 2002). Diversity in vegetation composition and structure within the plantations can slow down pathogen spread, thereby increasing yields (Fig. 13.2; McCracken et al. 2001; Mundt 2002). As a result, growers are today urged not to plant large areas of a single clone, even if that clone is currently known as resistant or less susceptible to *Melampsora* (McCracken and Dawson 1998). In the design of such mixtures, it is important to optimize not only the number of willow varieties, but also their differential resistance to specific pathotypes occurring in a particular region (McCracken and Dawson 1997; Hunter et al. 2002). Interestingly, the spatial design of willow short-rotation systems has a greater influence on rust than on beetle distribution. This may be due to the passive

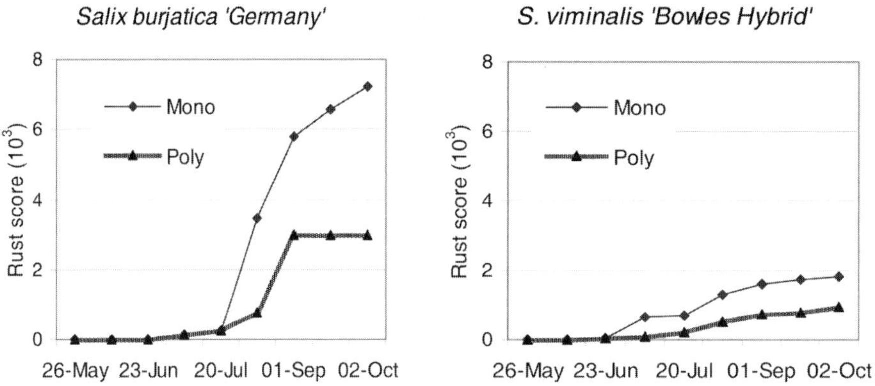

Fig. 13.2. *Melampsora* rust disease progress in two mono- and polyclonal willow stands (from McCracken and Dawson 1998; reproduced with kind permission of Blackwell Publisher). The rust score was calculated by multiplying the percentage number of leaves with rust by the mean rust cover per leaf

dispersal mechanism of airborne fungal pathogens (Boudreau and Mundt 1997). In cases where fungal pathogens are dispersed by insect vectors, as in the well-known example of Dutch elm disease (Brasier 1991), the beneficial effect of low host density is counteracted by the dispersal activity of the vector.

A lower density of susceptible hosts may work in the absence of vectors. In British Columbia, alternate rows of susceptible and less susceptible or even immune tree species were planted in 1968 after felling of a Douglas fir/lodgepole pine mixed forest severely affected by *Phellinus weirii* and *Armillaria ostoyae* root rots (Morrison et al. 1988). During the first 20 years, the spread of *Armillaria* was much reduced where the host trees were intermixed with non-susceptible species, probably due to a reduced chance of contact between diseased and undiseased root systems. Further assessments of this large-scale experiment are due. Gerlach et al. (1997) provided additional experimental evidence, albeit on a smaller scale, of the value of diversity (either species diversity or functional-group diversity) in diminishing damage by *Armillaria ostoyae*. The reduced disease incidence in diverse stands is explained by a lower density of the target species, which diminishes the chance of infection. This leads to the idea of a critical host population *density*, below which the disease will not occur (Burdon and Chilvers 1982; Newman 2000). Similarly, we propose the concept of a critical host *diversity*, a level of species richness which limits a particular disease to a silviculturally acceptable level.

13.2.2 Lucky Monocultures?

Despite the above evidence for diminished disease susceptibility of diverse communities, there are examples of monospecific or even single genotype stands that have shown little disease. These have been reported mainly from boreal and temperate forests; for example, large, apparently healthy monoclonal stands of *Populus tremuloides* are known in North America (Burdon 2001a, but see Hogg et al. 2002). Some examples are also known locally in humid tropical forests (Evans 1992); as in the case of dense, monospecific mangrove forests (Gilbert et al. 2002).

Monospecific plantations with apparently low disease susceptibility have rarely been studied in detail and any data may have been confused by the masking effects of protective measures. In some cases, stands may seem to be non-susceptible only because they have temporarily escaped disease (e.g., a monoclonal stand of willow or poplar; see Sect. 13.2.1) or because pathogens have not had time to build up. In the case of *Dothistroma* needle-cast in exotic monocultures of *Pinus radiata* in New Zealand, evidence of some stands remaining largely unaffected was reported (Chou 1981), but it is more typical to find severe disease and a requirement for large-scale copper fungicide treatment (Gibson 1975).

A situation where homogeneous monospecific forests initially remained largely disease-free, but were later affected by the buildup of a pathogen, is exemplified by the plantations of *Picea sitchensis* in Great Britain. In the 1930s, Peace (1938) evaluated butt rot as of minor importance in such plantations, but Greig (1962), Pratt (1979a,b), and Greig et al. (2001) later reported an increased frequency of decay by *Heterobasidion annosum*. The fungus had been infecting stumps left from harvesting, which thus became a source of increasing inoculum for standing trees via root contact. Currently, however, prevention is not focused on tree diversity, but on recognizing sites with characteristics conducive to disease development (Redfern et al. 2001). Such sites may be avoided for re-planting susceptible species, or may be designated as requiring preventive stump treatment, a method used since the 1960s. Despite the buildup of *H. annosum* in British stands of *P. sitchensis*, the overall load of diseases in these stands currently remains low.

Another example of an increasing incidence of a root and butt rot following harvesting is provided by *Pinus radiata* plantations established on former agricultural land in New Zealand, where losses due to *Armillaria* became progressively more troublesome after the first rotation (Chou 1991; Burdon 2002). Disease has been increasing also on sites cleared of the native, diverse forest (van der Pas 1981). It might be hoped that shortened rotations will limit the opportunity for this fungus to develop within trees. This does not, however, seem to be the way to avoid damage in the long term, perhaps because each rotation provides new food bases for the fungus. Indeed, when rotation periods of *Populus tremuloides* and *P. grandidentata* plantations in the Great

Lakes region were shortened, the incidence and extent of *Armillaria* root rot increased progressively (Stiell and Berry 1986; Stanosz and Patton 1987a,b; Kile et al. 1991). In mixed spruce/birch forests, thinning of overstory birches had no effect on the rate at which *Armillaria ostoyae* killed spruce (Simard and Hannam 2000). The observation period of 5 years may, however, have been too short for demonstrating a change in disease expression.

The above examples support the idea that a low incidence of disease is sometimes transient, while a pathogen may be absent or at a low incidence. Thus, although there are examples where low diversity seems to coexist with a non-susceptible state, such situations may be unsustainable.

13.2.3 Susceptibility Without Diversity

Despite exceptions, monocultures are generally associated with increased ecological risks. Korhonen et al. (1998b) thoroughly reviewed the debate concerning the relative incidence of *Heterobasidion annosum* s.l. root rot in mixed vs. pure stands under different climatic conditions (Table 13.1). Nine of the thirteen large-scale studies listed support the insurance hypothesis: a reduced incidence of butt rot is correlated with an increase in the number of tree species. Three studies did not show any difference between mixed and pure stands; indeed, one indicated an increase in susceptibility in mixed stands, but this may be explained by the combined influence of edaphic conditions and climate. It is difficult to find fully comparable pure and mixed stands, as noted long ago by Flury (1926). Moreover, *H. annosum* s.l. consists of three intersterility groups with distinctive host preferences (Korhonen et al. 1998a).

13.2.3.1 Unlucky Monocultures

Homogeneous monospecific or monoclonal plantations are particularly susceptible to epidemics (Gibson and Jones 1977; Hood et al. 1991; Hartley 2002) because they lack tree diversity, but not all cases follow this relationship. For instance, many epidemics affect exotic plantations of *Pinus radiata*, but this species coexists well with pathogens when planted within its natural range. Genetic data shows scope for breeding *P. radiata* varieties with adaptability to site types and resistance to particular pathogens (Burdon 2001a,b). The value of such varieties or of species diversity may, however, be limited in the face of an aggressive pathogen with a broad host range. For example, *Armillaria luteobubalina* has proved able to infect all eucalypt species exposed to it (Jarosz and Davelos 1995; Kile 2000). However, the susceptibility of different host species to attack by pathogenic *Armillaria* species appears to differ greatly (Wargo and Harrington 1991). Thus, the impact of any particular

Table 13.1. Susceptibility of pure vs. mixed forest stands to *Heterobasidion annosum* s.l. butt rot. (Data from the Russian literature have been kindly provided by K. Korhonen)

Key features of the study, location	Main tree species, and admixtured species	Effect of diversity on susceptibility[a]		Reference
8 stands in 4 regions (southern Italy)	*Abies alba*, pure and mixed with mainly *Fagus sylvatica*	–	Slightly lower infection with *H. abietinum* in mixed stands	Puddu et al. (2003)
122 Permanent experimental plots at 44 sites (southern Sweden)	*Picea abies*, pure and mixed with *Pinus sylvestris*	–	Lower proportion of butt rot incidence in mixed stands; most significant result with 50 % admixture	Lindén and Vollbrecht (2002)
Experimental stands planted on agricultural land in 1975; different mixture schemes (Lithuania)	*Pinus sylvestris*, pure and mixed with *Amorpha fruticosa*, *Betula pendula*, or *Robinia pseudoacacia*	–	Pure stands show the lowest productivity following increased disease incidence	Lygis et al. (2001)
Large-scale study of naturally and artificially regenerated forests, 60 plots (northern Switzerland)	*Picea abies*, pure and mixed with *Abies alba*	–	Mixtures with silver fir show a lower rot frequency	Graber (1994)
Nationwide survey of ca. 5000 stands (Norway)	*Picea abies*, pure and mixed with hardwood species	–	Admixture with other species significantly reduced rot frequency	Huse et al. (1994)
Predictive models for volume percentage affected by root rots (Belarus)	*Picea abies*, pure and mixed with *Pinus sylvestris* and hardwood species (30 %)	–	Incidence of rots is diminished by the presence of other species	Arnol'bik et al. (1990)
Comparison of 34 clear-cut pure and mixed stands of totally 35 ha (southern Finland)	*Picea abies*, pure and mixed with *Betula pendula* and *Pinus sylvestris* (up to 60 %)	–	Damage caused by *H. annosum* to spruce was slightly smaller in mixed stands than in pure ones	Piri et al. (1990)

Table 13.1. (*Continued*)

Key features of the study, location	Main tree species, and admixtured species	Effect of diversity on susceptibility[a]		Reference
Study of 50 plots in newly cleared areas (southeastern Norway)	*Picea abies*, pure and mixed with hardwood species	–	Frequency and extent of *H. annosum* in stumps decreased with increased proportion of deciduous trees	Enerstvedt and Venn (1979)
Assessment of 24 pure spruce and 10 mixed stands (Sweden)	*Picea abies*, pure and mixed with *Pinus sylvestris*	–	The average butt rot frequency in spruce was ca. 40 % in pure and 25 % in mixed stands	Rennerfelt (1946)
Advance regeneration study on 17 sample plots in 9 stands at 4 locations (southern Finland)	*Picea abies*, pure and mixed with *Betula pendula, Betula pubescens, Pinus sylvestris* (overstory), and *Betula* spp. and *Sorbus aucuparia* (regeneration)	=	No correlation was found between the proportion of admixed tree species and the incidence of *H. parviporum* root rot	Piri and Korhonen (2001)
Assessment of 15 pure and 25 mixed stands (southern Germany)	*Picea abies*, mixed with broadleaved trees and or with *Pinus sylvestris, Abies alba, Pseudotsuga menziesii*	=	Mixing neither decreased nor increased the percentage of spruce with butt rot	Siepmann (1984)
Large-scale study on ca. 1200 ha (eastern Russia)	*Picea obovata* and *Abies sibirica*, pure and mixed with deciduous trees	=	Infection also proved to be high in mixed broadleaf/conifer stands	Korotkov (1978)
Comparison of 35 pure spruce and 7 mixed stands (northern Germany)	*Picea abies*, pure and mixed with hardwood species	+	Butt rot presence proved to be higher in mixed than in pure stands	Kató (1967)

[a] – Diversity associated with lower susceptibility; = diversity not significantly related to affect susceptibility; + diversity associated with higher susceptibility

Armillaria sp. on a monoculture probably depends on the susceptibility of the host species concerned, but data on mortality rates due to *Armillaria* root disease in different community types remain rare (Kile et al. 1991). Equally, different *Armillaria* species, of which at least 40 are currently recognized worldwide, vary greatly in their pathogenicity (Gregory et al. 1991). A monoculture may thus be unlucky if the particular *Armillaria* species at the site happens to be particularly aggressive.

The example of *Armillaria ostoyae* root rot, a major concern of conifer forest management (Hagle and Shaw 1991), highlights another disease-promoting factor in monocultures, i.e., an enhanced buildup of inocula of host-specific pathogens. Seedlings of three conifer species established in recently logged sites in Minnesota showed *Armillaria*-induced mortality that increased with the proportion of conifers in the previous crop (Gerlach et al. 1997). Hence, selected diversity (mixtures of conifers with hardwoods rather than conifer monocultures) may reduce disease impact in the long term. Increased disease incidence has also been observed where paper birch is either removed or excluded from stands of susceptible conifers (Simard 1998). In general, the case of the unlucky monoculture clearly supports the insurance hypothesis (Sect. 13.2.1).

13.2.3.2 The Fate of the American Chestnut

The case of the American chestnut *Castanea dentata*, the "monarch of the eastern hardwood forests," at the beginning of the last century (Hepting 1974; Smith 2000) may be an example of susceptibility associated with a lack of intraspecific host diversity. Beginning with the introduction of the chestnut blight fungus *Cryphonectria parasitica* on imported Asiatic chestnut seedlings, the main story exemplifies the classic destabilizing effects of an exotic pathogen within a forest ecosystem (Stephenson 1986; Anagnostakis 1987; Burdon 1991; Oak 2002). In such events, the absence of co-evolution between the pathogen and its new host can largely explain the host's high susceptibility (Harper 1990; Hansen 1999). However, the narrow genetic diversity of the American chestnut, as compared to other *Castanea* species (Huang et al. 1998), could be a compounding factor in this instance. Thus, there is a hypothesis that the disease might not have reached devastating levels if host diversity had been greater.

Due to chestnut blight, *C. dentata* became almost extinct (Blanchard and Tattar 1997) and now survives almost only in the form of sprouting stumps (Tainter and Baker 1996). Yet, since sprouts usually fail to attain sexual maturity before they succumb to disease (Jarosz and Davelos 1995) or to competitive exclusion by shade-tolerant species (Schwadron 1995), there is a very limited chance that natural genetic variation may arise and perhaps offer resistance to the pathogen. In this situation, the high susceptibility of the

American chestnut is genetically perpetuated. The potential for co-evolution can, however, be envisaged in other cases where pathogens have encountered new hosts. Such events can occur naturally and may have played an evolutionary role long before man overrode continental divides, albeit at a more relaxed pace (Josephson Weddell 2002). These processes show the capacity of pathogens to shape forest diversity (Burdon 1991) and offer a chance to study such ecological interactions in real time (Dinoor and Eshed 1984; McDonald et al. 1998; Brasier 2001).

13.2.4 Susceptibility Despite Diversity

The above evidence shows that host diversity can be reasonably expected to play a key role in reducing the susceptibility of forests to fungal pathogens. In this section, however, we examine three pathosystems from three different continents where both diversity and susceptibility are present.

13.2.4.1 The Jarrah Forest Dieback

The introduction of the oomycete *Phytophthora cinnamomi,* a generalist and aggressive root pathogen (Dickman 1992; Hansen 1999; Prell and Day 2001) to which most native eucalypts proved to have little if any resistance, had severe consequences for the Jarrah (*Eucalyptus marginata*) forest of Western Australia (Newhook and Podger 1972; Jarosz and Davelos 1995; Shearer and Dillon 1995; Shearer and Smith 2000; Fitter 2001). In this case, in spite of a high functional diversity (not only of tree species but also of shrubs), the Jarrah forest was very susceptible to the introduced fungal pathogen (McDougall et al. 2002; Weste et al. 2002).

Here, the insurance hypothesis fails in the face of a non-specific pathogen. No other pathogen has killed so many different plants in different communities as has *P. cinnamomi* (Zentmyer 1980; May and Simpson 1997). A change toward a drier climate has been suggested as an alternative reason for Jarrah dieback (Florence 1996), but Wilson et al. (2000) could not show an association between the spatial disease pattern and specific hydrological features. There have been regional differences in disease development, but these may have been caused by the disappearance of a local *Acacia* species in the understory (Shea et al. 1979). The presence of legumes leads to an increase of available soil nitrogen, which in turn may result in a more diverse microflora, exerting antagonistic effects against the pathogen (Murray 1987; see also Walchhütter et al. 2000).

13.2.4.2 North American Forests

The North American forests display multifaceted interrelationships between tree diversity and pathogen activity. In this section, we focus on two pathogens with completely different dispersal strategies, an introduced rust (*Cronartium ribicola*) and an indigenous root rot fungus (*Phellinus weirii*). Both pathosystems show that tree diversity does not necessarily reduce forest susceptibility.

As with the chestnut canker pathogen, the causal species, *C. ribicola,* of white pine blister rust was introduced to both seaboards of North America at the beginning of the last century (Tainter and Baker 1996). A century later it is one of the most important diseases of white pines in both Eastern and Western America. It can affect all North American five-needled pine species, but susceptibility varies with species and age (Hoff and Hagle 1990; Burdon 2001a; Zeglen 2002). Host species richness seems to have very little effect on the disease within stands generally, but the presence and distribution of species within a stand are important in determining the severity of infection on individual trees. Here, it is functional diversity that matters. Yet, the severity of rust infection differs regionally, perhaps due to differences in climate and topographic position (Kendall and Keane 2001). In order to spread, *C. ribicola* needs the presence of the alternate host genus *Ribes* (Kimmey 1938), which for a time was subjected to attempted eradication. This control strategy has long since been abandoned in favor of developing resistance in white pines (Hoff and McDonald 1993). In future, disease might be controlled by influencing landscape connectivity, an approach suggested by a pioneering study of the related pathosystem of fusiform rust (*Cronartium quercuum*) in loblolly and slash pine forests of the southeastern United States (Perkins and Matlack 2002).

Western Oregon forests affected by the basidiomycete *Phellinus weirii* are another pathosystem showing a coexistence of tree diversity and susceptibility. Unlike blister rust, this pathogen is indigenous to the system (Holah et al. 1997). Root rot caused by *P. weirii* is one of the most important disturbances leading to long-term stand-replacing processes and an uneven-aged forest structure (Lewis and Lindgren 1999, 2000; Hansen and Goheen 2000). Tree diversity may retard mycelial advance (McCauley and Cook 1980), but *P. weirii* infects all the dominant conifer species and creates gaps in the forest overstory. The proportion of otherwise rare trees with resistance to root rot thereby increases (Burdon 1994; Jarosz and Davelos 1995; Ingersoll et al. 1996). Thus, *P. weirii* plays a key role in determining the overall diversity and community structure of the forests where it is active.

13.2.4.3 The Introduction of Lodgepole Pine

Diversity may not only co-exist with susceptibility; it may even be its cause (Garrett and Mundt 1999). For instance, if the alternate hosts of a rust pathogen are planted in the same forest, this augmentation of diversity clearly increases the susceptibility of the whole forest (Heybroek 1982; Mattila et al. 2001). However, disease may drive a co-evolutionary process selecting for increased resistance in the longer run. One example is provided by the *Pinus contorta–Cronartium comptoniae* pathosystem where hosts grown in the presence of the alternate host *Myrica gale* show a higher degree of resistance than those from areas at greater distance from the inoculum source (Hunt and van Sickle 1984).

A change in host diversity may promote disease also by driving the evolution of a pathosystem. Such a case may have begun with the introduction of a North American tree species, lodgepole pine (*Pinus contorta*) to Sweden. Despite an initial delay (Witzell and Karlman 2000), the large-scale plantations of *P. contorta* became epidemically affected by *Gremmeniella abietina*, a fungus which is largely absent from the native range of *P. contorta* (Karlman 2001). The fungus on *P. contorta* is different from the biotype which occurs naturally on the native Scots pine (*P. sylvestris*) and which is specially adapted to attack trees above snow cover (Hellgren and Barklund 1992; Hellgren 1995; Hellgren and Högberg 1995; Hellgren and Stenlid 1997). Lodgepole pine has, however, become a potential source of increased infection pressure on Scots pine, thus fueling the potential evolution of genotypes adapted to Scots pine (Engelmark et al. 2001). As new genotypes could be more aggressive (Ennos 2001), the introduction of the exotic lodgepole pine may represent a new threat of disease for the surrounding native forest (Karlman 2001).

13.3 Reversing the Terms

Figure 13.3 shows four hypotheses relating forest susceptibility to tree diversity (selected from a variety of hypothetical relationships between biodiversity and ecosystem processes, see Naeem et al. 2002). The null hypothesis represents no relationship, whereas the rivet hypothesis represents largely a linear relationship. The redundancy hypothesis, in which a modest amount of diversity is required for low susceptibility, largely equates with the insurance hypothesis (see Sect. 13.2.1). The latter is supported by frequent examples in which monocultures are highly susceptible (see Sect. 13.2.3) or have only a transient freedom from overt disease (see Sect. 13.2.2). However, the case studies where diversity does not prevent susceptibility (see Sect. 13.2.4) support the idiosyncratic response hypothesis, in which ecosystem functions are

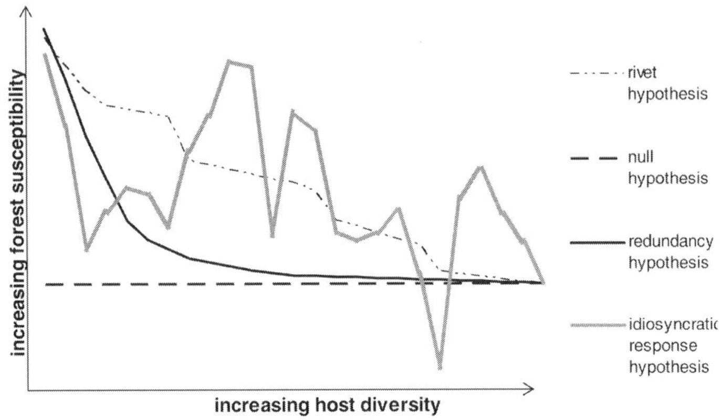

Fig. 13.3. Four hypothetical relations between forest susceptibility to fungal pathogens and tree diversity (adapted from Naeem et al. 1995, 2002; Naeem 2002b)

affected by variations in diversity, but in an unpredictable way in any given situation (Naeem et al. 1995; Garrett and Mundt 1999), at least when only two variables are considered. Accordingly, many other factors are relevant here, including the successional stage of the stand (Kimmins 1997b) and the spatial scale over which the effects are considered (Wiens 2000; Hemstrom 2001). Also, the specificity and aggressiveness of the pathogen, as well as its scale of dispersal and mode of action, are crucial elements for understanding disease impact in relation to host diversity and density (Strong and Levin 1975; Jeger 1999; Thrall and Burdon 2002). Both the idiosyncratic and redundant hypotheses may work, albeit in different contexts: diversity may play an active role against susceptibility within study sites, but the response may be idiosyncratic across sites, due to the variation of other significant factors (see Naeem 2002b).

Reversing the terms clarifies the issue. No longer asking whether forest susceptibility might be dependent on tree diversity, the research question becomes whether susceptibility leads to diversity (pathogen-driven forest diversity: van der Putten 2000; Singh 2002). It has indeed been shown that the impacts of antagonistic interactions, including diseases, lessen the ability of potentially dominant species to monopolize resources and may thus allow other species to subsist (van der Kamp 1991; Dobson and Crawley 1994; Dobson and Grenfell 1995; Fitter 2001; Chave et al. 2002). For instance, the coexistence of a large number of tree species in humid tropical forests may be facilitated by species-specific interactions between hosts and pathogens that affect tree fecundity, mortality, or general competitiveness (Augspurger 1990; Gilbert and Hubbell 1996; Ashton and LaFrankie 2000; Turner 2001; Wright 2002).

The Janzen-Connell hypothesis (which associates the diversity of tree species with the presence of specialist herbivores and pathogens, so that offspring of a tree species can establish only at a substantial distance from the parents; Janzen 1970; Connell 1978) has been widely tested. Thus, various studies, mostly focusing on soil pathogens in tropical rainforests, have corroborated the view that plant-pathogen interactions affect successional dynamics and species diversity (Augspurger 1984; Augspurger and Kelly 1984; Gilbert et al. 1994; Gilbert and de Steven 1996; Wills et al. 1997; Webb and Peart 1999; van der Putten 2000, 2001). There is also growing evidence of similar interactions in temperate communities (Whittaker et al. 2001; Gilbert 2002; Lambers et al. 2002). The existence of density-dependent mortality due to fungal soil pathogens was confirmed in North American *Prunus serotina* seedlings (Packer and Clay 2000) and in seeds of *Fagus crenata* in Japan (Tomita et al. 2002). Rare plant species seem to be more susceptible than common ones to such soil pathogens and may die out with a consequent reduction in diversity (Klironomos 2002, but see Blaney and Kotanen 2001). Even if dominant forest trees of the temperate zone have rarely been included in the experiments, there are hints from field observations that the regeneration of silver fir (*Abies alba*) may be controlled by the soil fungus *Cylindrocarpon destructans* (Kowalski 1980) in the presence of fir overstory. Experiments in mixed mountain forests in Bavaria corroborate this hypothesis (Mosandl and Aas 1986; Burschel et al. 1992).

Foliar pathogens in natural forests probably play an equally important role in stand dynamics as soil-based pathogens, but have received less attention (Benitez-Malvido et al. 1999; Roy and Kirchner 2000; García-Guzman and Dirzo 2001). Indeed, data on the effects of pathogens on tree diversity appear to be generally scarce (Kranz 1990). This may be due to the difficulties in reconstructing tree diversity data of a pre-epidemic population after the disease has reached the community (McDonald et al. 1998). Even if there is evidence that susceptibility enhances diversity, there remains a case for the reverse hypothesis, i.e., a decrease in diversity brought by susceptibility. Such an effect may result from the introduction of an exotic pathogen (as in the Jarrah forest case) but it has never been documented for native pathogens at larger scales (Hansen 1999).

13.4 Conclusions

This review shows that tree species diversity is tightly linked with fungal diseases. Tree species diversity may make forests less susceptible to fungal pathogens, thereby confirming the insurance hypothesis. Supporting evidence is provided by willow-cropping systems affected by *Melampsora* rust, *Heterobasidion annosum,* and *Armillaria* root rot in pure vs. mixed stands, chestnut blight (*Cryphonectria parasitica*), and various diseases in monocultures of *Eucalyptus* spp., *Picea sitchensis, Pinus radiata,* and *Populus* spp. In the Jarrah forest, however, tree species diversity has not impeded a dieback due to the generalist pathogen *Phytophthora cinnamomi.* Tree diversity coexists with susceptibility also in North American coniferous forests affected by the introduced rust *Cronartium ribicola*, and by native *Phellinus weirii* root rot. Moreover, the introduction of *Pinus contorta* from North America to Sweden, while representing a local increase in tree diversity, may enhance Scleroderris canker. Alternatively, by reversing the terms, susceptibility becomes the explanatory variable of diversity. Diversity may be increased by susceptibility, as the Janzen-Connell hypothesis predicts, or diminished, as the introduction of pathogens into previously unexposed ecosystems shows.

In the overall functioning of forest ecosystems, both tree diversity and tree diseases matter, although their effects at times may be confounded by other factors (Barthod 1994; Malmström and Raffa 2000; Heil 2001; Naeem 2002b). The effects of biodiversity on ecosystems vary strongly amongst the systems studied, and evidence is scarce, owing to the relative lack of studies in forests (Schläpfer and Schmid 1999). Environmental change, either involving single abiotic factors (Pfisterer and Schmid 2002), or when combined with pathogen and host introductions, may result in unprecedented effects (Lonsdale and Gibbs 1995; Coakley et al. 1999; Ayres and Lombardero 2000). It is still an open question to what extent data from artificial cropping systems are relevant to natural plant communities. Likewise, the relative importance of abiotic perturbations, animal predators, and herbivores versus pathogens has to be further investigated. Progress will come from species performance studies in existing forest ecosystems, combinatorial biodiversity experiments, and evolutionary epidemiology. A thorough understanding of the interrelations between diversity and disease in forest pathosystems is an essential prerequisite for sustainable ecosystem management. The role of pathogens in influencing tree diversity of forests, as well as the function of tree diversity on disease impact, undoubtedly deserve further research, both empirical and through modeling, and not only by plant pathologists.

Acknowledgments: We gratefully thank Paolo Capretti, Christoph Grünig, Everett Hansen, Gertrude Hirsch, Kari Korhonen, Christian Körner, Kathy J. Lewis, Vaidotas Lygis, Duncan Morrison, Lioba Paul, Hans Pretzsch, Katja Rodriguez, Michael Scherer-Lorenzen, Thomas Sieber, Bart Van der Kamp and an anonymous reviewer for commenting on the typescript, providing useful information, or discussing ideas contained here. Special thanks to Pat Johnson, Christian Körner and David Lonsdale for editorial and linguistic revision of the text.

References

Anagnostakis SL (1987) Chestnut blight; the classical problem of an introduced pathogen. Mycology 79:23–27

Arnol'bik VM, Smolyak Yu L, Fedorov VN (1990) Effect of root rot on the productivity of Norway spruce stands in Belarus. Lesovedenie i Lesnoe Khozyaistvo 25:95–100

Ashton PS, LaFrankie JV (2000) Patterns of tree species diversity among tropical rain forests. In: Kato M (ed) The biology of biodiversity. Springer, Berlin Heidelberg New York, pp 161–177

Augspurger CK (1984) Seedling survival of tropical trees: interactions of dispersal distance, light gaps, and pathogens. Ecology 65:1705–1712

Augspurger CK (1990) Spatial patterns of damping-off disease during seedling recruitment in tropical forests. In: Burdon JJ, Leather SR (eds) Pests, pathogens, and plant communities. Blackwell, Oxford, pp 131–144

Augspurger CK, Kelly CK (1984) Pathogen mortality of tropical tree seedlings: experimental studies of the effects of dispersal distance, seedling diversity, and light conditions. Oecologia 61:211–217

Ayres MP, Lombardero MJ (2000) Assessing the consequences of global change for forest disturbance from herbivores and pathogens. Sci Total Environ 262:263–286

Barthod C (1994) Sylviculture et risques sanitaires dans les forêts tempérées, 1re partie. Rev For Fr 46:609–628

Barthod Ch (1995) Sylviculture et risques sanitaires dans les forêts tempérées, 2e partie. Rev For Fr 47:39–53

Bengtsson J, Nilsson SG, Franc A, Menozzi P (2000) Biodiversity, disturbances, ecosystem function and management of European forests. For Ecol Manage 132:39–50

Benitez-Malvido J, García-Guzmán G, Kossmann-Ferraz I (1999) Leaf-fungal incidence and herbivory on tree seedlings in tropical rainforest fragments: an experimental study. Biol Cons 91:143–150

Blanchard RO, Tattar TA (1997) Field and laboratory guide to tree pathology. Academic Press, San Diego

Blaney CS, Kotanen PM (2001) Effects of fungal pathogens on seeds of native and exotic plants: a test using congeneric pairs. J Appl Ecol 38:1104–1113

Boudreau MA, Mundt CC (1997) Ecological approaches to disease control. In: Rechcigl J, Rechcigl N (eds) Environmentally safe approaches to disease control. CRC Press, Boca Raton, pp 33–62

Brasier CM (1991) *Ophiostoma-novo-ulmi* sp-nov, causative agent of current Dutch elm disease pandemics. Mycopathologia 115:151–161

Brasier CM (2001) Rapid evolution of introduced plant pathogens via interspecific hybridization. Bioscience 51:123–133

Burdon JJ (1987) Diseases and plant population biology. Cambridge Univ Press, Cambridge

Burdon JJ (1991) Fungal pathogens as selective forces in plant populations and communities. Aust J Ecol 16:423–432

Burdon JJ (1993) The structure of pathogen populations in natural plant communities. Annu Rev Phytopathol 31:305–323

Burdon JJ (1994) The role of parasites in plant populations and communities. In: Schulze E-D, Mooney HA (eds) Biodiversity and ecosystem function. Springer, Berlin Heidelberg New York, pp 165–179

Burdon JJ, Chilvers GA (1982) Host density as a factor in plant disease ecology. Annu Rev Phytopathol 20:143–166

Burdon JJ, Wennström A, Ericson L, Müller WJ, Morton R (1992) Density-dependent mortality in *Pinus sylvestris* caused by the snow blight pathogen *Phacidium infestans*. Oecologia 90:74–79

Burdon JJ, Wennström A, Müller WJ, Ericson L (1994) Spatial patterning in young stands of *Pinus sylvestris* in relation to mortality caused by the snow blight pathogen *Phacidium infestans*. Oikos 71:130–136

Burdon RD (2001a) Genetic diversity and disease resistance: some considerations for research, breeding, and deployment. Can J For Res 31:596–606

Burdon RD (2001b) *Pinus radiata*. In: Last FT (ed) Tree crop ecosystems. Elsevier, Amsterdam, pp 99–161

Burdon RD (2002) *Pinus radiata* D. Don. Forestry compendium (compiled from) Pines of Silvicultural Importance. CABI, Wallingford, pp 359–379

Burschel P, El Kateb H, Mosandl R (1992) Experiments in mixed mountain forests in Bavaria. In: Kelty MJ, Larson BC, Oliver CD (eds) The ecology and silviculture of mixed-species forests. Kluwer, Dordrecht, pp 183–215

Castello JD, Leopold DJ, Smallidge PJ (1995) Pathogens, patterns, and processes in forest ecosystems. Bioscience 45:16–24

Chave J, Muller-Landau HC, Levin SA (2002) Comparing classical community models: theoretical consequences for patterns of diversity. Am Nat 159:1–23

Chou CKS (1981) Monoculture, species diversification and disease hazards in forestry. NZ J For 26:20–36

Chou CKS (1991) Perspectives of disease threat in large-scale *Pinus radiata* monoculture – the New Zealand experience. Eur J For Pathol 21:71–81

Clay K, Kover PX (1996) The red queen hypothesis and plant/pathogen interactions. Annu Rev Phytopathol 34:29–50

Coakley SM, Scherm H, Chakraborty S (1999) Climate change and plant disease management. Annu Rev Phytopathol 37:399–426

Connell JH (1978) Diversity in tropical rain forests and coral reefs. Science 199:1302–1310

Courtecuisse R (2001) Current trends and perspectives for the global conservation of fungi. In: Moore D, Nauta MM, Evans SE, Rotheroe M (eds) Fungal conservation. Issues and solutions. Cambridge Univ Press, Cambridge, pp 7–18

Dìaz S, Cabido M (2001) Vive la différence: plant functional diversity matters to ecosystem processes. Tr Ecol Evol 16:646–655

Dickman A (1992) Plant pathogens and long-term ecosystem changes. In: Carroll GC, Wicklow DT (eds) The fungal community. Its organization and role in the ecosystem. Dekker, New York, pp 499–520

Dinoor A, Eshed N (1984) The role and importance of pathogens in natural plant communities. Annu Rev Phytopathol 22:443–466

Dobson AP, Crawley M (1994) Pathogens and the structure of plant communities. Tr Ecol Evol 9:393–398

Dobson AP, Grenfell BT (1995) Introduction. In: Grenfell BT, Dobson AP (eds) Ecology of infectious diseases in natural populations. Cambridge Univ Press, Cambridge, pp 1–19

Eble GJ (1998) The role of development in evolutionary radiations. In: McKinney ML, Drake JA (eds) Biodiversity dynamics. Turnover of populations, taxa and communities. Columbia Univ Press, New York, pp 132–161

Edmonds RL, Agee JK, Gara RI (2000) Forest health and protection. McGraw-Hill, Boston

Ehrlich PR (1994) Foreword. Biodiversity and ecosystem function: need we know more? In: Schulze E-D, Mooney HA (eds) Biodiversity and ecosystem function. Springer, Berlin Heidelberg New York, pp vii–xi

Enerstvedt LI, Venn K (1979) Decay in mature *Picea abies (*L.) Karst. stands. A study on clear-cuttings in Ovre Eiker, Norway. Rep Norw For Res Inst 35:241–264

Engelmark O, Sjöberg K, Andersson B, Rosvall O, Ågren GI, Baker WL, Barklund P, Björkman C, Despain DG, Elfving B, Ennos RA, Karlman M, Knecht MF, Knight DH, Ledgard NJ, Lindelöw Å, Nilsson C, Peterken GF, Sörlin S, Sykes MT (2001) Ecological effects and management aspects of an exotic tree species: the case of lodgepole pine in Sweden. For Ecol Manage 141:3–13

Ennos RA (2001) The introduction of lodgepole pine as a major forest crop in Sweden: implications for host-pathogen evolution. For Ecol Manage 141:85–96

Evans J (1992) Plantation forestry in the tropics, 2nd edn. Clarendon Press, Oxford

Finckh MR, Wolfe MS (1998) Diversification strategies. In: Jones DG (ed) The epidemiology of plant diseases. Kluwer, Dordrecht, pp 231–239

Fitter AH (2001) Specificity, links and networks in the control of diversity in plant and microbial communities. In: Press MC, Huntley NJ, Levin S (eds) Ecology: achievement and challenge. Blackwell Science, Oxford, pp 95–114

Florence RG (1996) Ecology and silviculture of eucalypt forests. CSIRO, Collingwood, Victoria, Australia

Flury P (1926) Über Zuwachs und Ertrag reiner und gemischter Bestände. Schweiz Z Forstwes 77:337–342

García-Guzmán G, Dirzo R (2001) Patterns of leaf-pathogen infection in the understorey of a Mexican rain forest: incidence, spatiotemporal variation, and mechanisms of infection. Am J Bot 88:634–645

Garrett KA, Mundt CC (1999) Epidemiology in mixed host populations. Phytopathology 89:984–990

Gerlach JP, Reich PB, Puettman K, Baker T (1997) Species, diversity, and density affect tree seedling mortality from *Armillaria* root rot. Can J For Res 27:1509–1512

Gibson IAS (1975) Impact and control of *Dothistroma* blight of Pines. Eur J For Pathol 4:89–100

Gibson IAS, Jones T (1977) Monoculture as the origin of major forest pests and diseases. In: Cherrett JM, Sagar GR (eds) Origins of pests, parasite, disease and weed problems. Blackwell, Oxford, pp 139–161

Gilbert GS (2002) Evolutionary ecology of plant diseases in natural ecosystems. Annu Rev Phytopathol 40:13–43

Gilbert GS, de Steven D (1996) A canker disease of seedlings and saplings of *Tetragastris panamensis* (Burseraceae) caused by *Botryosphaeia dothidea* in a lowland tropical forest. Plant Dis 80:684–687

Gilbert GS, Hubbell SP (1996) Plant diseases and the conservation of tropical forests. Bioscience 46:98–106

Gilbert GS, Hubbell SP, Foster RB (1994) Density and distance-to-adult effects of a canker disease of trees in a moist tropical forest. Oecologia 98:100–108

Gilbert GS, Mejia-Chang N, Rojas E (2002) Fungal diversity and plant disease in mangrove forests: salt excretion as a possible defense mechanism. Oecologia 132:278–285

Glatzel G (1991) The impact of historic land-use and modern forestry on nutrient relations of Central-European forest ecosystems. Fert Res 27:1–8

Graber D (1994) Die Fichtenkernfäule in der Nordschweiz: Schadenausmass, ökologische Zusammenhänge und waldbauliche Massnahmen. Schweiz Z Forstwes 145:905–925

Gregory SC, Rishbeth J, Shaw CG III (1991) Pathogenicity and virulence. In: Shaw CG III, Kile GA (eds) *Armillaria* root disease. Agriculture handbook no 691. USDA, Washington, DC, pp 76–87

Greig BJW (1962) *Fomes annosus* (Fr.) Cke. and other root-rotting fungi in conifers on ex-hardwood sites. Forestry 35:164–182

Greig BJW, Gibbs JN, Pratt JE (2001) Experiments on the susceptibility of conifers to *Heterobasidion annosum* in Great Britain. For Pathol 31:219–228
Grogan RG (1987) The relation of art and science of plant pathology for disease-control. Annu Rev Phytopathol 25:1–8
Haack RA, Byler JW (1993) Insects & pathogens – regulators of forest ecosystems. J For 91(9):32–37
Hagle SK, Shaw CG III (1991) Avoiding and reducing losses from *Armillaria* root disease. In: Shaw CG III, Kile GA (eds) *Armillaria* root disease. Agriculture handbook no 691. USDA, Washington, DC, pp 157–173
Hamilton WS (1980) Sex versus non-sex versus parasite. Oikos 35:282–290
Han ZM, Yin TM, Li CD, Huang MR, Wu RL (2000) Host effect on genetic variation of *Marssonina brunnea* pathogenic to poplars. Theor Appl Genet 100:614–620
Hansen A, Rotella J (1999) Abiotic factors. In: Hunter ML Jr (ed) Maintaining biodiversity in forest ecosystems. Cambridge Univ Press, Cambridge, pp 161–209
Hansen EM (1999) Disease and diversity in forest ecosystems. Australas Plant Pathol 28:313–319
Hansen EM, Goheen EM (2000) *Phellinus weirii* and other native root pathogens as determinants of forest structure and process in Western North America. Annu Rev Phytopathol 38:515–539
Harper JL (1977) The population biology of plants. Academic Press, London
Harper JL (1990) Pests, pathogens, and plant communities: an introduction. In: Burdon JJ, Leather SR (eds) Pests, pathogens, and plant communities. Blackwell, Oxford, pp 3–14
Hartley MJ (2002) Rationale and methods for conserving biodiversity in plantation forests. For Ecol Manage 155:81–95
Hatcher PE (1995) 3-way interactions between plant-pathogenic fungi, herbivorous insects and their host plants. Biol Rev 70:639–694
Heil M (2001) Induced systemic resistance (ISR) against pathogens – a promising field for ecological research. Perspect Plant Ecol 4:65–79
Hellgren M (1995) Comparison of *Gremmeniella abietina* isolates from *Pinus sylvestris* and *Pinus contorta* in terms of conidial morphology and host colonization. Eur J For Pathol 25:159–168
Hellgren M, Barklund P (1992) Studies of the life cycle of *Gremmeniella abietina* on Scots pine in southern Sweden. Eur J For Pathol 22:300–311
Hellgren M, Högberg N (1995) Ecotypic variation of *Gremmeniella abietina* in Northern Europe: disease patterns reflected by DNA variation. Can J Bot 73:1531–1539
Hellgren M, Stenlid J (1997) Diseases of conifers caused by *Gremmeniella abietina*. In: Hansen EM, Lewis K (eds) Compendium of forest pathology. APS Press, St Paul, pp 43–45
Hemstrom MA (2001) Vegetative patterns, disturbances, and forest health in eastern Oregon and Washington. Northwest Sci 75:91–109
Hepting GH (1974) Death of the American chestnut. J For Hist 18:60–67
Heybroek HM (1982) Monoculture versus mixture: interactions between susceptible and resistant trees in a mixed stand. In: Heybroek HM, Stephan BR, von Weissenberg K (eds) Resistance to disease and pests in forest trees. Pudoc, Wageningen, pp 326–341
Hoff RJ, Hagle S (1990) Diseases of whitebark pine with special emphasis on white pine blister rust. In: Schmidt WC, McDonald KJ (eds) Proceedings – Symposium on Whitebark pine ecosystems: ecology and management of a high-mountain resource. INT-GTR 270. USDA FS, Ogden, UT, pp 179–190
Hoff RJ, McDonald GI (1993) Variation of virulence of white pine blister rust. Eur J For Pathol 23:103–109

Hogg EH, Brandt JP, Kochtubajda B (2002) Growth and dieback of aspen forests in northwestern Alberta, Canada, in relation to climate and insects. Can J For Res 32(5):823–832

Holah JC, Wilson MV, Hansen EM (1997) Impacts of a native root-rotting pathogen on successional development of old-growth Douglas fir forests. Oecologia 111:429–433

Holdenrieder O (1991) Der Forstschutz – Objekte, Probleme, Strategien. Schweiz Z Forstwes 142:795–807

Holling CS, Meffe GK (1996) Command and control and the pathology of natural resource management. Conserv Biol 10:328–337

Hood IA, Redfern DB, Kile GA (1991) *Armillaria* in planted hosts. In: Shaw CG III, Kile GA (eds) *Armillaria* root disease. Agriculture handbook no 691. USDA, Washington, DC, pp 88–101

Huang HW, Dane F, Kubisiak TL (1998) Allozyme and RAPD analysis of the genetic diversity and geographic variation in wild populations of the American chestnut (Fagaceae). Am J Bot 85:1013–1021

Hunt RS, van Sickle GA (1984) Variation in susceptibility to sweet fern rust among *Pinus contorta* and *P. banksiana*. Can J For Res 14:672–675

Hunter T, Peacock L, Turner H, Brain P (2002) Effect of plantation design on stem-infecting form of rust in willow biomass coppice. For Pathol 32:87–97

Huse KJ, Solheim H, Venn K (1994) Stump inventory of root and butt rots in Norway spruce cut in 1992. Rapport-fra-Skogforsk, no 23, 26 pp

Ingersoll CA, Wilson MV, Thies WG (1996) Effects of *Phellinus weirii* gaps on early successional vegetation following timber harvest. Can J For Res 26:322–326

Ingram DS (1999) Biodiversity, plant pathogens and conservation. Plant Pathol 48:433–442

Janzen DH (1970) Herbivores and the number of tree species in tropical forests. Am Nat 104:501–528

Jarosz AM, Davelos AL (1995) Effects of disease in wild plant populations and the evolution of pathogen aggressiveness. New Phytol 129:371–387

Jeger MJ (1999) Improved understanding of dispersal in crop pest disease management: current status and future directions. Agric For Meteorol 97:331–349

Jones CG, Lawton JH, Shachak M (1997) Positive and negative effects of organisms as physical ecosystem engineers. Ecology 78:1946–1957

Josephson Weddell B (2002) Conserving living natural resources in the context of a changing world. Cambridge Univ Press, Cambridge

Karlman M (2001) Risks associated with the introduction of *Pinus contorta* in northern Sweden with respect to pathogens. For Ecol Manage 141:97–105

Kató F (1967) Auftreten und Bedeutung des Wurzelschwammes (*Fomes annosus* [Fr.] Cooke) in Fichtenbeständen Niedersachsens. Schr Forstl Fak Univ Göttingen 39:33–120

Kendall KC, Keane RE (2001) Whitebark pine decline: infection, mortality, and population trends. In: Tomback DF, Arno SF, Keane RE (eds) Whitebark pine communities. Island Press, Washington, DC, pp 221–242

Kile GA (2000) Woody root rots of eucalypts. In: Keane PJ, Kile GA, Podger FD, Brown NB (eds) Diseases and pathogens of eucalypts. CSIRO, Collingwood, Victoria, Australia, pp 293–306

Kile GA, McDonald GI, Byler JW (1991) Disease in natural forests. In: Shaw CG III, Kile GA (eds) *Armillaria* root disease. Agriculture handbook no 691. USDA, Washington, DC, pp 102–121

Kimmey JW (1938) Susceptibility of *Ribes* to *Cronartium ribicola* in the West. J For 36:312–320

Kimmins JP (1997a) Balancing act. Environmental issues in forestry, 2nd edn. UBC Press, Vancouver

Kimmins JP (1997b) Biodiversity and its relationship to ecosystem health and integrity. For Chron 73:229–232

Kirchner JW, Roy BA (2000) Evolutionary implications of host-pathogen specificity: the fitness consequences of host life history traits. Evol Ecol 14:665–692

Klironomos JN (2002) Feedback with soil biota contributes to plant rarity and invasiveness in communities. Nature 417:67–70

Knops JMH, Tilman D, Haddad NM, Naeem S, Mitchell CE, Haarstad J, Ritchie ME, Howe KM, Reich PB, Siemann E, Groth J (1999) Effects of plant species richness on invasion dynamics, disease outbreaks, insect abundances and diversity. Ecol Lett 2:286–293

Korhonen K, Capretti P, Karjalainen R, Stenlid J (1998a) Distribution of *Heterobasidion annosum* intersterility groups in Europe. In: Woodward S, Stenlid J, Karjalainen R, Hüttermann A (eds) Heterobasidion annosum. Biology, ecology, impact and control. CABI, Wallingford, pp 93–104

Korhonen K, Delatour C, Greig BJW, Schönhar S (1998b) Silvicultural control. In: Woodward S, Stenlid J, Karjalainen R, Hüttermann A (eds) Heterobasidion annosum. Biology, ecology, impact and control. CABI, Wallingford, pp 283–313

Korotkov GP (1978) *Heterobasidion annosum* infection in spruce/fir stands. Lesnoe Khozyaistvo 6:75–78

Kowalski S (1980) *Cylindrocarpon destructans* (Zins.) Scholt., sprawca zamierania samosiequ jodly (*Abies alba* Mill.) w niektorych drzewostanach gorskich poludniowej polski. Acta Agrar Silv Ser Silv 19:57–73

Kranz J (1990) Fungal diseases in multispecies plant communities. New Phytol 116:383–405

Lacy GH, Stromberg EL (2001) Susceptibility. In: Maloy OC, Murray TD (eds) Encyclopedia of plant pathology, vol II. Wiley, New York

Lambers JHR, Clark JS, Beckage B (2002) Density-dependent mortality and the latitudinal gradient in species diversity. Nature 417:732–735

Lewis KJ, Lindgren BS (1999) Influence of decay fungi on species composition and size class structure in mature *Picea glauca x engelmannii* and *Abies lasiocarpa* in subboreal forests of central British Columbia. For Ecol Manage 123(2/3):135–143

Lewis KJ, Lindgren BS (2000) A conceptual model of biotic disturbance ecology in the central interior of B.C.: how forest management can turn Dr. Jekyll into Mr. Hyde. For Chron 76:433–443

Lindén M, Vollbrecht G (2002) Sensitivity of *Picea abies* to butt rot in pure stands and in mixed stands with *Pinus sylvestris* in southern Sweden. Silva Fenn 36:767–778

Lively CM (2001) Parasite-host interactions. In: Fox CW, Roff DA, Fairbairn DJ (eds) Evolutionary ecology. Concepts and case studies. Oxford Univ Press, Oxford, pp 290–302

Lonsdale D, Gibbs JN (1995) Effects of climate change on fungal diseases of trees. In: Frankland JE, Magan N, Gadd GM (eds) Fungi and environmental change. British Mycological Society Symp vol XX, pp 1–19

Loreau M, Naeem S, Inchausti P, Bengtsson J, Grime JP, Hector A, Hooper DU, Huston MA, Raffaelli D, Schmid B, Tilman D, Wardle DA (2001) Biodiversity and ecosystem functioning: current knowledge and future challenges. Science 294:804–808

Lundquist JE (1995) Pest interactions and canopy gaps in Ponderosa pine stands in the Black Hills, South-Dakota, USA. For Ecol Manage 74:37–48

Lygis V, Vasiliauskas R, Stenlid J, Vasiliauskas A (2001) Preliminary evaluation of Scots pine plantations "resistant" to *Heterobasidion annosum*. In: Laflamme G, Bérubé JA, Bussières G (eds) Proceedings of the 10th International conference on root and butt rots of forest trees, Québec City, Canada, September 16–22, 2001. Laurentian Forestry Centre, Sainte-Foy, Information Report LAU-X-126, pp 362–365

Malmström CM, Raffa KF (2000) Biotic disturbance agents in the boreal forest: considerations for vegetation change models. Global Change Biol 6 [Suppl 1]:35–48

Maloney PE, Rizzo DM (2002) Pathogens and insects in a pristine forest ecosystem: the Sierra San Pedro Martir, Baja, Mexico. Can J For Res 32:448–457

Mattila U, Jalkanen R, Nikula A (2001) The effects of forest structure and site characteristics on probability of pine twisting rust damage in young Scots pine stands. For Ecol Manage 142:89–97

May TW, Simpson JA (1997) Fungal diversity and ecology in eucalypt ecosystems. In: Williams JE, Woinarski JCZ (eds) Eucalypt ecology. Individuals to ecosystems. Cambridge Univ Press, Cambridge, pp 246–277

McCann KS (2002) The diversity-stability debate. Nature 405:228–233

McCauley KJ, Cook SA (1980) *Phellinus weirii* infestation of two mountain hemlock forests in the Oregon Cascades. For Sci 26:23–29

McCracken AR, Dawson WM (1997) Growing clonal mixtures of willow to reduce effect of *Melampsora epitea* var. *epitea*. Eur J For Pathol 27:319–329

McCracken AR, Dawson WM (1998) Short rotation coppice willow in Northern Ireland since 1973: development of the use of mixtures in the control of foliar rust (*Melampsora* spp.). Eur J For Pathol 28:241–250

McCracken AR, Dawson WM, Watson S, Allen CY (2000) Pathotype composition in *Melampsora epitea* populations occurring on willow (*Salix*) grown in mixed and monoculture plantations. Eur J Plant Pathol 106:879–886

McCracken AR, Dawson WM, Bowden G (2001) Yield responses of willow (*Salix*) grown in mixtures in short rotation coppice (SRC). Biomass Bioenerg 21:311–319

McDonald BA, Bellamy BK, Zhan J, Appel DN (1998) The effect of an oak wilt epidemic on the genetic structure of a Texas live oak population. Can J Bot 76:1900–1907

McDougall KL, Hobbs RJ, Hardy GES (2002) Vegetation of *Phytophthora cinnamomi*-infested and adjoining sites in the northern Jarrah (*Eucalyptus marginata*) forest of Western Australia. Aust J Bot 50:277–288

Mitchell CE, Tilman D, Groth JV (2002) Effects of grassland plant species diversity, abundance, and composition on foliar fungal disease. Ecology 83:1713–1726

Morrison DJ, Wallis GW, Weir LC (1988) Control of *Armillaria* and *Phellinus* root diseases: 20-year results from the Skimikin stump removal experiment. Canadian Forestry Service, Pacific Forestry Centre, Victoria, Information Report BC-X-302

Mosandl R, Aas G (1986) Vorkommen und Bedeutung von Keimlingspilzen im Bergmischwald der ostbayerischen Kalkalpen. Forst Holzwirt 41:471–475

Mundt CC (2002) Use of multiline cultivars and cultivar mixtures for disease management. Annu Rev Phytopathol 40:381–410

Murray DIL (1987) Rhizosphere microorganisms from the Jarrah forest of Western Australia and their effect on vegetative growth and sporulation in *Phytophthora cinnamomi* Rands. Aust J Bot 35:567–580

Naeem S (2002a) Biodiversity equals instability? Nature 416:23–24

Naeem S (2002b) Ecosystem consequences of biodiversity loss: the evolution of a paradigm. Ecology 83:1537–1552

Naeem S, Thompson LJ, Lawler SP, Lawton JH, Woodfin RM (1995) Empirical evidence that declining species diversity may alter the performance of terrestrial ecosystems. Philos Trans R Soc Lond 347:249–262

Naeem S, Loreau M, Inchausti P (2002) Biodiversity and ecosystem functioning: the emergence of a synthetic ecological framework. In: Loreau M, Naeem S, Inchausti P (eds) Biodiversity and ecosystem functioning – synthesis and perspectives. Oxford Univ Press, Oxford, pp 3–11

Newhook FJ, Podger FD (1972) The role of *Phytophthora cinnamomi* in Australian and New Zealand forests. Annu Rev Phytopathol 10:299–326

Newman EI (2000) Applied ecology and environmental management, 2nd edn. Blackwell Science, Oxford

Oak SW (2002) Native diseases and insects that impact oaks. In: McShea WJ, Healy WM (eds) Oak forest ecosystems. Ecology and management for wildlife. Johns Hopkins University Press, Baltimore, pp 80–99

Packer A, Clay K (2000) Soil pathogens and spatial patterns of seedling mortality in a temperate tree. Nature 404:278–281

Peace TR (1938) Butt rot of conifers in Great Britain. Q J For 32:81–104

Peacock L, Hunter T, Turner H, Brain P (2001) Does host genotype diversity affect the distribution of insect and disease damage in willow cropping systems? J Appl Ecol 38:1070–1081

Pei MH, Royle DJ, Hunter T (1996) Pathogenic specialization in *Melampsora* var. *epitea* on *Salix*. Plant Pathol 45:679–690

Perkins TE, Matlack GR (2002) Human-generated pattern in commercial forests of southern Mississippi and consequences for the spread of pests and pathogens. For Ecol Manage 157:143–154

Perry DA (1998) The scientific basis of forestry. Annu Rev Ecol Syst 29:435–466

Perry DA, Amaranthus MP (1997) Disturbance, recovery, and stability. In: Kohm KA, Franklin JF (eds) Creating a forestry for the 21st century. The science of ecosystem management. Island Press, Washington, DC, pp 31–56

Petchey OL, Gaston KJ (2002) Functional diversity (FD), species richness and community composition. Ecol Lett 5(3):402–411

Pfisterer AB, Schmid B (2002) Diversity-dependent production can decrease the stability of ecosystem functioning. Nature 416:84–86

Piri T, Korhonen K (2001) Infection of advance regeneration of Norway spruce by *Heterobasidion parviporum*. Can J For Res 31:937–942

Piri T, Korhonen K, Sairanen A (1990) Occurrence of *Heterobasidion annosum* in pure and mixed spruce stands in southern Finland. Scand J For Res 5:113–125

Pratt JE (1979a) *Fomes annosus* butt rot of Sitka spruce. I. Observations on the development of butt-rot in individual trees and stands. Forestry 52:11–29

Pratt JE (1979b) *Fomes annosus* butt rot of Sitka spruce: III. Losses in yield and value of timber in diseased trees and stands. Forestry 52:113–127

Prell HH, Day PR (2001) Plant-fungal pathogen interaction. A classical and molecular view. Springer, Berlin Heidelberg New York

Puddu A, Luisi N, Capretti P, Santini A (2003) Environmental factors related to damage by *Heterobasidion abietinum* in *Abies alba* forests in southern Italy. For Ecol Manage 180(1–3):37–44

Rajora OP, Mosseler A (2001) Challenges and opportunities for conservation of forest genetic resources. Euphytica 118:197–212

Ramstedt M (1999) Rust disease on willows – virulence variation and resistance breeding strategies. For Ecol Manage 121:101–111

Ramstedt M, Hurtado S, Åström B (2002) Pathotypes of *Melamspora* rust on *Salix* in short-rotation forestry plantations. Plant Pathol 51:185–190

Rao MR, Nair PKR, Ong CK (1997) Biophysical interactions in tropical agroforestry systems. Agroforest Syst 38:3–50

Redfern DB, Pratt JE, Gregory SC, MacAskill GA (2001) Natural infection of Sitka spruce thinning stumps in Britain by spores of *Heterobasidion annosum* and long-term survival of the fungus. Forestry 74:53–71

Rennerfelt E (1946) Om rotrötan (*Polyporus annosus* Fr.) I Sverige. Dess utbredning och sätt att uppträda. Meddel. Statens Skogsforskningsinst 35(8), 88 pp

Roll-Hansen F (1989) *Phacidium infestans*. A literature review. Eur J For Pathol 19:237–250

Roy BA, Kirchner JW (2000) Evolutionary dynamics of pathogen resistance and tolerance. Evolution 54:51–63

Roy BA, Kirchner JW, Christian CE, Rose LE (2000) High disease incidence and apparent disease tolerance in a North American Great Basin plant community. Evol Ecol 14:421–438

Schläpfer F, Schmid B (1999) Ecosystem effects of biodiversity: a classification of hypotheses and exploration of empirical results. Ecol Appl 9:893–912

Schmidt RA (1978) Diseases in forest ecosystems: the importance of functional diversity. In: Horsfall JG, Cowling EB (eds) Plant disease. An advanced treatise, vol II. How disease develops in populations. Academic Press, New York, pp 287–315

Schowalter T, Hansen E, Molina R, Zhang Y (1997) Integrating the ecological roles of phytophagous insects, plant pathogens, and mycorrhizae in managed forests. In: Kohm KA, Franklin JF (eds) Creating a forestry for the 21st century. The science of ecosystem management. Island Press, Washington, DC, pp 171–189

Schwadron PA (1995) Distribution and persistence of American chestnut sprouts, *Castanea dentata* [Marsh] Borkh., in northeastern Ohio woodlands. Ohio J Sci 95:281–288

Shea SR, McCormick J, Portlock CC (1979) The effects of fires on regeneration of leguminous species in the northern Jarrah (*Eucalyptus marginata*) forest of Western Australia. Aust J Ecol 4:195–105

Shearer BL, Dillon M (1995) Susceptibility of plant-species in *Eucalyptus marginata* forest to infection by *Phytophthora cinnamomi*. Aust J Bot 43:113–134

Shearer BL, Smith IW (2000) Diseases of eucalypts caused by soilborne species of *Phytophthora* and *Pythium*. In: Keane PJ, Kile GA, Podger FD, Brown NB (eds) Diseases and pathogens of eucalypts. CSIRO, Australia, pp 259–291

Shurtleff MC, Averre CW III (1997) Glossary of plant-pathological terms. APS Press, St Paul

Siepmann R (1984) Stammfäuleanteile in Fichtenreinbeständen und in Mischbeständen. Eur J For Pathol 14:234–240

Siitonen J (2001) Forest management, coarse woody debris and saproxylic organisms: fennoscandian boreal forests as an example. Ecol Bull 49:11–41

Simard SW (1998) Intensive management of young mixed forests: effects on forest health. In: Slurrock R (ed) 45th Western International Forest Disease Work Conference. Canadian Forest Service, University of Northern British Columbia, pp 48–54

Simard SW, Hannam KD (2000) Effects of thinning overstory paper birch on survival and growth of interior spruce in British Columbia: implications for reforestation policy and biodiversity. For Ecol Manage 129:237–251

Singh JS (2002) The biodiversity crisis: a multifaceted review. Curr Sci 82:638–647

Smith DM (2000) American chestnut: ill-fated monarch of the eastern hardwood forest. J For 98:12–15

Spies TA, Turner MG (1999) Dynamic forest mosaics. In: Hunter ML Jr (ed) Maintaining biodiversity in forest ecosystems. Cambridge Univ Press, Cambridge, pp 95–160

Stanosz GR, Patton RF (1987a) *Armillaria* root rot in Wisconsin aspen sucker stands. Can J For Res 17:995–1000

Stanosz GR, Patton RF (1987b) *Armillaria* root rot in aspen stands after repeated short rotations. Can J For Res 17:1001–1005

Stephenson SL (1986) Changes in a former chestnut-dominated forest after a half century of succession. Am Midl Nat 116:173–179

Stiell WM, Berry AB (1986) Productivity of short-rotation aspen stands. For Chron 62:10–15

Strong DR Jr, Levin DA (1975) Species richness of the parasitic fungi of British trees. Proc Natl Acad Sci USA 72:2116–2119

Tainter FH, Baker FA (1996) Principles of forest pathology. Wiley, New York

Thrall PH, Burdon JJ (2002) Evolution of gene-for-gene systems in metapopulations: the effect of spatial scale of host and pathogen dispersal. Plant Pathol 51:169–184
Tokeshi M (1999) Species coexistence. Ecological and evolutionary perspectives. Blackwell, Oxford
Tomita M, Hirabuki Y, Seiwa K (2002) Post-dispersal changes in the spatial distribution of *Fagus crenata* seeds. Ecology 83:1560–1565
Turner IM (2001) The ecology of trees in the tropical rain forest. Cambridge Univ Press, Cambridge
Twery MJ, Gottschalk KW (1996) Forest health: another fuzzy concept. J For 94(8):20
Van der Kamp BJ (1991) Pathogens as agents of diversity in forested landscapes. For Chron 67:353–354
Van der Maarel E (1993) Some remarks on disturbance and its relations to diversity and stability. J Veg Sci 4:733–736
Van der Pas JB (1981) A statistical appraisal of *Armillaria* root-rot in New Zealand plantations of *Pinus radiata*. NZ J For Sci 11:23–36
Van der Putten WH (2000) Pathogen-driven forest diversity. Nature 404:232–233
Van der Putten WH (2001) Interactions of plants, soil pathogens and their antagonists in natural ecosystems. In: Jeger MJ, Spence NJ (eds) Biotic interactions in plant-pathogen associations. CABI, New York, pp 285–305
Walchhütter T, Weste G, Guest D (2000) Regeneration after dieback due to *Phytophthora cinnamomi* – are suppressive soils involved? In: Hansen E, Sutton W (eds) *Phytophthora* diseases of forest trees. IUFRO Working Party. Oregon State University, Corvallis, pp 40–43
Wargo PM, Harrington TC (1991) Host stress and susceptibility. In: Shaw CG III, Kile GA (eds) *Armillaria* root disease. Agriculture handbook no 691. USDA, Washington, DC, pp 88–101
Webb CO, Peart DR (1999) Seedling density dependence promotes coexistence of Bornean rain forest trees. Ecology 80:2006–2017
Weste G, Brown K, Kennedy J, Walshe T (2002) *Phytophthora cinnamomi* infestation – a 24 year study of vegetation change in forests and woodlands of the Grampians, Western Victoria. Aust J Bot 50:247–274
Whittaker RJ, Willis KJ, Field R (2001) Scale and species richness: towards a general, hierarchical theory of species diversity. J Biogeogr 28:453–470
Wiens JA (2000) Ecological heterogeneity: an ontogeny of concepts and approaches. In: Hutchings MJ, John EA, Stewart AJA (eds) The ecological consequences of environmental heterogeneity. Blackwell, Oxford, pp 9–31
Wills C, Condit R, Foster RB, Hubbell SP (1997) Strong density- and diversity-related effects help to maintain tree species diversity in a neotropical forest. Proc Natl Acad Sci USA 94:1252–1257
Wilson BA, Aberton J, Cahill DM (2000) Relationships between site factors and distribution of *Phytophthora cinnamomi* in the Eastern Otway Ranges, Victoria. Aust J Bot 48:247–260
Wingfield MJ, Slippers B, Roux J, Wingfield BD (2001) Worldwide movement of exotic forest fungi, especially in the tropics and the Southern Hemisphere. Bioscience 51:134–140
Witzell J, Karlman M (2000) Importance of site type and tree species on disease incidence of *Gremmeniella abietina* in areas with a harsh climate in North Sweden. Scand J For Res 15:202–209
Wohlgemuth T, Bürgi M, Scheidegger C, Schütz M (2002) Dominance reduction of species through disturbance – a proposed management principle for central European forests. For Ecol Manage 166:1–15
Wolfe MS (2000) Crop strength through diversity. Nature 406:681–682

Wright SJ (2002) Plant diversity in tropical forests: a review of mechanisms of species coexistence. Oecologia 130:1–14

Yachi S, Loreau M (1999) Biodiversity and ecosystem productivity in a fluctuating environment: the insurance hypothesis. Proc Natl Acad Sci USA 96:57–64

Zeglen S (2002) Whitebark pine and white pine blister rust in British Columbia, Canada. Can J For Res 32:1265–1274

Zentmyer GA (1980) *Phytophthora cinnamomi* and the diseases it causes. St Paul, APS

Zhu Y, Chen H, Fan J, Wang Y, Li Y, Chen J, Fan JX, Yang S, Hu L, Leungk H, Mewk TW, Tengk PS, Wang Z, Mundtk CC (2000) Genetic diversity and disease control in rice. Nature 406:718–722

14 Implication of Forest Diversity in Resistance to Strong Winds

J.-F. Dhôte

14.1 Introduction

14.1.1 Scales of Interest and Associated Problems

The impact of strong winds on forest ecosystems is a complex phenomenon, involving many time and spatial scales.

- There are instantaneous impacts. At the tree level, storms affect survival, of course. At the stand level, storms may disrupt silvicultural plans by creating gaps of various sizes, from single-tree gaps to group or mass destruction (Quine et al. 1999). At the ownership level, windthrow affects various aspects of management planning, e.g., harvest schedule; income; sustainability of goals, including yield, soil protection, species or ecosystem conservation, and esthetics.
- There are also delayed effects (over the 10–30 years after storm), impacting further tree and stand dynamics. First, tree health may be affected: wind-induced wounds are entry points to various pathogens, which may start a process of canopy replacement in natural forests (tree death >gap creation >regeneration; Korpel 1995). Another concern for silviculturists is the affect on tree and wood quality: for example, in common beech (*Fagus sylvatica*), the "red heartwood" discoloration is known to depend on crown wind exposure. With the increasing devotion of European silviculturists to "single-tree management," the whole silvicultural system is based on the estimation of single trees' future expected value, and thinning efforts are largely concentrated to promote growth of a small number of high-quality trees. Windthrow is a specific risk in these systems. At the stand level, storms modify regeneration processes, species mix, and soil and amenity recovery is not immediate. At larger scales, resource imbalance, production losses, and market disturbance may need 15–20 years to recover after major events like the 1990 or 1999 storms in northern Europe.

Ecological Studies, Vol. 176
M. Scherer-Lorenzen, Ch. Körner, and E.-D. Schulze (Eds.)
Forest Diversity and Function: Temperate and Boreal Systems

When addressing forest diversity with regard to strong winds, many different issues can be considered:

- The resistance of trees: how does the physical stability of trees depend on species, tree dimension and shape, soil and site factors, canopy characteristics?
- The resilience of the ecosystem: how fast are ecological functions and processes restored; toward what new equilibrium do the dynamics lead after disturbances? similar to beforehand? qualitatively different? small differences in species proportions or soil processes?
- The resilience of the management (or production) system: how quickly can the sustainability criteria be restored (evenness of the regeneration area, distribution of timber supply by diameter classes, tree quality and species)?
- The optimization of management under risk: what is the impact of specified levels of risk on the optimal rotation age?; what thinning strategies are appropriate to reduce wind damages? is an uneven-aged system more profitable, or less at risk, than an age-class system?
- The various attitudes of forest owners toward risk: is risk to be avoided through adapted management practices, or confronted by taking into account storm damage probabilities in the design of silvicultural scenarios, or by choosing adequate economic insurance options (Gardiner and Quine 2000)?
- The social impact of technical solutions: what is the importance of amenities in silvicultural system conception?; what is the acceptability of increasing clear-felled areas and short-rotation forestry?; is there a specific concern for heavy windthrow in semi-natural forests or forest reserves?

In this chapter, I will only consider the physical stability of trees, i.e., resistance, as dependent on a small number of factors related to stand cover. In particular, neither the prominent role of wind disturbances in the natural dynamics of temperate lowland nor unmanaged montane forests will be discussed (for this, see e.g., Falinski 1986; Korpel 1995; Smejkal et al. 1995; Pontailler et al. 1997; Emborg et al. 2000). In the context of these "natural forests", one may consider that tree stability is not a "problem" per se, since windthrow is just a stimulus that contributes to forest dynamics. It is when speaking about managed forests that windthrow is regarded as a risk. Sometimes, on the other hand, it may also be an opportunity: for example, Norway spruce plantations on labile soils are highly susceptible, and the restoration of such forests after heavy storm damages can serve as an opportunity to adopt more appropriate management practices.

14.1.2 Different Aspects of Forest Diversity

How shall we define forest diversity as related to strong wind impact? Not only species richness or composition is important, but more generally forest stand structure should be considered (Schütz 1990). Stand structure, in turn, is very diverse:

- Single-layer mixtures: one main vegetation story comprising several species. For example, beech–spruce mixtures are discussed by Lüpke and Spellmann (1997).
- Vertical stratification: different stories made up of different species. For example, many oak forests develop into complex, stratified systems, with oak dominating the upper story, while shade-tolerant (or semi-tolerant) species grow in the understory (*Fagus, Carpinus, Fraxinus, Acer, Tilia*). A vertical stratification can also be found in pure stands with different age-related cohorts.
- Size (or age) unevenness within stories: trees belonging to the same vegetation story may be almost even aged, or spread over a large range of ages. For example, in many regions of France, oak-based coppices with standards and oak high forests have similar vertical structures and species composition (main story consisting of oak, understory consisting of other species), but they differ mainly in age evenness in each story; oak standards are sparsely distributed and largely of uneven age. These age-related factors imply different levels of canopy roughness (Schütz et al. 2001).
- Horizontal diversity, involving the spatial distribution of different species and size (or age) classes: the gradient here is from intimate, single-tree association of all sizes and species (selection forests, *Plenterwald, futaie jardinée*), through patch or group associations (100–1,000 m^2), to regular canopies over large areas (1 ha or more).

Due to limited material for analysis, this chapter focuses on single-layer mixtures, selection forests, and coppice-with-standard systems. Is there a significantly better resistance to strong winds by trees in these systems, compared to the standard situation of pure, even-aged forests?

14.1.3 Factors Contributing to Tree Stability

The effect of storms on tree stability involves many interacting factors (Savill 1983; Bouchon 1987; Ruel 1995), which may be classified into five categories.

- Climatic factors. Many analyses of storm damages have stressed the importance of wind speed and gustiness (Gardiner and Quine 2000; Schütz et al. 2001), duration of the event, wind orientation (compared to the dominant wind direction), difference between winter storms and summer thunderstorms. Most European statistics refer to winter storms, to which evergreen conifers are most susceptible (see below). Also, the amount of rainfall in the preceding weeks, as well as additional snow, increases storm damages (Bock and Duplat 2001).

- Site factors. Soil depth, texture, and water content on the one hand; topography, elevation, and wind exposition on the other hand, are major risk factors (Miller 1985; Gardiner and Quine 2000).
- Species traits. Both root characteristics (such as resistance to uprooting, depth, type of rooting, root-ball weight, root health) and stem mechanical properties and defects (such as resistance to breaking and leafy state) are determinants of tree stability.
- Tree size. Tree height (lever arm), crown-exposed area (wind drag force), and bole taper (resistance to breaking) are important (Bock and Duplat 2001).
- Silviculture and stand structural effects. Canopy roughness (existence of recent thinnings, size evenness, presence of internal or external edges) and the opposition between single-tree stability and mass stability (Schütz et al. 2001) have to be considered. However, also soil preparation techniques, choice of genetic material (selection for vigor generally impacts root/shoot proportions), and fertilization (which affects root/shoot ratios) influence tree stability.

The first three categories of factors are general, and not linked to forest diversity itself, although one may adapt species choice and silvicultural methods to minimize risk under some specific climatic or site conditions. This is the case, for example, with the so-called Atlantic silviculture of conifer plantations in Great Britain (Savill 1983; Gardiner and Quine 2000; Quine 2000).

Tree size is directly influenced by silviculture. For example, coppicing broad-leaved trees implies that total height never reaches a stage where wind risk becomes significant. Similarly, standards in coppice-with-standards systems are definitely different from high forest trees (larger exposed area, but smaller and more robust stems).

Even the pure, even-aged stands in the age-class system are not homogeneous with respect to the stability criteria mentioned above. Intensively managed conifer plantations are often based on large spacings, heavy early thinnings, artificial pruning, and early rotation age, any or all of which factors may somehow compensate for the higher windthrow susceptibility. This contrasts with the extensive silviculture of broad-leaved trees (oaks, common beech), with very long rotations and light thinnings.

As a consequence, when trying to estimate the impact of forest diversity on tree stability, one has to separate (1) the effects of tree size from (2) the specific role of stand structure and composition (canopy characteristics). It is also important to consider a number of confounding factors, and to try to filter them out (Ford 1978).

14.1.4 Randomness of Wind Damage and Analysis Methods

We have mentioned that tree stability is influenced by several factors. However, storm damage remains merely a random event. Hence, there are no purely deterministic modeling approaches of tree and stand stability. For example, *ForestGALES* (Gardiner and Quine 2000) is a "hybrid model", combining deterministic components (e.g., critical loads to uproot or break a tree, measured by pulling experiments) and random components (statistical models of wind speed occurring at regional and local levels).

The notion of *storm return interval* (the average number of years between two storms on the same site, i.e., the reverse of storm annual probability) is essential in many forestry applications. For example, Schütz et al. (2001) argue that, because the return interval of major storms (200 to 300 years) is larger than the rotation ages of most species, forest managers should not shorten rotations. This statement is largely disputed, especially in France, but the important thing here is that the probabilistic features of storms have to be taken into account in forest management.

Another methodological consequence of this randomness is that most of our knowledge issues not from controlled-conditions experimentation, but from analysis of historical damage reports. A number of such reports are now available, and may be classified into three categories:

- Large-scale statistics, computed at the country, state or regional levels (typically 10^6 ha; Laiho 1987; Schmidt-Haas and Bachofen 1991; Anonymous 1994; Hesse 1994; König et al. 1995; Schreiner et al. 1996; Piton 2002).
- Forest impact studies (10^4 ha): summaries of damage analyses in a whole forest estate, without control of silviculture, or site–species interactions (Kuner 1967; Rodenwald 1973; Heinrich 1991; König 1995).
- Local impact studies: summaries of damage reports on existing networks of permanent plots with controlled silviculture and/or site conditions (Lüpke and Spellmann 1997).

This scientific situation creates specific opportunities and problems, which must not be overlooked, since it may account for some contradictions or uncertainties in the literature. When working with permanent plots with known silviculture, usually the number of plots is too low for accurately estimating a low probability risk (1 to 5 %). Or the phenomenon is locally so heterogeneous in space that the response is saturated. For example, all plots in a thinning trial have been swept down, whether stable or not, since wind speed was simply too high on that stand; but a similar trial, 10 km distant, escaped devastating wind gusts, and is still intact. In this case, it is impossible to draw conclusions on the relationship between thinning methods and wind risk.

Region-wide statistics theoretically provide the large data bases needed to estimate probabilities. However, problems still arise with such data due to confounding factors: species are not randomly distributed, with regard to the wind-risk influencing factors; silvicultural techniques have quickly evolved in time, and are thus statistically linked to stand age (Ford 1978); some kinds of stand structure occupy a very small part of the existing forest area (e.g., selection forests in Germany or France with only 1 to 5 % of the area), and hence are more likely to have been exposed to different levels of wind speed as compared to the remaining stands. As a first conclusion, assessing the importance of several factors in the susceptibility to wind damage is a difficult task, and needs thorough knowledge of forest resources, ecology, forest practices, sampling strategies, and statistical methods.

In the remaining part of the chapter, I shall:

- recall the main factors of tree and stand stability and examine additional information provided by the ongoing analyses of the December 1999 storms in France ("Lothar" and "Martin");
- analyze whether unstable species (like Norway Spruce, *Picea abies* Karst.) are improved when mixed with more stable species, based on the German experience after the 1990 storms;
- analyze tree stability in more complex canopies such as selection forests or coppice-with-standards.

14.2 Species-Specific Susceptibility to Wind Damage

The range of species susceptibility is quite large. After the 1987 storm in the Massif Central (mountainous region of central France, wind blowing from the south, whereas the prevailing winds are from the west), the following qualitative ranking of conifers was established, from the lowest to the highest risk (Bouchon 1987):

European larch <Corsican pine <Douglas fir <Sitka spruce <European fir <Scots pine <Norway spruce.

Statistics for the 1982–1994 period from the Finnish National Forest Inventory were computed by Jalkanen and Mattila (2000). The following figures represent the percentage of plots in northern Finland showing signs of wind damage (whatever the severity of damage):

Birch spp.: 0.5 % <Scots pine: 2.2 % <Norway spruce: 4.8 %

State-level analyses in three German states were computed after the 1990 storms and compared by Lüpke and Spellmann (1997). The figures in Table 14.1 are based on damage volume (percent of the existing resource), and values have been normalized, with 1 being the most stable species.

However, there are some contradictions regarding species ranking among different studies. In the same German statistics, European fir (*Abies alba* Mill.) is systematically two to three times more stable than Norway spruce, in contrast to the results of Polge (1960). Furthermore, although Scots pine is two to three times more stable than Norway spruce in the above-mentioned

Table 14.1. Relative sensitivity of four important species in Germany during the 1990 storm. From Lüpke and Spellmann (1997). Numbers are relative to oak damage (oak = 1)

Species	Land		
	Bayern	Baden-Württemberg	Hessen
Norway spruce (*Picea abies* Karst.)	8	12	10
Scots pine (*Pinus sylvestris* L.)	3.5	4	4
Common beech (*Fagus sylvatica* L.)	2	3	2.5
Sessile oak (*Quercus petraea* Liebl.)	1	1	1

figures from Finland and Germany (see also Schreiner et al. 1996), both species had similar susceptibilities in the 1999 French statistics (Renaud 2001; Piton 2002).

Recent analyses have assessed wind damages by logistic regression, using a number of predictors (mainly wind speed, stand height, proximity of a thinning, site factors, rooting depth). These results have shed new light on the frequent contradictions in the literature regarding species ranking. Canham et al. (2001) presented logistic curves of windthrow probability for six broadleaved and one conifer species: species-specific risk is largely diverging at intermediate storm severity, the shape of susceptibility curves is quite different, and the relative ranking of species may change from intermediate to severe storms.

Renaud (2001) reported that the proportion of windthrown trees decreases sharply with rooting depth in Norway spruce and common beech, but much less in Scots pine and European fir. However, this study was based on 321 French plots of the European Network for Forest Damage Monitoring, with a very large geographical distribution, and the author suspected that there might be "hidden variables" contributing to this differential species response to rooting depth.

Piton (2002) applied the logistic regression method to four species in a small region, the Département du Haut-Rhin (Alsace valley and Vosges mountains), using field observations on 1,029 plots of the National Forest Inventory. Damages in fir and Norway spruce were related to wind speed, dominant height, recent thinning intensity, and stand exposition (west/east), whereas Scots pine responded only to thinning, common beech to height and exposition.

Due to the synthesis of several sensitivity factors in the logistic regression approach, these recent results carry a more complete evaluation of species sensitivity, e.g., by highlighting that species react differently to factors like height or rooting depth. For example, the decision to promote spruce or beech should consider rooting depth, so that some site types are avoided, or that silvicultural interventions are adopted to minimize stand height on these sites.

14.3 Effect of Location, Developmental Stage, and Canopy Closure

14.3.1 Increased Risk on Labile Soils and Exposed Locations

Shallow "physiological depth" (i.e., the effective depth explored by root systems) systematically implies high damage proportion (König 1995). In the study by Renaud (2001), damage proportion is reduced by a factor of 3 when rooting depth increases from 20 to 60 cm, for Norway spruce and common beech. The analysis of Bock and Duplat (2001) was based upon a 145-plot sample in common beech high forests, evenly distributed according to soil types in northeastern France. They reported three main factors of susceptibility: wind speed, dominant height, and soil depth. Shallow soils (less than 50 cm) had almost the same sensitivity, whether the limit was set by bedrock or a compact clay horizon. Deep soils (more than 50 cm) were 1.5 to 3 times more stable.

In the mountainous regions, conifer damage was clearly linked to exposition and proximity to the nearest edge (Jacquemin 2001). As reported by Piton (2002), 30-m-high spruce and fir stands were eight times more at risk on western than on eastern slopes (the 1999 storm came from the west).

14.3.2 Increasing Risk with Stand (or Tree) Height

Generally, the proportion of wind damage increases sharply, and non-linearly with stand height. In the logistic regression analysis of König (1995), four factors were found significant: height, site quality (soil "physiological depth"), presence of a thinning in the last 5 years, and wind speed.

Bock and Duplat (2001) showed that the sensitivity to wind damage increased abruptly with height above the threshold value of 23 m. According to Renaud (2001), the proportion of damage in the 100–120 km/h wind-speed class increased from 0.4 to 0.8 in European fir when dominant height increased from 20 to 35 m; damage ro common beech increased from 0.15 to 0.65.

In some studies, height is replaced by substitutes like stand age or dominant diameter (Jalkanen and Mattila 2000; Jacquemin 2001). Damage proportion usually increases strongly from young stands to pole stage and mature stands. Except in the study of Jalkanen and Mattila (2000), there seems to be a difference between pines and other conifer species (spruce, Douglas fir, fir). In the stands analyzed by König et al. (1995), 20–40 % of Scots pine damage occurs in the 20–60 year age classes, but 11–13 % for Norway spruce. In contrast, Piton (2002) showed that there is no significant effect of age or height on the proportion of damage in northeastern Scots pine or southwestern maritime pine (*Pinus pinaster*).

14.3.3 Increasing Risk with Recent Canopy Opening

All reviewed studies consistently report of an increasing risk with a recent canopy opening. In König et al. (1995), 44 % of the damage is located in open stands, 46 % in stands with light cover, 10 % in closed canopies. Piton (2002) considered not only the presence/absence of thinning in the five preceding years, but also the percentage of volume removed. This variable was significant in four out of five different species analyzed in three remote areas of France. Maritime pine stands with 30 % volume recently removed were twofold more at risk than unthinned stands. In European fir stands analyzed by Jacquemin (2001), both the duration since the last thinning and the proportion of volume removed were found significant; altogether, these two factors explain 15 % of the damage variability.

As a partial conclusion, it appears that

- the risk is considerably higher on shallow soils and in exposed locations;
- almost all species exhibit a sharp and nonlinear increase in risk as stand height increases; however, several independent observations from different regions in Europe suggest that pines do not behave as the other conifers, showing a more even distribution of damage between height classes; this deserves more attention;
- the risk is considerably higher during the 3–5 years after canopy opening (thinning), and seems to depend also on the percentage of volume removed (when more volume has been removed, the stand remains unstable for a longer period). Thinnings temporarily increase canopy roughness, which is a drawback in even-aged stands (in these stands, individual trees are not very stable by themselves, but the stand may be stable, or metastable, due to the dispersion of the wind energy by intercrown contacts).

14.4 Analyzing Stability in Complex Forest Structures

To study the specific role of forest diversity, two methods have been applied:

- Comparison of situations which are reasonably similar with regard to the factors described above, and which differ only by forest structure or mixture. In this case, statistics compiled from many forest stands are directly compared.
- Inclusion of forest structure or mixture factors into a general modeling approach, taking all other factors that might influence stand stability into account. Usually, logistic regressions are applied in this approach.

The second method has received more interest in the recent analyses.

14.4.1 Stability of Species in Single-Layered Mixtures

This part of the review focuses mainly on approximate single-layered mixtures (as far as this can be assessed in the reviewed papers), and on the issue of conifer–broadleaf mixtures. The general idea is that more sensitive species like evergreen conifers might be stabilized when mixed with less sensitive species.

Three large-scale studies investigated the impact of conifer dominance. For the 1990 storm in Switzerland, Schmidt-Haas and Bachofen (1991) found that stands with more than 90 % conifers were largely over-represented in the population of stands having single-tree damage, compared to their proportion in the whole forested land. The study by Zindel (1991) reported that Norway spruce is stabilized when mixed in small groups inside common beech stands (only single-tree damage is mentioned). In the Finnish National Forest Inventory statistics for the 1982–1994 period (Jalkanen and Mattila 2000), there was no clear effect of conifer dominance, once location (climate), age, and diameter were entered into the logistic model. The authors stressed that conifer dominance is largely confounded with elevation in the Finnish data, and elevation in turn is associated with faster winds.

The statistics for Bavaria in Germany in 1990 (König et al. 1995) allow a more thorough examination of damage, since a classification into stand types was available (Table 14.2).

In terms of susceptibility, only the third column of Table 14.2 needs to be considered. Here, it is obvious that species identity is much more important than the difference between pure and mixed stands. Pure stands of Scots pine

Table 14.2. Distribution of damage according to stand type from the 1990 storm in Bavaria (König et al. 1995). Stand types: main species first, then secondary species. Statistics for heavily damaged stands (*Flächenwürfe*)

Stand type	Percentage of this type in the total damaged area	Percentage of damaged area in the area of this type
Norway spruce/Scots pine	27.4	5.7
Pure Norway spruce	39.8	3.2
Scots pine/Norway spruce	10.1	1.8
Norway spruce/broad-leaved species	15.8	1.6
Common beech/other broad-leaved species	3.8	0.6
Pure Scots pine	2.1	0.5
Scots pine/broad-leaved species	0.5	0.3
Pure oaks	0.6	0.2

or oaks were quite stable, and all stands based on Norway spruce were unstable, whether pure or mixed. The good stability of Scots pine does not seem to be profitable to Norway spruce. The authors discussed two hypotheses that could account for the bad performance of pine/spruce mixtures: (1) a concentration of these stand types on labile soils (high water table); (2) a difference in stand classification between the damage assessment campaign and the former forest inventory. Lüpke and Spellmann (1997) commented on this study and added that damage in mixtures of species with different susceptibilities occurs more often than single-tree or small-group windthrow, which do not appear in the Table 14.2 data.

On a smaller-scale, Kuner (1967) examined damage in a 6,000-ha forest in Switzerland, with deep brown soils on moraines. Only 1:1 group mixtures of broad-leaved trees and conifers were studied (European fir, Norway spruce, Douglas fir). Heavy damage was recorded in conifers. Lüpke and Spellmann (1997) reviewed a series of reports with control of site conditions. Spruce stability did not differ between pure and mixed stands; the proportion of damage mainly depended on the proportion of spruce. Clearly, site conditions were the most important factors. Schütz (1990) claimed that common beech develops deeper root systems in mixtures, but Lüpke and Spellmann (1997) considered that root anchorage of Norway spruce in mixtures is not significantly improved. The question of root systems in pure vs. mixed stands, and its effect on both stability and nutrition, certainly merits more attention.

The last element to take into account for wind risk in mixtures is related to the different competitiveness and growth patterns of constituent species. Beech/spruce mixtures have been intensively studied in central Europe, especially in Germany. Spruce outgrows beech in situations where temperature is lower, soil more acid, and water supply more favorable (Kramer 1988). Because these environmental conditions change from northern Germany to the plateaus of Switzerland and northeastern France, the canopy structure of these mixtures changes, too (Lüpke and Spellmann 1997). It turns out that spruce is much more at risk where it is relatively more productive. Rapid height growth is a twofold drawback, since height is an absolute risk factor and, furthermore, a conifer outgrowing a broad-leaved species develops more exposed crowns.

In conclusion, although the available material is limited, sometimes difficult to interpret, and with contradictions among sources, it does not seem that the physical stability of sensitive species is improved when mixed with stable species. However, this does not mean that mixing is useless: mixing seems to prevent massive damage, and implies more single-tree windthrow, so that the canopy cover is not completely disrupted by storms, which is favorable in many aspects (organization of logging, soil protection, light attenuation, and effect on regeneration processes).

14.4.2 Stability of Selection Versus Even-Aged Forests

In this case, I tried to summarize a series of factors that could theoretically contribute to a differential stability of trees in finely irregular (selection system) versus regular forest stands. These factors are compiled in Table 14.3.

From this summary, one cannot assign a clear superiority of one system over the other. Advantages and drawbacks are distributed – with a slight qualitative advantage for selection forests – and the economic resilience is largely unpredictable (adverse effects may compensate for beneficial ones). Hence, qualitative rationales are not sufficient, and quantitative elements are necessary. Two types of arguments may be used: (1) simulation approaches, combining forest growth, fluid mechanics, and tree biomechanics models; (2) analyses of damage reports.

Simulation models are not nearly mature enough for addressing such complex problems. Research groups involved in this field have recently started research programs for even-aged situations (Gardiner et al. 2000; Lee 2000).

Damage reports are also difficult to analyze in this respect. In Europe, a very low fraction of the forest area is actually treated by the selection method: 2 % in Germany, 5 % in France, 8 % in Switzerland (Schütz 1999). It is remarkable that even Switzerland, which has a long-lasting commitment to selection

Table 14.3. Theoretical advantages (+) and drawbacks (–) of even-aged vs. selection forests with regard to different criteria of wind stability

Category	Factor	Even-aged forest	Selection forest
Canopy effects	Canopy roughness	+	–
	Mass stability[a]	+	–
	Edge frequency[b]	–	+
Tree stability	Crown-exposed area	+	–
	Lever arm	–	+
	Bole taper	–	+
	Single-tree acclimation to wind	–	+
Recovery dynamics	Advance regeneration	–	+
	Systemic damage of mature trees on poles[c]	+	–
System resilience	Ecological resilience	–	+
	Economical resilience	+/– ?	+/– ?

[a] Wind energy is dissipated by mechanical contacts between swinging crowns of similar heights

[b] Edges are produced by the mosaic of age classes

[c] In selection forests, windthrow of mature trees may break high-quality poles in the understory, whereas these categories are spatially separated in age-class forests

forestry, was not completely successful in this effort. Reasons for this were analyzed by Schütz (1999). Whereas the selection system works well in the conifer/beech mixtures of mountainous sites (Jura, pre-Alps), the effort to extend the method to lowland or piedmont broad-leaved forests allowed shade-tolerant species (beech and fir) to dominate in the regeneration story, even on sites where they were not desirable species. It is probably necessary, for other broad-leaved species to survive, to maintain quite low levels of standing volume.

A second difficulty lies in the comparison method. Since selection and even-aged systems are management options and not only silvicultural practices, comparative analyses should consider (1) selection forests and (2) a balanced, whole range of ages in similar even-aged forests. Populations of stands for comparison should also be reasonably similar with respect to site conditions and species.

This lack of similarity may explain why case studies are hardly conclusive (see Dvorak and Bachmann 2001 for a recent analysis). Experienced silviculturists like Schütz therefore do not insist so much on wind stability, among various arguments for adopting selection forestry (Schütz et al. 2001).

14.4.3 Stability of Coppice-with-Standards Versus High Forest Trees

Another form of uneven-aged stands is provided by the coppice-with-standards method. Although this method is partly artificial, it produces very irregular canopies, and is still largely widespread in many regions of France (26 % of the forest area).

Piton (2002) analyzed 720 plots from the French National Forest Inventory, in high forests and coppice-with-standards dominated by oaks (493 *Quercus petraea* and 227 *Q. robur*) in the Département de l'Yonne (central France). Storm damage in 1999 was explained by a logistic model comprising three factors: wind speed, dominant height, and the coefficient of variation of tree heights. For wind speed of ca. 160 km/h, the proportion of damage increased from 0.15–0.2 to 0.45–0.55 when dominant height increased from 20 to 30 m. An interesting result was the role of height unevenness. When the coefficient of variation of individual heights increased from 10 to 20 %, the proportion of damage increased by 20–35 %. In this case, more irregular canopies seem to be more at risk, at a given height.

Bock and Duplat (2001) analyzed 78 plots in stands dominated by common beech in a small region of Lorraine (Plateau de Haye). Sampling was strictly controlled for site conditions and balanced between coppice-with-standards and high forests. During the 1999 storm, the climatic conditions in this small region were extreme: 140–160 km/h winds, rainfall in December 1999 twice the 30-year mean. The probability of a tree being damaged was modeled by logistic regression, with factors soil depth, crown radius (exposed area), and mean crown height (lever arm of drag forces). Interestingly, this study showed that:

- Populations of trees from coppice-with-standards or high forests are almost completely separated (the former having smaller heights and larger crowns).
- Individual tree windthrow probability combines crown radius and crown height, which compensate for each other. Both populations are equally at risk (probability between 0.3 and 0.8), although for different reasons: coppice-with-standards trees because they have a large exposed area, high forest trees because they are high and slender.
- However, due to the extreme conditions, it is still possible that intermediate-severity storms might reveal some differences that are not apparent here.

For the same reason as in the analysis of selection forests, these raw results should not be interpreted alone. Coppice-with-standard systems need to be compared to high forest systems over appropriate time intervals. Since the former have lower height growth, but faster diameter growth, commercial maturity is reached sooner, and stands spend less time in sensitive situations (driven by height). Hence, a complete comparison of the management systems should include growth curves, canopy roughness, and rotation ages in risk and benefit assessments.

14.5 Conclusions

From a critical survey of damage reports and of theoretical and applied modeling approaches, the most important and general factors controlling tree response to strong winds are found to be climatic conditions (wind speed and gustiness, rainfall and snow), site characteristics (rooting depth, soil moisture and texture, topography, exposition), tree size (height, crown exposed area, slenderness), and canopy roughness (time elapsed since and intensity of thinnings, height irregularity, edge frequency).

Empirically, species exhibit a very large range of susceptibility to wind damage. Evergreen conifers are, on average, more endangered by winter storms than broad-leaved trees. However, there are large differences within these groups. The relative contribution of crown architecture, foliage distribution, stem mechanical properties, and root systems to this species-specific susceptibility is not yet fully understood. However, most modeling approaches reveal that species respond in quite different ways to similar risk factors. Combined with the fact that forest management and climate–ecological conditions are largely variable over Europe, this may account in part for the contradictions in the literature. For example, in central Europe and Scandinavia, Norway spruce is an abundant species (whether natural or planted) and occupies a large range of sites, including unstable situations (shallow soils, high water tables); in contrast, in France, it is confined to the eastern mountain sites (Vosges, Jura, Alps). Hence, spruce sensitivity appears more pronounced in German or Scandinavian studies, as does the better resistance of Scots pine.

In addition to these general factors, there is no clear improvement in tree stability in single-layered mixtures or in complex, uneven-aged canopies (selection forests, coppice-with-standards), in comparison to even-aged monocultures. For wind resistance, species identity matters more than just tree species richness or forest structural diversity. However, a comprehensive evaluation of management systems with respect to wind damage has not yet been achieved. Such a study would need to consider not only the physical stability of individual trees, but a number of other factors influenced by storms, or by management itself:

- The persistence of a partial cover after storms is an advantage of mixtures or uneven-aged stands. In the former, regeneration of shade-tolerant (or intermediate) species may start in better conditions under partial cover; in the latter, a continuous regeneration flow is inherent to the system, and contributes to a faster recovery after disturbance. The price to pay for these advantages is either to restrain the development of light-demanding species, or to induce secondary damage to the advance regeneration (windthrow of mature trees breaking poles).
- Management systems based on low levels of growing stocks (broad-leaved uneven-aged forests, e.g., coppice-with-standards or selection forests) highly modify the morphology and growth rates of individual trees, compared to dense high forests. Although wind risk seems to increase with canopy roughness, height growth is somehow slower, and diameter reaches maturity more rapidly. Trees here can be expected to spend a smaller fraction of their life span in situations at risk, which may be an important advantage for large-scale forest owners.
- A comprehensive evaluation of alternative management options can be envisaged, provided that appropriate simulation models are made available. For research purposes, mechanistic models of soil–canopy–wind interactions could bring some insight into the physical issues of wind sensitivity. For management purposes, forest growth and yield simulators, including risk and microeconomics components, could be used to calculate risk-minimizing silvicultural options, or to evaluate optimal strategies according to various owners' attitudes toward risk.

References

Anonymous (1994) Dokumentation des Sturmschäden 1990. Schrift der LFV Bd Württemberg, Freiburg, 190 pp

Bock J, Duplat P (2001) Influence des paramètres sylvicoles et stationnels sur l'intensité des dégâts provoqués par la tempête du 26 décembre 1999 dans les hêtraies du quart Nord-Est de la France. Rapport Contrat ONF-ECOFOR no 2000–39, ed. Office National des Forêts, Fontainebleau (France), 60 pp

Bouchon J (1987) Etat de la recherche relative aux dégâts forestiers dûs aux tempêtes. Rev For Fr 39:301–312

Canham CD, Papaik MJ, Latty EF (2001) Interspecific variation in susceptibility to windthrow as a function of tree size and storm severity for northern temperate tree species. Can J For Res 31:1–10

Dvorak L, Bachmann P (2001) Sturmschäden in ungleichförmigen Beständen in Abhängigkeit vom Bestandesaufbau. ETHZ, Professur Forsteinrichtung und Waldwachstum, Zürich (Switzerland), 87 pp

Emborg J, Christensen M, Heilmann-Clausen J (2000) The structural dynamics of Suserup Skov, a near-natural temperate deciduous forest in Denmark. For Ecol Man 126:173–189

Falinski JB (1986) Vegetation dynamics in temperate lowland primeval forests. Ecological studies in Bialowieza forest. Junk, Dordrecht, 537 pp

Ford ED (1978) An ecological basis for predicting the growth and stability of plantation forests. In: Ford ED, Malcolm DC, Atterson J (eds) The ecology of even-aged forest plantations. Forestry Comm, Cambridge, UK, pp 147–174

Gardiner BA, Quine CP (2000) Management of forests to reduce the risk of abiotic damage – a review with particular reference to the effects of strong winds. For Ecol Manage 135:261–277

Gardiner BA, Peltola H, Kellomäki S (2000) Comparison of two models for predicting the critical wind speeds required to damage coniferous trees. Ecol Modelling 1:1–23

Heinrich J (1991) Ursachen von Sturmschäden in Buchenbeständen (*Fagus sylvatica*) aus bodengeographischer Sicht. Allg Forst J Z 162:145–149

Hesse (1994) Orkanschäden 1990 und ihre Bewältigung bis 1993. Dokumentation. Hess Minist für Landesentw, Wohnen, Landw, Forsten und Natursch, 117 pp

Jacquemin B (2001) Stabilité du Sapin en Auvergne face aux vents forts. Rapport Interne Office National des Forêts, Clermont-Ferrand (France), 41 pp

Jalkanen A, Mattila U (2000) Logistic regression models for wind and snow damage in northern Finland based on the National Forest Inventory data. For Ecol Manage 135:315–330

König A (1995) Sturmgefährdung von Beständen im Altersklassenwald. Sauerländer, Frankfurt/Main, 194 pp

König A, Mössmer R, Bäumler A (1995) Waldbauliche Dokumentation der flächigen Sturmschäden des Frühjahrs 1990 in Bayern und meteorologische Situation zur Schadenszeit. Berichte aus der Bayerischen Landesanstalt für Wald und Forstwirtschaft, 2, Munich, 396 pp

Korpel S (1995) Die Urwälder der Westkarpaten. Fischer, Stuttgart, 310 pp

Kramer H (1988) Waldwachstumlehre. Parey, Hamburg, 374 pp

Kuner M (1967) Die Windstürme der letzten Jahre. J For Suisse 7:435–445

Laiho O (1987) Susceptibility of forest stands to windthrow in southern Finland. Folia For 706:23–24

Lee X (2000) Air motion within and above forest vegetation in non-ideal conditions. For Ecol Manage 135:3–18

Lüpke B, Spellmann H (1997) Aspekte der Stabilität und des Wachstums von Mischbeständen aus Fichte und Buche als Grundlage für waldbauliche Entscheidungen. Forstarchiv 68:167–179

Miller KF (1985) Windthrow hazard classification. For Comm Booklets 85:14

Piton B (2002) Facteurs de sensibilité au vent des peuplements. Etude à partir des données de l'IFN relatives aux tempêtes de décembre 1999. Mémoire de fin d'études FIF. ENGREF, Nancy (France), 153 pp

Polge H (1960) Sensibilité relative du sapin pectiné et de l'épicéa commun aux coups de vent. Rev For Fr 10:637–642

Pontailler JY, Faille A, Lemée G (1997) Storms drive successional dynamics in natural forests: a case study in Fontainebleau forest (France). For Ecol Manage 98:1–15

Quine CP (2000) Estimation of mean wind climate and probability of strong winds for wind risk assessment. Forestry 73:249–257

Quine CP, Humphrey JW, Ferris R (1999) Should the wind disturbance patterns observed in natural forests be mimicked in planted forests in the British uplands? Forestry 72:337–358

Renaud JP (2001) Première évaluation de la sensibilité des peuplements forestiers aux tempêtes, à partir des dommages subis par le Réseau Européen. Les Cahiers du DSF, Paris, pp 1–4

Rodenwald U (1973) Eine reviergeschichtliche Studie. Allg Forst J Z 144:125

Ruel JC (1995) Understanding windthrow: silvicultural implications. For Chron 71:434–445

Savill PS (1983) Silviculture in windy climates. For Abstr 44:473–488

Schmidt-Haas P, Bachofen M (1991) Die Sturmgefährdung von Einzelbäumen und Beständen. J For Suisse 142:477–504

Schreiner M, Aldinger E, Bantle P (1996). Standort und Sturmwurf 1990 – dargestellt am Östlichen Odenwald und Nordöstlichen Schwarzwald. Mitt Ver Forst Standortskd Forstpflanzenztg 38:27–36

Schütz JP (1990) Sylviculture 1. Principes d'éducation des forêts. Pr Polytechn Univ Romandes, Lausanne (Switzerland), 243 pp

Schütz JP (1999) Close-to-nature silviculture : is this concept compatible with species diversity? Forestry 72:359–366

Schütz JP, Ranger J, Guinard D, Birot Y, Dhôte JF, Jacquemin B, Planchais I, Pilard-Landeau B, Bastien Y, Lacombe E (2001) Analyse et propositions concernant la stabilité sylvicole et écologique des forêts face à des tempêtes de type Lothar. Rapport au Cons Scient de l'ONF, 14/8/2001 Office National des Forêts, Paris (France), 39 pp

Smejkal GM, Bindiu C, Visoiu-Smejkal DV (1995) Banater Urwälder. Mirton, Temeswar (Roumanie), 198 pp

Zindel U (1991) Die Sturmschäden in den hessischen Forstämtern Frankenberg, Langen und Schlüchtern nach den Stürmen vom Februar 1990 – Ergebnisse einer Luftbildauswertung. Sturmschäden des Frühjahrs 1990 in Hessen. Forschungsber Hess Forstl Versuchsanstalt 12:41–90

15 Fire Regime and Tree Diversity in Boreal Forests: Implications for the Carbon Cycle

C. Wirth

15.1 Introduction

Although the tree component of boreal forest ecosystems hardly deserves the attribute "diverse" from a species point of view, its functional diversity – the diversity of species' traits (Tilman and Lehman 2001) – may still be regarded as high. This shall be illustrated by an extreme example: 95 % of the vast Eurasian boreal forest covering an area of about 500 million ha is dominated by just nine forest-forming tree species belonging to only six genera: *Abies sibirica, Picea obovata, Pinus sibirica, Pinus sylvestris, Larix sibirica, Larix gmelinii, Betula pubescens, Betula pendula,* and *Populus tremula*. However, this small assembly of species hosts representatives of four of five distinct plant functional types (PFT) with respect to fire adaptation in trees ("fire PFTs": namely resisters, avoiders, invaders, and endurers – cf. Sect. 15.3; Rowe 1983; Agee 1998). This functional diversity gives rise to a high diversity of fire regimes ranging from superficial non-lethal surface fires to devastating crown fires (Heinselman 1981; Shvidenko and Nilsson 2000a). The very nature of the prevailing fire regime characterized by the frequency and intensity of fires obviously has serious implications for biogeochemical cycling, in general, and carbon cycling, in particular.

In fact, functional redundancy of tree species is minimal in boreal forests as the following consideration may show that the strategy of fire adaptation is just one of many traits relevant for ecosystem functioning. It may thus appear reasonable to subdivide the four groups further according to just two very basic features, namely, leaf habitus (evergreen/deciduous) and shade tolerance (high/low). If we apply this simple scheme, we arrive immediately at the point where almost each species, or at least each genus, forms a distinct functional group (Fig. 15.1). Here, the distinction between functional and species diversity becomes obsolete. Thus, instead of searching for effects of species diversity per se, it is more instructive to ask: what is the effect of the presence or absence of a given plant functional type on biogeochemical cycles? Since

Ecological Studies, Vol. 176
M. Scherer-Lorenzen, Ch. Körner, and E.-D. Schulze (Eds.)
Forest Diversity and Function: Temperate and Boreal Systems

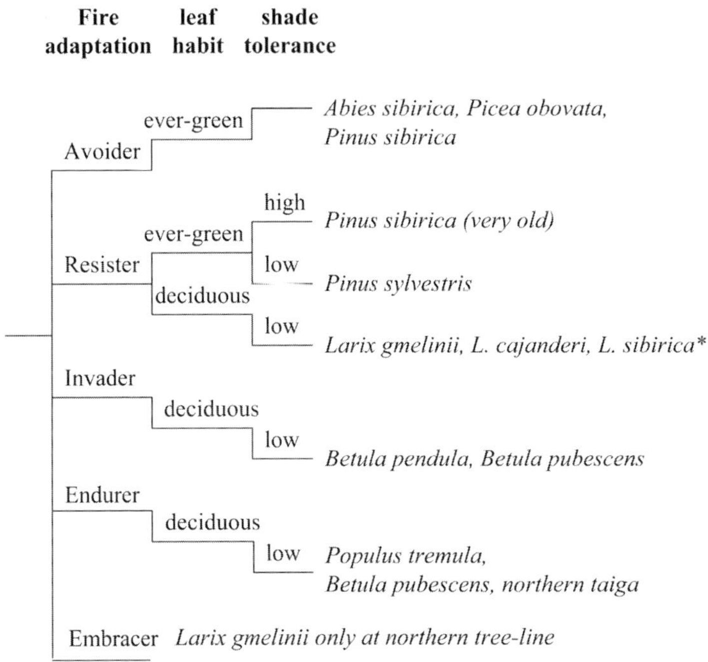

Fig. 15.1. Classification of the major forest-forming tree species of Siberia according to three basic morphological and physiological features that are relevant for ecosystem functioning. This exercise is done to illustrate the case that functional redundancy is very low in boreal forest ecosystems and that almost every individual species may be regarded as a separate functional group whose presence or absence matters to the functioning of system. Changes in the fire adaptation strategy may be induced by the environment or the ontogeny. * The distribution ranges of *Larix gmelinii*, *L. cajanderi* and *L. sibirica* are disjunct

this chapter focuses on fire, we may refine this central question and add the clause "as mediated through its associated fire regime".

After reviewing the basic strategies of fire adaptation in boreal tree species, I will explore patterns of tree diversity, functional diversity, fire regime, and biomass based on a meta-data set that has been compiled for this specific purpose. In the second part, I will briefly review ecosystem responses to different fire regimes. Finally, the circumboreal distribution of fire PFTs and likely consequences for selected biogeochemical cycles at the biome level are analyzed, based on results of large-scale forest inventories and case studies.

15.2 Methods

The literature was searched for boreal forest ecosystem studies reporting (1) the mean fire return interval (FRI), (2) the average intensity or type of each fire, as well as (3) data on species composition. Since high-elevation forests are very similar to boreal forest ecosystems, data on the former were also included in the database. In addition, important geographical details, climate data, and the methods of fire return interval determination were recorded. FRI based on the following methods were considered: age-class analysis at the landscape level (van Wagner 1978; Johnson and Miyanishi 2001), analysis of regeneration epochs within a stand (Cogbill 1984), dendrochronological analysis of fire scars (e.g., Arbatskaya and Vaganov 1997; Wirth et al. 1999), dating of char coal layers in peat (Cwynar 1987). FRI determined as the inverse of annual area burned from fire statistics were not included. This is because annual area burned is highly variable and most statistics cover periods too short to obtain reliable estimates of FRI. While publications focusing on fire history usually lack a detailed description of the vegetation, publications from the field of vegetation science often provide only general information of the fire regime. Compiling a large enough data set was therefore only possible by taking semi-quantitative and, to some extent, qualitative statements into consideration. However, a qualitative description of the FRI (e.g., "frequent", "occasionally") was not accepted. If only minimum and maximum FRI were given, the central value [(max. + min.)/2] was used. In most studies, the fire regime/intensity was described qualitatively. This information was categorized according to a three-level ordinal scale indicating fire intensity (1 = light surface fire, 2 = severe surface fire/partial crown fire, 3 = severe crown fire). Instead of considering just the number of tree species, diversity was described by the well-known Shannon-Wiener function measuring the information content H' of a sample (Krebs 1999)

$$H' = \sum_{i=1}^{s} p_i(\log_2 p_i)$$

where s denotes the number of species and p_i the proportion of total sample belonging to the ith species. Optimally, p_i could be approximated at the stand level by the basal area or density fraction of species occurring in a mixture. In many cases, however, the semantics of qualitative statements needed to be translated into quantitative proportions. This was done according to the following scheme: In monospecific stands p_i obviously equals unity. In mixtures of n species without any ranking reported, the species were assigned equal proportions (p_i=1/n). If a "dominant" fraction comprising d species was indicated, each dominant species was assigned a value of p_{id}=0.7/d (e.g., "dominated by species a and b" translates into p_a=p_b=0.35). The remaining fraction of 0.3 was equally distributed to *c* species listed in the context of phrases such as "co-occurring", "with admixtures of", "forming a subcanopy", etc., (p_{ic}=0.3/*c*). Case-specific adjustments were made if additional information was available from other regional studies.

In forest communities dominated by stand-replacing fires, the mean fire return interval can only be reconstructed using age-class analysis or fire records (Johnson and Miyanishi 2001). This necessarily refers to larger spatial units such as catchments, mountain ridges, altitudinal vegetation belts, etc., where several successional stages of serial communities coexist in the landscape. In those cases either quantitative or qualitative information on successional trends was processed separately for the three stages (early, mid- and late successional) and averaged to capture the tree diversity at the landscape level, i.e., γ-diversity is addressed rather than α-diversity. In contrast, fire scar analysis has usually been performed at the stand level. To match the spatial scales for comparison I thus averaged FRI data from fire scars from several stands representing a specific forest community in a study region. Forest communities were classified as belonging to either of five fire PFTs (see below) according to the fire PFT represented by the dominant species. If no fire PFT clearly dominated, the forest community was classified as "mixed." In doubtful cases, the assignment of species to a fire PFT was done based on the autecological record as given in the fire effects information site (FEIS; http://www.fs.fed.us/database/feis/). To account for climate as an important driver for fire return intervals, the average growing season temperature (av T_{GS}) and precipitation ($\sum P_{GS}$) was taken as the average and sum, respectively, of the months May to August. The monthly data were taken from the closest climate station at comparable altitude and exposition (http://www.wordclimate.com). A simple index was calculated as the ratio $\sum P_{GS}$/av T_{GS}, where low values indicate a climate favoring fires. The database contains 151 entries (for their origins see legend to Fig. 15.5).

To analyze the imprint of the circumboreal distribution of fire PFTs on the carbon cycling at the biome scale, data of macro-regional inventories as well as from case studies were taken from the literature. In addition, data from an IGBP (International Geosphere–Biosphere Programme) high-latitude transect intercomparison were used that were compiled by a group of international experts (McGuire et al. 2002). In the case of Russian transects, those data were extracted from the national databases hold at the International Institute for Applied Systems Analysis in Laxenburg, Austria (IIASA).

15.3 Patterns of Functional Diversity and Fire Regime

15.3.1 Five Basic Strategies for Coping with Fire

Evolution has created distinct strategies of plants to ensure the completion of their life cycle in the face of fire (Gill 1981). Five such strategies are commonly realized in the boreal and high elevation tree species (Rowe 1983). These strategies, which I will term "fire PFTs," can be broadly categorized as being directed to optimize either survival or dispersal (Table 15.1).

Beyond a certain size threshold *resisters* are able to survive surface fires of low to medium intensity. To insulate their cambial sheath from lethal temperatures they build a thick bark around the stem base (Gill 1981; Ryan and Reinhardt 1988). In line with its function as heat shield, the thermal diffusivity of bark is low compared with other tissue types (Hare 1965; Vines 1968; Gutsell and Johnson 1996). Bark is poorly inflammable due to the high temperatures needed to volatilize its major chemical constituents, such as lignins (up to 50% of dry weight, Hakkila 1989) and the lipid-derived suberins (Rundel 1981; Hengst and Dawson 1994). Since the construction costs of these substances are high compared to cellulose (Poorter and Villar 1997), and since bark is also enriched in nutrients in comparison with wood (Hakkila 1989; Wirth et al. 2002b), trees are rarely protected completely with a thick layer of bark. Instead, bark thickness decreases rapidly with tree height (cambial age) and is minimal around branches (Fig. 15.2). Fires reaching the crown therefore lead to immediate cambial damage and the degree of crown scorch, in turn, is inversely related to survival (van Wagner 1972). To minimize the risk of crown fires, resisters tend to quickly shed dead branches (self-pruning). Low reaching dead branches potentially act as a fuel ladder connecting the forest floor and the canopy (Rowe 1983). As a consequence of their survival strategy individual resisters can attain high ages (Schulze et al. 1995; Wirth et al. 1999; Anthoni et al. 2002). Most resisters such as various species of the genus *Pinus* and *Larix* are shade intolerant and require freshly burned areas with exposed mineral soil for successful regeneration (Sannikov and Goldammer 1996; Green et al. 1999). The following species in the database are classified as resisters: The Siberian Larix species (*L. sibirica, L. gmelinii, L. cajanderi*), *Larix occidentalis, Pinus ponderosa, Pinus resinosa, Pinus sylvestris,* and *Pseudotsuga menziesii.*

Unlike resisters the second fire-PFT specialized on survival, the *endurers*, survive only below ground. They possess the capacity to resprout from below ground adventitious buds reaching down to 20 cm soil depth (Brown and Debyle 1987). While common in herbs and shrubs this fire PFT is less common in forest-forming trees and in the boreal zone is mainly represented by species of the genus *Populus.* Although not classified as typical endurers, some *Betula* species and among the conifers *Picea mariana, Pinus banksiana,*

Table 15.1. Fire-related and successional traits of five plant functional types related to fire adaptation (fire PFTs) common to boreal and high-elevational forest ecosystems

		Resister	Endurer	Embracer	Invader	Avoider
	Fire-related traits					
Survival	Thick bark	+	–	–	–	–
	Self-pruning	+	±	–	±	–
	Re-sprouting	–	+	–	±	–
Dispersal	Flammability	–	–	+	–	–
	Serotiny	–	–	+	–	–
	Early reproduction	±	±	+	±	–
	Long-distance dispersal	±	±	–	+	–
	Successional traits					
	Shade tolerance	–	–	–	–	+
	Longevity	+	–	–	–	+
	Successional status	Both	Early	Early	Early	Late

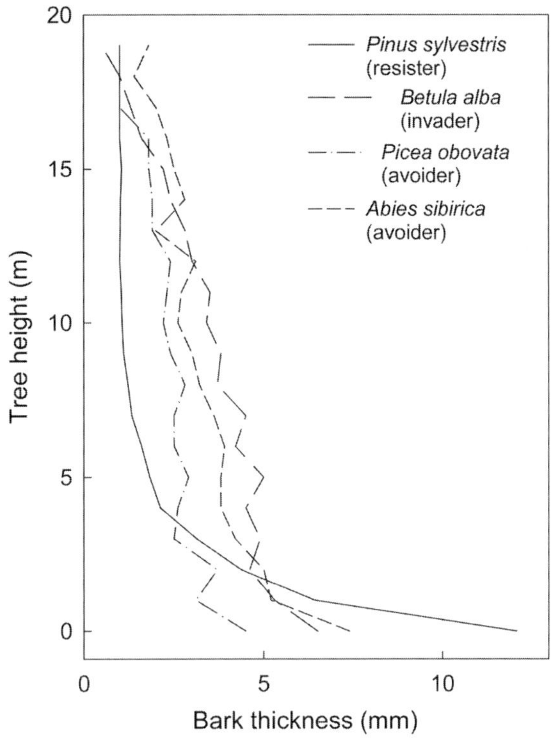

Fig. 15.2. Height profiles and bark thickness of four boreal tree species of Siberia. The four species represent the three fire PFTs "resister", "invader", and "avoider". With a thick bark near the ground the resister species *Pinus sylvestris* (*solid line*) is adapted to survive surface fires. The invader species *Betula alba* and the avoider species *Abies sibirica* and *Picea obovata* do not survive surface fires

and *Thuja occidentalis* are to some extent capable of resprouting (Nikolov and Helmisaari 1992). The vegetative regeneration of aspen by "suckers" leads to the development of polycormones consisting of several genetically identical shoots being closely connected through root grafting. In addition, species of aspen are highly effective long-distance dispersers and colonize freshly burnt areas. The following species in the data-base are classified as endurers: *Populus balsimifera, Populus tremuloides,* and *Populus tremula.*

Embracers (sometimes also termed "evaders") form a canopy seed bank. Seeds accumulate over decades in serotinous cones sealed by resin plugs that require fire temperatures to melt and release the seeds (Keeley and Zedler 1998). The thick cone scales also protect seeds from heat during the melting process. Because high temperatures are essential for successful dispersal, embracers have developed traits to promote the development of high intensity crown fires. The development of crown fires is facilitated by the retention of low-reaching dead branches that serve as fuel ladders, especially if covered with pendent lichens. In addition, dry dead branches ignite easily and produce the heat needed to drive out the moisture of living canopy fuels above. The flammability of canopy fuels is further increased by loosely packed needles on thin twigs (Schwilk and Ackerly 2001). Since embracers hardly invest in protective structures such as a thick bark, they usually do not survive fires (Jackson et al. 1999). This has imposed selection for an early onset of reproduction to minimize the risk of failing to complete the life-cycle within the fire return interval. The following species in the data-base are classified as embracers: *Pinus banksiana,* and *Pinus contorta var. latifolia. Picea mariana* takes an intermediate position between embracers and avoiders (see below) and was therefore given the status of a separate group.

Invaders are killed by even light burns but have specialized on re-colonizing burnt areas from outside. They are prolific producers of wind-dispersed seeds that cover the landscape in a dense seed rain (on average 40,000 seeds m^{-2}; Zasada et al. 1992). As shade-intolerant pioneer species they rely on open areas for successful regeneration (Nikolov and Helmisaari 1992). Further, they depend on nutrient-rich soils (as they result from the thermal mineralization of organic matter during a fire) to realize the high intrinsic growth rates necessary to overtop evergreen late-successional competitors (Horn 1974). Among boreal tree species invaders are primarily found in the genus *Betula*, and to a lesser extent in *Populus* species. *Betula papyrifera* was the only invader species for which independent information on the fire regime was found.

Avoiders are also easily killed by fire, and avoider-dominated forests sustain intense crown fires. With few exceptions this fire PFT is represented by shade-tolerant species, which in the course of secondary succession gain dominance over pioneer species on rich soils with high water supply. They do not possess specific traits associated with fire. As late-successional species avoiders can become very old and regenerate with gap-phase dynamics given the absence of fire. The flammability of forest floors is low because except during extreme droughts the microclimate close to the ground is humid due to a dense canopy and a lush understory vegetation. The following species in the database are classified as avoiders: *Abies balsamifera, Abies concolor, Abies lasiocarpa, Abies sibirica, Acer saccharum, Picea abies, Picea engelmanii, Picea glauca, Picea obovata,* and *Pinus strobus.*

These fire PFTs represent distinct, stable evolutionary strategies resulting from defense/growth or defense/reproduction trade-offs. Comparing 38 North American species of the genus *Pinus*, Schwilk and Ackerly (2001) could clearly demonstrate that traits supporting the above strategies to cope with fire are highly inter-correlated (Fig. 15.3). The degree of serotiny was negatively correlated with the minimum reproductive age, needle density, twig thickness, and self-pruning ability. Thus, serotinous species reach reproductive maturity early and exhibit a more flammable crown with a high surface-to-volume ratio and a fuel ladder of dead branches. On the other hand, bark thickness and self-pruning ability as fire-protective traits were positively correlated with minimum reproductive age, mature height, needle density, and with each other. Negative correlation between serotiny and bark thickness has also been observed at the ecotypic level in *Pinus contorta* and *Pinus*

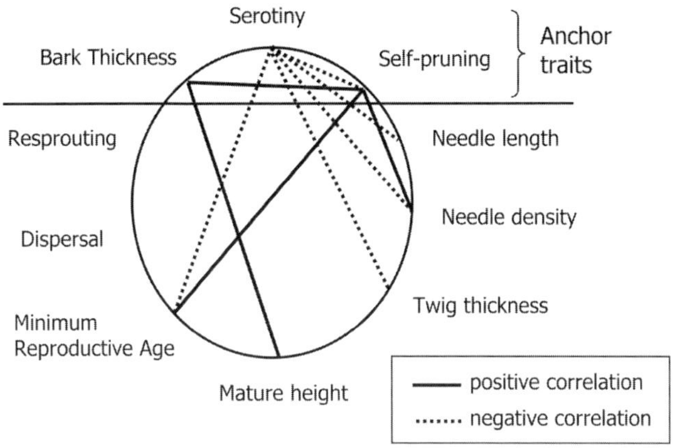

Fig. 15.3. Significant pairwise correlations of species trait values in 27 North American pine species of the subgenus *Pinus* (after Schwilk and Ackerley 2001). *Solid lines* indicate significant positive correlations and *dashed lines* significant negative correlations. To preserve power, only pairwise correlations among the important anchor traits (bark thickness, serotiny, and self-pruning) and between each anchor trait and the other seven traits were tested

banksiana (Gill 1981). Jackson et al. (1999) compared the ontogenetic development of bark thickness (*B*) and tree diameter (*D*) for North American pines. The allometric coefficient *b* of the underlying allometric relationship $B=a*D^b$ is used to indicate how allocation priorities change over time. With $b < 1$, bark thickness is favored over diameter growth early at young ages (negative bark allometry, convex shape of function = "safety first"), while in species with $b > 1$ diameter growth is initially prioritized (positive bark allometry, concave shape of function = "growth first"). Resisters growing in habitats with frequent fires exhibit negative bark allometry, embracers and avoiders positive bark allometry (Fig. 15.4). This may be interpreted as a typical defense/growth trade-off in the sense that the development of protective structures occurs at the expense of overall growth. However, since bark thickness can also be controlled by the rate of bark shedding, this conclusion remains speculative.

15.3.2 Fire PFTs and Their Associated Fire Regimes

Upon pooling data from all regions, it was found that resister communities formed by pine species generally support light surface fires (fire intensity value close to 1 out of 3 on the ordinal scale). The distribution of FRI was skewed with a median of 28 years and the 15th and 85th percentiles (pct) at 12 and 55 years, respectively (Fig. 15.5). Resister communities formed by larch

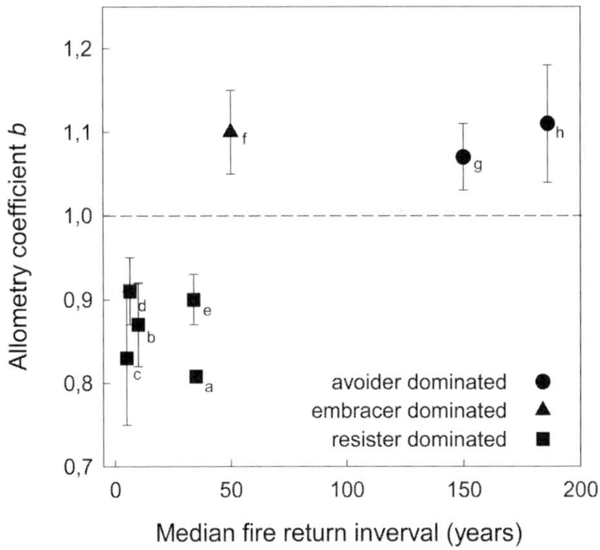

Fig. 15.4. Relationship between the median fire return interval in North American pine species and their allometric coefficient (*b*), scaling bark thickness (*B*), to tree diameter at breast height (*D*), according to the allometric equation $B=a\times D^b$ (after Jackson et al. 1999); $b<1$ indicates a "safety first strategy", whereas $b>1$ represents a "growth-first strategy" (see text for explanation). The individual data points show groups of 32 pine species occurring in different forest communities and were further categorized according to the dominating fire PFT within the groups. *a* Boreal resister pines, *b* eastern dry mixed-species forest, *c* eastern park-like forest/savanna, *d* western park-like forest/savanna, *e* western dry mixed-species forest, *f* embracer-dominated monospecific forests, *g* woodland communities, *h* moist mixed-species forest

forests exhibit a longer FRI of 54 years (pct 30 and 90 years) where fire intensities are somewhat higher (1.6±0.8 dimensionless). It should be noted that mean FRIs for resister communities are usually derived from fire scar analysis, recording only the history of surface fires. However, most resister communities form a fine-grained mosaic of age classes, as such a pattern typically results from stand-replacing fires (Sannikov and Goldammer 1996; Wirth et al. 1999). In fact, the fire regime in resister-dominated communities has to be regarded as bimodal, with a fast cycle of surface fires being overlaid by a long cycle of intense crown fires (Kilgore 1981).

The other extreme is represented by avoider-dominated communities. Here, the mean FRIs are about five times longer than those found in resister communities, exhibiting a median of 150 years (pct 99 and 300 years). Throughout, the intensity of the fires were classified as extremely high, i.e., severe, stand-replacing crown fires prevail (fire intensity value 2.7±0.5). Embracer communities formed by *Pinus banksiana* and *P. contorta* take an intermediate position with fires of moderate intensity (fire intensity value

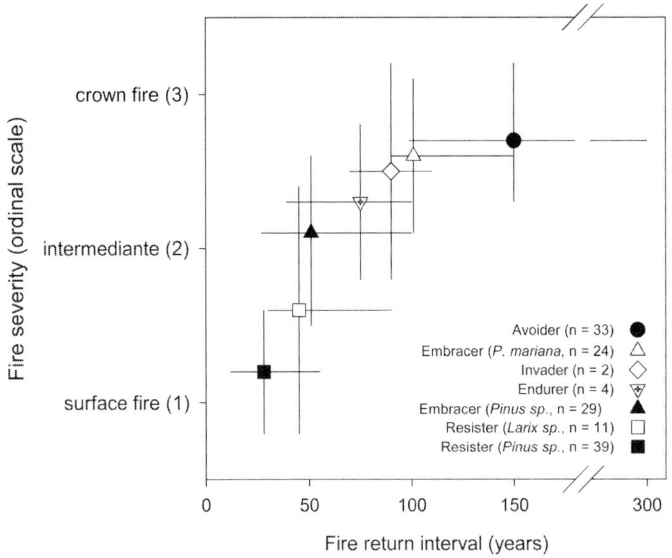

Fig. 15.5. Ordination of forest communities dominated by a specific fire PFT according to median fire return interval and fire intensity (ordinal scale 1–3, *1* surface fire, *2* high intensity surface/low intensity crown fire, *3* crown fire). *Vertical error bars* represent ±1 SD, whereas *horizontal error bars* show the distance between the 15th and 85th -percentile to account for the right-skewed distribution of the fire return intervals. The underlying data base contains 151 entries taken from the following original works and reviews: Weaver (1959); Utkin (1965); Sneck (1970); Day (1972); Frissell (1973); Houston (1973); Viereck (1973); Henry and Swan (1974); Johnson and Rowe (1975); Rowe et al. (1975); Arno (1976); Gabriel (1976); Maikawa and Kershaw (1976); Tande (1977, 1979); Zackrisson (1977); Hawkes (1979); Johnson (1979); Arno (1980); Black and Bliss (1980); Yarie (1980); Heinselman (1981); Romme and Knight (1981); Carrol and Bliss (1982); Romme (1982); Wright and Bailey (1982 p. 233); Cogbill (1984); Foster (1985); Foster and King (1986); Anderson et al. (1987); Engelmark (1987); Bergeron and Dubuc (1989); Payette et al. (1989); Agee et al. (1990); Bergeron and Brisson (1990); Johnson et al. (1990); Masters (1990); Taylor and Fonda (1990); Bergeron (1991); Johnson and Larsen (1991); Lynham and Stocks (1991); Bradshaw and Hannon (1992); Loope and Gruell (1973); Abaimov and Sofronov (1996); Lehtonen et al. (1996); Snytkin (1996); Vaganov et al. (1996); Valendik (1996); Arbatskaya and Vaganov (1997); Larsen (1997); Ivanova (1998/1999); Sheppard and Lassoie (1998); Wirth et al. (1999); Stuart and Salazar (2000); Bergeron et al. (2001); Donnegan et al. (2001); Furyaev et al. (2001); Pitkäninen and Grönlund (2001); Mollicone et al. (2002)

2.1±0.5) arriving rather early after 51 years (pct 27 and 100 years). *Picea mariana* as a semi-serotinous species is positioned halfway between avoiders and embracers. Fires recur earlier than in typical avoider communities (FRI 101 years; pct 90 and 150 years) but are just as intense (fire intensity value 2.6±0.5). Invader and endurer species usually form pioneer stages in longer-term successions toward avoider-dominated forest communities. In some cases where endurers and invaders form stable communities supporting their own fire cycle, this succession is halted after about 75 and 90 years, respectively, by fires of intermediate intensity.

Climate is an important determinant of the fire regime (Swetnam 1993; Campbell and Flannigan 2000). Under warm and dry conditions, the water content of the surface fuels is reduced, increasing its flammability (van Wagner 1983). The above differences in FRI could thus simply result from climatic differences between the geographic ranges in which the fire PFTs were sampled. Although, indeed, large differences between PFT sites exist for mean annual temperature ranging from –8.3 °C for larch to 1.2 °C for embracer pines, precipitation sum and average temperature during the growing season are quite similar (Table 15.2). To exclude the climatic influence on fire return intervals from the comparison of fire PFTs, I conducted an analysis of covariance with the ratio of growing season precipitation and growing season temperature average (P_{GS}/av T_{GS}) as covariate (Table 15.2). High values of this ratio indicate humid conditions during the growing season and consequently low fire danger. Both the covariate, P_{GS}/av $T_{GS,}$ and the categorical variable "fire PFT" were highly significant (P <0.001) and no significant interactions could be detected (data not shown).

It should be noted at this point that one important driver of fire regime was not considered, namely, site quality (Keeley and Bond 2001), because information on texture and water status of soils was rarely reported in the available

Table 15.2. Climate data for sites dominated by certain fire PFTs plus results of the ANCOVA comparing fire return intervals of the different fire PFTs with the ratio of growing season precipitation and temperature as covariate

Fire PFT	av T_{yr} [a] (°C)	Σ P_{yr} [b] (mm)	av T_{GS} [c] (°C)	Σ P_{GS} [d] (mm)	Mean FRI at covariate mean[e] (years)
Avoider	1.1±3.7	732±533	12.8±2.2	244±114	166.8[f]
Embracer *Pinus*	1.2±3.7	483±162	12.5±2.2	215±82	58.3
Embracer *Picea*	–2.4±4.3	561±337	12.5±2.4	231±116	124.1[f]
Endurer/invader	–0.8±3.1	648±289	13.3±1.3	283±117	56.5
Resister Larix	–8.3±3.2	342±137	11.6±2.2	187±69	58.6
Resister Pine	0.5±4.6	456±135	12.4±2.1	201±66	36.3[ref]

[a] Mean annual temperature
[b] Annual precipitation sum
[c] Average growing season temperature (May through August)
[d] Precipitation sum during growing season
[e] FRI, Fire return interval (years); the ratio Σ P_{gs}/ av T_{GS} was used as covariate
[f] Significantly different from resister pines set as zero reference (ref) in the dummy coding of fire PFTs

literature. Avoiders dominate on fertile soils with a good water supply. The humid conditions on the forest floor prevailing in such dense forest communities are not suitable for the development and spread of fire, and FRIs are therefore high, irrespective of the climate. Similarly, the difference between pine and larch within the resister fire PFT emerges most likely because resister pines (*P. sylvestris* and *P. resinosa*) in the boreal zone occur on edaphically drier sites more prone to fire than typical larch sites with more fine-textured soils and underlying permafrost.

15.3.3 Fire Regime, Tree Diversity and Biomass

The intermediate disturbance theory predicts highest species diversity to occur at intermediate levels of disturbance (Conell 1979; Roberts and Gilliam 1995), resulting in a bell-shaped course of diversity along any disturbance axis. A high disturbance frequency creates a hostile environment allowing only specialized species to exist, whereas a low disturbance frequency leaves enough time for competitive exclusion and thus gives way to the dominance of only a few strong competitors. Indeed, species diversity H' for all available forest types initially increased with increasing FRI. But instead of decreasing again, H' saturated at a level of 1.4 after about 200 years (Fig. 15.6A). Further, there was substantial scatter. While most forest types in the high frequency range between 8 and 40 years FRI were monocultures, diverse systems with H' >2 also occurred.

The scatter, however, resolved into distinct clusters if data points were grouped according to fire PFTs (Fig. 15.6B). The following section discusses qualitatively why the fire PFTs occupy different positions in the diversity–disturbance space. To help interpret the position of the fire PFT clusters in diversity–disturbance space, it is useful to imagine a central site quality axis between resister pines and avoiders (dashed line in Fig. 15.6B). Although the database lacks edaphic information this interpretation is justified by the fact that in Eurasia these two forest types coexist in the landscape under identical climatic conditions, and their occurrence is almost completely controlled by site quality (resister pines on poor soils, avoiders on rich soils; see above). The resister communities among the pines grew mostly as monocultures (median H'=0.2). The explanation for this uniformity is straightforward. By favoring recurring surface fires resisters indirectly exclude non-resistant species and "not yet" resistant younger offspring from under-canopy regeneration (Agee 1998; Wirth et al. 1999). To give an example, on sandy soils of intermediate water supply and fertility in Siberia secondary succession starts with a mixture of *Pinus sylvestris* and *Betula* sp., but with the first surface fire the birch component usually disappears (Furyaev et al. 2001). On the other hand, in typical avoider communities an average of five species coexist, resulting in a high H'. While there are indeed shifts in the species composition with time-

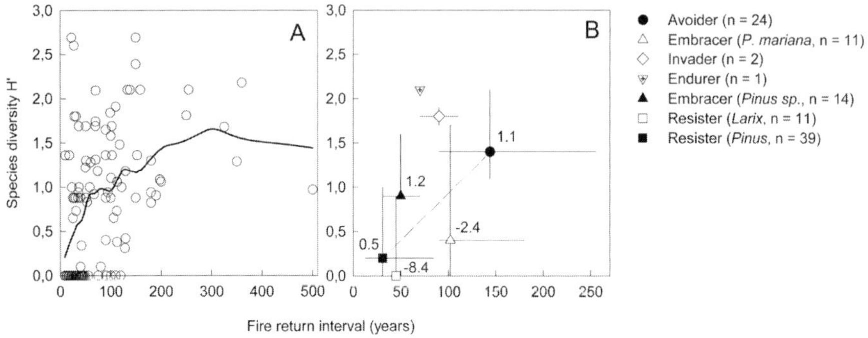

Fig. 15.6. **A** Shannon-Wiener index of tree species diversity H′ for all forest communities in the database against their fire return interval fitted by a LOESS regression. **B** Ordination of the forest communities in the diversity–disturbance space separated according to the dominating fire PFT. *Data points* represent the median. *Error bars* show the distance between the 15th and 85th percentile to account for the skewed distributions of the two quantities shown. For fire PFTs represented by more than two entry numbers next to data points indicate mean annual temperatures as given in Table 15.2. The *dashed line* connecting resister pines and avoiders represents a site quality axis and serves to discuss the relative position of fire PFTs in the diversity–disturbance space (see text)

since-fire, competitive exclusion rarely happens. This is supposedly because the persistence of successional niches is long in comparison with the median FRI of 144 years. Further, avoider communities are also prone to small scale disturbances other than fire (Furyaev et al. 1983; Vygodskaya et al. 2002).

If we accept that resister pines and avoiders are separated along a site-quality axis (dashed line in Fig. 15.6B), the departure of the remaining fire PFTs from this axis in the diversity–disturbance space points to factors other than site quality. This shall be explored in the following for those fire PFTs which are represented by more than two data points. While forests formed by the resister species *Larix* sp. and the embracer species *Picea mariana* fall below the axis (i.e., comparatively lower diversity at longer fire intervals) the embracer pines are well above it (i.e., comparatively higher diversity at shorter fire intervals). All resister larch forests in the data base were monocultures (median H′=0). Interspecific exclusion by fire as in the case of resister pines also occurs in Siberian larch forests, but permafrost adds to the prevention of other species' invasion into larch stands. North American embracer communities formed by *Picea mariana* occur in the forest tundra ecotone. Although close to the arctic tree line *Picea mariana* indeed forms monocultures, further south it is frequently accompanied by *Picea glauca*, *Larix laricina*, and deciduous pioneer species (median H′=0.4). For both resister larch and embracer *Picea mariana* forests the comparatively cold climate (av T_{yr} <<0 in both cases) indirectly leads to longer FRI as compared to their pine counterparts (resister pine 35 years ⇔ resister larch 45 years, embracer pine 50 yrs ⇔ embracer spruce 102 years). In both cases low temperatures go along with low rates of potential evapotranspiration and growth on a substrate formed by a poorly drained active layer above the permafrost. Under such conditions the moisture of the surface fuel is less likely to drop below the so-called moisture-of-extinction at which fires can spread. The high diversity of embracer pine communities (median H′=0.9) as compared to resister pine communities can hardly be explained by site-specific differences. This is because *P. sylvestris* as well as *P. banksiana* and *P. contorta* are commonly found on dry, sandy soils and climatic differences are small (Table 15.2). One explanation could be that the longer fire interval of 50 years as opposed to 31 years in resister pines allows non-resistant species to reach the reproductive age.

Studies on fire frequency either refer to larger landscape units or integrate over long time periods. Therefore, data on biomass of the respective spatial or temporal unit are difficult to retrieve and are thus rarely reported. To analyze whether fire PFTs forming communities with a higher diversity also carry a higher landscape-level biomass, different sources of information had to be combined, and the analysis was only possible for aggregated data. To this end, I grouped species-specific data on landscape-level biomass, as reported in Shepashenko et al. (1998) for Russia and Yarie and Billings (2002) for North America, according to the fire PFTs. There was no obvious relationship between aboveground biomass at the landscape level as an indicator of ecosystem function and species diversity H′ for the different fire PFTs (Fig. 15.7).

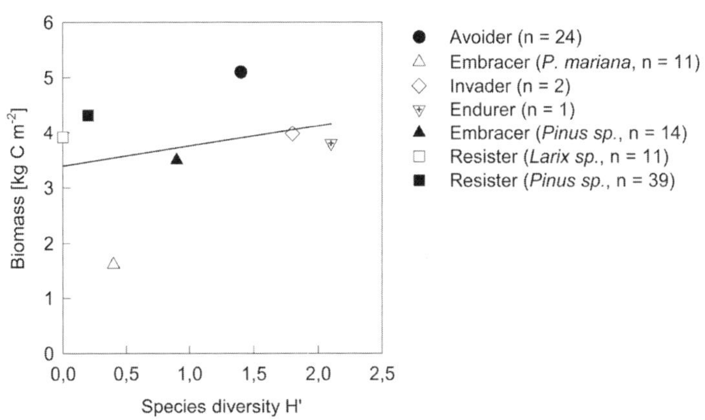

Fig. 15.7. Relationship between tree biomass and species diversity H′. Data for species diversity are averages for fire PFTs as in Fig. 15.6. Biomass data are based on species-specific data on landscape-level biomass as reported in Shepashenko et al. (1998) for Russia and Yarie and Billings (2002) for North America, grouped according to the fire PFTs. There is no obvious relationship between landscape-level biomass and species diversity H′ (*P*=0.55)

15.3.4 Circumboreal Distribution of Fire PFTs

The most striking feature of the distribution of fire PFTs in the boreal zone is the complete absence of embracer species from the whole Eurasian boreal forest on the one hand and the very limited occurrence of resister species in boreal North America on the other (Table 15.3). Embracers such as the serotinous pines *Pinus banksiana* and *Pinus contarta* var. *latifolia* and the semi-serotinous *Picea mariana* occupy 42 and 59 % of the forested areas in Alaska (Yarie and Billings 2002) and Canada (Bourgeau-Chavez et al. 2000; National Forestry Database Program; http://nfdp.ccfm.org), respectively. In contrast, no embracer species occur in the Eurasian boreal forest. While the resisters *Pinus sylvestris*, *Larix sibirica*, and *Larix gmelinii* constitute 63 % of the boreal forest in central Siberia and even 85 % in eastern Siberia (Shvidenko and Nilsson 1994), the only resister species occurring in boreal North America, *Larix laricina*, occupies marginal peatlands sites and dominates not more than 0.4

Table 15.3. Percentage cover and average biome-specific fire return interval of tree species grouped according to fire PFTs in the boreal forest of North America (Alaska and Canada) and Siberia (western/eastern Siberia and far eastern Siberia)

	Boreal North America			Boreal Siberia		
	Alaska[a] (%)	Canada[b] (%)	FRI[c] (years) mean ± SD	Western and eastern Siberia[d] (%)	Far eastern Siberia[d] (%)	FRI (years) mean ± SD
Avoider (av)	38	34.5	147±46	23.1	8.9	265±85
Embracer *Picea* (epc)	42.1	37.1	126±85	--[e]	–[e]	–[e]
Embracer *Pinus* (epn)	–[e]	21.3	48±23	–[e]	–[e]	–[e]
Invader (in)	11.8	1.6	90±20	15.0	5.5	–[c]
Endurer (en)	7.7	5	54±16	3.3	0.5	–[c]
Resister *Larix* (rl)	0.4	0.5	–[c]	37.6	79.5	54±24
Resister *Pine* (rp)	–[e]	–[e]	–[e]	21.0	5.6	28±11

[a] Yarie and Billings (2002) av: *Picea glauca*; epc: *Picea mariana*; in: *Betula papyfera*; en: *Populus tremoloides, P. balsamifer, P. trichocarpa*; rp: *Larix laricina*

[b] Bourgeau-Chavez (2000) and National Forestry Database (2002) av: *Picea glauca, Abies balsamifera, A. lasiacarpa*; epc: *Picea mariana*; epn: *Pinus contorta* var. *latifolia, Pinus banksiana*; in: *Betula* spp.; en: *Populus* spp.; rp: *Larix laricina*

[c] No biome-specific data available

[d] Shvidenko and Nilsson (1994) based on the Russian Forest State Account, av: *Picea obovata, Abies sibirica, Pinus sibirica, Pinus korajensis*; in: *Betula pubescens ssp. alba, Betula pendula*; en: *Populus tremula*; rl: *Larix sibirica, Larix gmelinii*; rp: *Pinus sylvestris*

[e] Fire PFT does not occur in respective biome

and 0.5 % of the forested area in Alaska and Canada, respectively. Despite their different fire adaptation strategies, North American embracer pines and Eurasian resisters have similar edaphic requirements and exhibit a similar degree of drought tolerance, allowing them to dominate dry, coarse-textured upland soils on both continents. Avoiders are more prominent in boreal North America than in Eurasia. Their limited occurrence in Siberia (most pronounced in eastern Siberia with just 9 % cover) is most likely due to the continental climate and permafrost favoring larch on most sites (Archibold 1995). On both continents invaders and endurers together comprise between 5 and 20 % of the area, mostly as early successional stages of typical avoider communities and to a lesser extent as persistent vegetation types.

15.3.5 Comparing Boreal Siberian and North American Fire Regimes

Fire regimes of the boreal forest of Siberia and North America have been reported to differ markedly with respect to the prevailing fire type (Table 15.4). While in Siberia in a typical year some 75 % of the forest area burned is by surface fires (Kolchugina and Vinson 1995; Conard and Ivanova 1997; Shvidenko and Nilsson 2000a), crown fires dominate in the boreal forest of North America (Heinselman 1981; Kasischke 2000). One major driver for this difference becomes obvious if one combines the information about how fire regimes are associated with the occurrence of a specific fire PFT (Fig. 15.5) and about how these fire PFTs are distributed within the two biomes (Table 15.3). The importance of surface fires in Siberia is associated with the dominance of resister-dominated forest types, whereas the crown fires are favored in Canada and Alaska by the dominance of embracer- and avoider-dominated forest types. The partitioning between surface and crown fires is therefore largely a species effect.

A comparison of the extent of burning in Siberia and boreal North America based on official statistics is difficult, because the Russian records before 1988 purposely underestimated fire areas for political reasons (Stocks et al. 1996). Unfortunately, reliable long-term observations are indispensable for any comparison since the interannual variability in fire incidence is extremely high (Murphy et al. 2000). Even after 1988 satellite-derived estimates for specific years were higher than ground-based fire records (Shvidenko and Nilsson 2000b) indicating that in Russia between 1.5 and 12 million ha of forest burned annually, depending on the weather (Cahoon et al. 1994, 1996). Using remote sensing data from 1998, Conard et al. (2002) estimated an area burned of 13.3 million ha, which exceeded the official estimate based on airplane observations by a factor of 5. Reviewing satellite data, Kasischke (2000) concluded that the percentage area burned annually in boreal North America and Siberia is broadly similar. A different picture emerged when I used the database to arrive at an independent estimate of the area annually burned. Com-

Table 15.4. Fire return intervals (FRI) at the regional level. These are calculated as the average of FRI of biome-specific fire PFTs weighted for their areal extent within the given region. The fractional area annually burned was estimated as the reciprocal of FRI

	Forest area (10^6 ha)	FRI (years)	Proportion of surface fires (%)	Estimated fractional area annually burned (%)	Estimated forest area annually burned (10^6 ha)
Western and eastern Siberia	317.8[a]	56.8	74.8[d]	1.76	5.59
Far eastern Siberia	279.4[a]	56.4		1.77	4.94
Alaska	17.2[b]	114.7	<10[e]	0.87	0.15
Canada	404.2[c]	94.1		1.09	4.41

[a] Forested area as given by Nilsson et al. (2000) – these numbers exclude the forest of European Russia
[b] After Yarie and Billings (2002)
[c] After Kurz and Apps (1999)
[d] After Shvidenko and Nilsson (2000a) for fire protected territory of the Russian forest fund during the period 1971–1995
[e] After a qualitative statement in Kasischke (2000)

bining the data on biome- and PFT-specific fire return intervals and areal extent of fire PFTs within the regions (Table 15.3), it is possible to estimate mean regional FRIs and, by taking the reciprocal, the percentage area burned annually. FRIs of western/eastern Siberia and far eastern Siberia (57 and 56 years, respectively) were found to be shorter than those of Canada and Alaska (94 and 115 years, respectively). In central and eastern Siberia 1.8 % and in Canada and Alaska only 1.1 and 0.9 %, respectively, of the forest area is expected to burn per year. This is in line with results from a recent comparison of the IGBP high-latitude transects based on corrected fire statistics (McGuire et al. 2002) but contradicts the conclusion of Kasischke (2000). My FRI-based analysis suggests that in Siberia 10.6 million ha burn annually (Table 15.4), which exceeds the satellite-based average approximately by a factor of 2. One possible reason for this discrepancy is the circumstance that FRI data derived from fire scars and age-class distribution integrate over periods of several centuries and therefore extend back to periods without fire suppression. Another important reason may result from the difference in the dominating fire types mentioned above: surface fires prevail in Siberia. These spread below the canopy and are hard to detect with satellite imagery, their thermal signal being too low for hot-spot detection. Moreover, they leave no fire scars in the landscape that could be retrieved by satellite. In contrast, crown fires are comparatively easy to spot by both methods. It may therefore well be that a larger proportion of the Siberian forest burns annually, although partly unobserved by satellites.

15.4 The Significance of Fire PFTs for Carbon Cycling

15.4.1 How Crown and Surface Fires Affect Ecosystem Functioning

Without any doubt the occurrence of a severe crown fire in a forest ecosystem represents a catastrophic disturbance that alters ecosystem functioning substantially and sustainably. At the shortest time scale, (1) matter is lost though volatilization, particulate loss in smoke, or leaching (Lorbert and Warnatz 1993; Neary et al. 1999), (2) matter is also redistributed between ecosystem compartments (Wirth et al. 2002b), and (3) organisms of all trophic levels may be killed (DeBano et al. 1998). The functions these organisms have carried out such as net primary production (NPP) and decomposition are halted or at least reduced. Rather quickly the processes recover, but their relative importance and magnitude in relation to the pre-fire levels remain altered over decades (Olsen 1981; Wirth et al. 2002b). Decomposition recovers first and exceeds pre-fire levels, since the remnants of killed but not burnt trees form masses of new substrate for rapidly growing microbial communities. These flourish in an environment that is usually warmer (due to higher levels of radiation reaching the forest floor; Amiro 2001), moister (due to reduced transpiration and interception; DeBano et al. 1998), and enriched in available nutrients (due to thermal mineralization by the fire, increased microbial mineralization, and reduced nutrient acquisition by the vegetation; McLean et al. 1983; Raison 1979). Although such an environment is equally favorable for plant growth, the recovery of net primary production, and even more so biomass, usually lags behind. This delay has demographic and biogeochemical reasons. Seeds of trees need to be dispersed into the disturbed area (Greene and Johnson 2000) and safe sites for successful recruitment of trees may be limited by competition with herbaceous post-fire vegetation or by habitat degradation (Abrams et al. 1985; Shvidenko and Nilsson 2000a). After successful regeneration the build-up of a support structure able to sustain high LAI, and therefore high NPP of the tree canopy, needs additional time.

Unlike crown fires, surface fires consume only part of the forest floor fuels (between 30 and 70% depending on the intensity; Gorbachev and Popova 1996; Wirth et al. 2002a, b) and hardly any canopy fuels. In surface fires, subcanopy regeneration is suppressed and trees that lack sufficient thermal protection due to their thin bark are selectively killed (Bond and van Wilgen

1996; Waldrop and Brose 1999; Wirth et al. 1999). However, the impact this has on NPP is small since these subpopulations usually contribute less than 10% to overall net primary production of trees (Dekort 1993). Growth depressions of surviving canopy trees as a consequence of partial cambial damage and mortality of superficial fine roots may reduce NPP by about 20% over a period of 10 years (Wirth et al. 2002 c). On the other hand, recurring surface fires regularly release nutrients that would otherwise be locked-up in the forest floor under the boreal climate (Gower et al. 1996). The increased nutrient availability may sustain productivity until high-stand age is reached to outweigh the initial detrimental effects of fire (DeBano et al. 1998). Most importantly, recurring surface fires keep the load of surface fuels low, therefore reducing the risk of crown fires and prolonging the cycle of stand replacement.

15.4.2 Regional Carbon Inventories and Effects of Species–Fire Regime Interactions

Given what was said above, the prevalence of either surface or crown fires should generate distinct patterns in the structure of boreal forest ecosystems. In summary, Siberia represents a biome dominated by resister communities subject to a high-frequency regime of surface fires and a low-frequency regime of crown fires. Boreal North America, on the other hand, is dominated by avoider and embracer communities subject to a crown fire regime of intermediate frequency. Consequently, I hypothesize for Siberia (1) a higher average stand age, (2) a higher tree biomass, (3) lower carbon stocks in the organic layer (i.e., surface fuel load), and (4) lower emissions per average fire than for boreal North America. In the following I explore results of macro-regional carbon inventories and case studies of the circumboreal forest for the hypothesized patterns.

Comparing age-class distributions between Siberia and boreal North America is somewhat complicated by the fact that the Russian statistics operate with relative age classes (Alexeyev and Birdsey 1996), while the Canadian and Alaskan forest inventories use absolute age classes. Further, the Russian forest is characterized by a significant component of uneven-aged forests. A direct comparison is therefore only possible if broad categories are used (Fig. 15.8). As expected Siberian forest stands attain higher ages than those in boreal North America. In Siberia, 54% of all stands are older than 80 years, whereas only 41% of all stands in boreal North America fall into this category (FAO TBFRA of UN-ECE/FAO 2000; Yarie and Billings 2002). In addition, the low percentage cover of middle-aged stands (41–80 years) in Siberia (21%) suggests a very "flat" age-class distribution with the right tail extending to very high ages. This is corroborated by Shvidenko and Nilsson (2002) who estimated an average age of the total forest

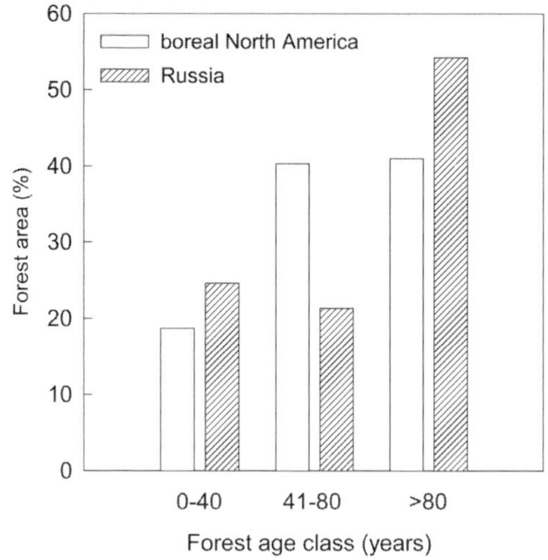

Fig. 15.8. Age-class distribution with respect to forest cover in boreal North America and Russia. Data sources are Yarie and Billings (2002) for Alaska and FAO TBRFA (2000) for Canada and Russia. The data shown for boreal North America represent the average of Alaska and Canada weighted for their respective forest areas

of 100 years (115 years in the coniferous forest and about 47 years for the soft-deciduous forest).

According to several different inventories, the carbon stored in tree biomass per unit area is higher by 50 to 100 % in Siberian as compared with boreal North American forest ecosystems. Siberian upland forests in the central taiga biome within the IGBP high-latitude transects store about twice as much biomass as their North American counterparts (Fig. 15.9A; McGuire et al. 2002): Whereas biomass per area in the Alaskan and Canadian forests is low at 2.6 and 2.0 kg C m^{-2}, respectively, high densities of 4.9 and 3.8 kg C m^{-2} are reported for eastern and far eastern Siberia, respectively. A comparable pattern emerges in three other large-scale forest inventories for Russia and Canada as a whole. Very similar to the transect inter-comparison, the FAO TBFRA 2000 reports 4.4 and 2.9 kg C m^{-2} for Russia and for Canada, respectively (Fig. 15.9B, a). National inventories quantify 4.4 and 3.4 kg C m^{-2} for Russia (Shvidenko and Nilsson 2002) and Canada (Bourgeau-Chavez et al. 2000), respectively (Fig. 15.9B, b). According to Dixon et al. (1994), the difference is even more pronounced, namely, 8.3 vs. 2.8 kg C m^{-2} (Fig. 15.9B, c).

The transect comparison of McGuire et al. (2002) also provides estimates of organic layer area densities of upland forests (Fig. 15.9A). Here, organic layer includes all topsoil organic horizons with a carbon concentration of more than 17 %. The organic layer density is highest in the Canadian transect (4.3 kg C m^{-2}) where thick mats of humus, mosses, and litter may accumulate,

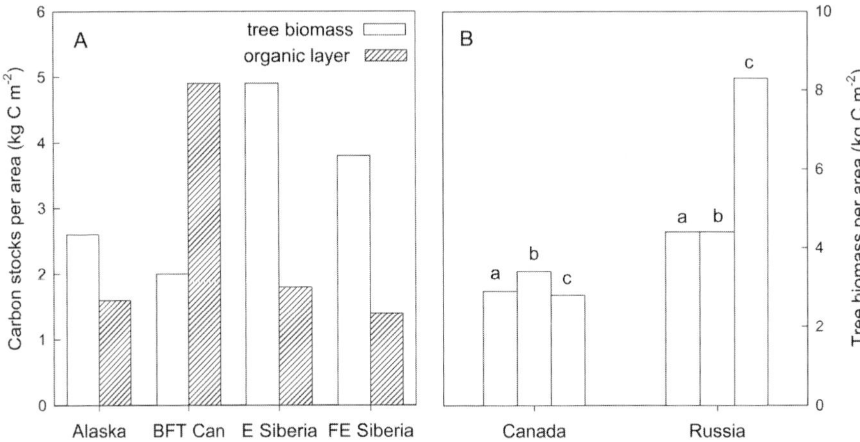

Fig. 15.9. Carbon density of tree biomass (above- and belowground) and organic layer in the boreal forest zone of Alaska, Canada, eastern Siberia, and far eastern Siberia. The organic layer includes all organic soil horizons with a carbon concentration of more than 17 %. **A** Data from the IGBP high-latitude transect comparison project (McGuire et al. 2002). In the case of Russian transects, these data were extracted from the national databases held at the International Institute for Applied Systems Analysis in Laxenburg, Austria (IIASA). *BFT Can* Boreal forest transect (BOREAS) case study in Canada, *E Siberia* eastern Siberia, *FE Siberia* far eastern Siberia. **B** Average tree biomass quantified by large-scale inventories for Canada and Russia: *a* FAO TBFRA; *b* national inventories: for Russia, the most recent estimate of Shvidenko and Nilsson (2002) is presented and for Canada the aboveground biomass given in Bourgeau-Chavez et al. (2000) was multiplied by 1.23 to obtain whole-tree biomass based on expansion factors in Shepashenko et al. (1998); *c* Dixon et al. (1994)

while in all other biomes, including Alaska, the densities range between 1.4 and 1.8 kg C m^{-2}.

Field observations clearly show that carbon emission rates are much higher during crown fires (3.7±2.5 kg C m^{-2}) than during surface fires (0.7±0.4 kg C m^{-2}). The data situation reflects well the relative importance of fire types in the two biomes (Table 15.5). For boreal North America, only data on crown fires were available, whereas about half the data points for Eurasia represented surface fires. By far the most data points existed for forests formed by the embracer *Picea mariana*, where 4.4±2.6 kg C m^{-2} were lost per crown fire (Harden et al. 2000; Ottmar et al. data from the FROSTFIRE experiment, unpubl.; Kasischke et al. 2000; Michalek et al. 2000). Other forest types revealed lower carbon emission rates during crown fires (*Picea glauca* 2.5±0.6 kg C m^{-2}, *Populus* sp. 1.5±0.1 kg C m^{-2}, and 1.2±0.6 kg C m^{-2} in embracer communities dominated by *Pinus banksiana*; Stocks 1989; Harden et al. 2000; Kasischke et al. 2000). For Eurasia, only two case studies report emission rates for crown fires. Wang et al. (2001) estimated a loss of 5.0±3.3 kg C m^{-2} per crown fires and 2.2±2.6 kg C m^{-2} per partial crown fires

Table 15.5. Absolute and relative carbon loss rates through direct combustion by fire in different forest communities in boreal North America and Siberia. Data are based on experimental studies comparing pre- and post-fire levels of carbon stocks (± SD)

	Type of fire[a]	Carbon emission per fire (kg C m^{-2})	Number of studies	Percentage C loss[b] (%)	Source
Boreal North America					
Pinus banksiana	Crown	1.2±0.81	2	34	Stocks (1989); Harden et al. (2000)
Picea glauca	Crown	2.5±0.6	2	n.d.	Kasischke et al. (2000)
Picea mariana	Crown	4.4±2.6	16	42±27	Kasischke et al. (2000); Michalek et al. (2000); Ottmar et al., unpubl. results from the FROSTFIRE experiment, http://www.fsl.orst.edu/fera/Fire2000_ffposters/
Populus sp.	Crown	1.5±0.1	2	n. d.	Kasischke et al. (2000)
Siberia					
Pinus sylvestris	Crown	1.9	1	17	FIRESCAN (1996)
	Mixed	1.5	1	20	McRae et al. (2004)
	Surface	0.8±0.4	8	12±6	McRae et al. (2004); Wirth et al. (2002a)
Larix gmelinii	Crown	5.0±3.3	3	63±19	Wang et al. (2001)
	Mixed	2.2±2.6	4	27±25	Wang et al. (2001)
Averages					
	Crown	3.7±2.5	19	43±26	
	Mixed	3.1±2.4	3	34±15	
	Surface	0.7±0.4	9	11±7	

[a] Crown, Intense crown fire; mixed, intense surface fire/low intensity crown fire; surface, low intensity surface fire
[b] All carbon pools except the mineral soil carbon

in larch forests in northeastern China. The FIRESCAN Science team (1996) measured a carbon loss of 1.9 kg C m^{-2} following a crown fire in a Scots pine forest. In contrast, carbon emissions rates of typical surface fires are much lower at about 0.8±0.4 kg C m^{-2} (Wirth et al. 2002a; McRae et al. 2004). I am not aware of any data for carbon emissions from surface fires in larch forests and from crown fires in typical Siberian avoider communities. The relative carbon loss in relation to pre-fire levels of available fuel in the biomass, the soil organic layer in coarse woody debris was on average by a factor of four higher during crown fires than during surface fires (11 and 43 %, respectively).

For Russia the average emission rates of 1.3 kg C m^{-2} assumed in regional carbon budget models (Shvidenko and Nilsson 2000b) agree well with the measurements and reflect the prevalence of surface fires. On the other hand, Amiro et al. (2001) estimated average emission rates for Canada of 1.2 kg C m^{-2}, which appears way too low in the light of the available data on crown fires.

15.5 Discussion

The biome-scale patterns are indeed broadly consistent with the hypotheses based on species/fire regime interaction. They suggest that indeed the presence or absence of certain fire PFTs has a significant influence on the cycling of carbon in boreal forest ecosystems at the biome level. This functional significance was mediated through the fire regime associated with the fire PFT (Agee 1998). However, the causal relationship between fire regime and forest composition resembles very much a hen-and-egg question. It was often emphasized that life history strategies and morphological traits of boreal and high-elevation tree species have evolved in the face of fire (Keeley and Zedler 1998; Keeley and Bond 2001). However, the causality works in both directions: trees having evolved certain adaptations in turn alter their own fire regime. For example, a regime of surface fires selects for species that are able to resist fires, but at the same time the presence of resisters favors light surface fires.

15.5.1 Mixed Strategies

Not always can tree species unambiguously be assigned to a specific fire PFT. In such cases the classification in Section 15.3.1 rather reflects the central tendency. At the ecotypic level or under certain environmental condition some species exhibit a combination of traits that are characteristic for two or more different fire PFTs. It was mentioned earlier that *Picea mariana* takes an intermediate position between embracers and avoiders. While *P. mariana* indeed accumulates a canopy seed bank and behaves as a true embracer on upland sites, it is frequently found in moist habitats which are hardly prone to fire (Bourgeau-Chavez et al. 2000). On such sites, *P. mariana* may be regarded as an avoider. Siberian *Betula* species very efficiently invade disturbed areas by seeds but also possess the ability to resprout from stumps – a typical endurer trait. Its resprouting strategy also allows *Betula pubescens* to form stable populations in the zone of continuous permafrost, where it coexist with *Larix sp.* at the landscape level (Abaimov et al. 1998). *Larix gmelinii* which dominates on continuous permafrost in northeastern Siberia represents a special case. Although potentially a strong resister, the harsh environmental conditions force the species to adopt an embracer/invader strategy. Under the extreme growing conditions in the northern taiga, *L. gmelinii* rarely exceeds a height of 8 m and a breast-height diameter of 12 cm. Given this "premature" stature the bark hardly ever grows thick enough (>5 to 7 mm) to protect the cambium from heat damage, and crowns are generally low-reaching (Tsvetkov 1996). It has even been observed that in northern ecotypes the bark peels off in stripes forming a fuel ladder (Tsvetkov, pers. com). As a consequence, even light surface fires cause almost 100 % mortality. On the other hand, *L. gemelinii* is the only boreal *Larix* species that exhibits a modest degree of serotiny: it stores a fraction of its seeds in the cone for a period of 3 to 4 years. Dispersal is therefore to some extent decoupled from the cycle of years of good seed crops. *L. sibirica* and *L. cajanderi*, which dominate further south and east, lose their seeds within 2 to 3 weeks (Abaimov et al. 1998).

15.5.2 The Contemporary Distribution of Fire PFTs: A Unique Biogeographical Outcome?

The contemporary distribution of fire PFTs poses an interesting question: does the absence of embracer species from boreal Eurasia, and the minor importance of resister species in boreal North America, represent a unique and accidental biogeographical outcome? Profound consequences lie herein regarding dynamic global vegetation models, assuming as they do climate and soils to be the main drivers for the distribution of plant functional types (Prentice et al. 1992). In fact, the Siberian resister *Pinus sylvestris* and the North American embracers *Pinus contor*ta and *Pinus banksiana* are the pine species with the highest ecological amplitude and therefore the largest ranges of all contemporary pine species. This may reflect their evolution during the Eocene, when pine populations were fragmented and displaced to southern and northern refugia (Millar 1998). In North America the lineages of the subsection *Contortae* leading to *Pinus contorta* and *P. banksiana* had such northern origins, whereas in Eurasia the same happened to lineages of the subsection *Pinus* represented by the extant *P. sylvestris*. Have intercontinental differences in climate, through differences in fire regimes, selected for different fire adaptation strategies during the Eocene? After the last glacial maximum these three species were those that were able to quickly reinvade the boreal zone at a pace of 100 to 300 m/year. *P. contorta* and *banksiana* reinvaded from southern refugia, whereas *P. sylvestris* could spread from refugia within the boreal forest zone (MacDonald et al. 1998). Climate and fire adaptation are not necessary linked. This is illustrated by the fact that *Pinus contorta* introduced in Scandinavia is highly competitive (Sykes 2001). However, the dominant role of the resister *Larix* in Siberia is certainly due to climatic reasons since the genus *Larix* hosts species that are particularly adapted to the extreme continental climate in far eastern Siberia.

15.5.3 Fire, Climate, and Biomass

The transition between the forest–tundra and the boreal forest in Canada is located about 5° further south than the Russian taiga, and the boreal forest belt itself is narrower than in Russia. Both growing-season temperature and photosynthetically active radiation are higher in the North American boreal forest (McGuire et al. 2002), and in eastern Canada the average annual precipitation exceeds 1,000 mm and is thus higher than anywhere in Siberia. In contrast to large parts of the Russian taiga, the boreal forest of Canada grows outside the region of continuous permafrost. Although the climate is obviously colder and dryer in the Siberian taiga as compared to the boreal forest in North America, inventories indicate that tree biomass area densities are higher in Siberia. This has been attributed to the abundance of resisters in Siberian forests, which attain higher stand ages and therefore exhibit long periods of biomass carbon accumulation under a regime of low intensity surfaces fires. Recurring surface fires induce repeated selective mortality, slowing down biomass accumulation rates (Yevdokimenko 1996; Wirth et al. 1999, 2002b). The longevity effect seems to overcompensate the reduction of biomass accumulation rates caused by surface fires.

15.5.4 Crown Versus Surface Fire: How Much Carbon Is Lost?

In central and eastern Siberia 1.8 % and in Canada and Alaska only 1.1 and 0.9 %, respectively, of the forest area is expected to burn every year (Table 15.4). On the other hand, the fire regime in boreal forest in North America is dominated by crown fires with markedly higher carbon emission rates. Combining the continent-specific partitioning in surface and crown fires given in Table 15.4 (ratio of crown to surface fire area: 25:75 in Russia and 90:10 in boreal North America) and the typical carbon emission rates for the two fire types (Table 15.5), the average fire in Russia emits 1.5 kg C m^{-2}, while in boreal North America the carbon loss is 3.4 kg C m^{-2}, over twice as much. It follows that the annual losses of carbon by fire per unit forest area should be roughly similar for the two boreal regions. With forest areas in Siberia, Canada, and Alaska of 597, 245, and 17 million ha, respectively, the average annual carbon emission is expected to be on the order of 0.16, 0.09, and 0.005 Pg C $year^{-1}$, respectively. These numbers are similar to the mean estimates by French et al. (2000) for boreal North America (0.053 Pg C $year^{-1}$) and the estimate for the 1980s by Kolchugina and Vinson (1995) for Siberia (0.137 Pg C $year^{-1}$). Lower estimates for Russian forests are reported by Shvidenko and Nilsson (2000b), to wit, 0.018 Pg C $year^{-1}$. This discrepancy is due to a very low estimate of the annual area burned in Russia of 0.23 % per year based on 5 years of data from 1988 through 1992. The forest composition with respect to fire PFTs obviously does not change the absolute loss of carbon, but only the partitioning between surface and crown fires. While irrelevant for the average carbon loss, this partitioning might still influence nutrient losses. The strong convective movement of air masses during an intense crown fire injects fire plumes into higher layers of the convective boundary layer, or even into the free troposphere, from where gases and aerosols may be transported over long distances (Cooke and Wilson 1996; Liousse et al. 1996; FIRESCAN science team 1996). This may be particularly pronounced in the presence of multiple stratified atmospheric layers which restrict vertical transport and force plumes into a rapid horizontal movement. Under these conditions fire-generated particles and gaseous products can travel 1,000 km in just 10 h (Kaufmann et al. 1996; Laursen and Radke 1996). In contrast, plumes generated by surface fires remain close to the surface. From there smoke contents may be recycled in the vicinity of the source through wet and dry deposition. Following Vitousek's notion that repeated fires may cause nitrogen limitation in terrestrial ecosystems (Vitousek and Howarth 1991) it may thus be hypothesized that crown-fire-dominated ecosystems lose more N per unit land area than surface-fire-dominated ecosystems, with the former becoming more likely N limited.

15.6 Conclusions

The number of forest-forming tree species in the circumboreal zone is rather low and so is tree diversity at any given location. However, the functional diversity with respect to strategies of fire adaptation is high, leading to a diversity of fire regimes. No obvious relationship between tree diversity per se and biomass accumulation appeared. Instead, the functional significance was mediated through the fire regime associated with the fire PFT. It emerged that the regional mix of fire PFTs may have demographic and biogeochemical consequences at the biome scale. The following broad pattern emerged:

- Russian boreal forests are dominated by resister communities associated with a bimodal fire regime with a short cycle of frequent low intensity surface fires being superimposed by a long cycle of stand-replacing crown fires. Since resisters usually survive surface fires they attain high stand ages. The frequent surface fires feed primarily on the organic layer and consume biomass only lightly. Therefore, the average organic layer densities (1.5 kg C m^{-2}) and the emission rates per fire (1.5 kg C m^{-2}) are low, while the biomass densities are high (4.4 kg C m^{-2}).
- The boreal forests of North America are dominated by embracers and avoiders, two fire PFTs that favor high intensity crown fires that occur at an intermediate frequency. Since usually all trees are killed by the fire, stand development is halted before maximum biomass densities can develop in old-growth stages, and average biomass levels are low (3–3.5 kg C m^{-2}). Fire intervals are long enough to allow for substantial accumulation of carbon in the organic layer (1.5–4.8 kg C m^{-2}). Since both the forest floor and the canopy fuels are consumed, emission rates per fire are high (3.4 kg C m^{-2}).

Contradicting recent remote sensing studies, the new data assembled here suggest that in Russia on average a higher fraction of forest burns annually than in boreal North America (1.8 vs. 1.0 %). However, since the emission rates are lower in Russia due to a dominance of surface fires, the carbon losses per year and unit area are comparable for the two regions. In summary, the mix of fire PFTs obviously does not effect the absolute loss of carbon but only the partitioning between surface or crown fires. This partitioning in turn may have implications for nutrient cycling and long-term transport of gases and aerosols.

References

Abaimov AP, Sofronov MA (1996) The main trends of post-fire succession in near-tundra forests of Central Siberia. In: Goldammer JG, Furyaev VV (eds) Fire in ecosystems of Boreal Eurasia, Forestry Sciences, vol 48. Kluver, Dordrecht, pp 372–386

Abaimov AP, Lesinski JA, Martinsson O, Milyutin LI (1998) Variability and ecology of Siberian larch species. Swedish University of Agricultural Sciences, Department of Silviculture. Reports No 43, Umeå, Sweden

Abrams MD, Sprugel DG, Dickmann DI (1985) Multiple successional pathways on recently disturbed Jack pine sites in Michigan. For Ecol Manage 10:31–48

Agee JK (1998) Fire and pine ecosystems. In: Richardson DM (ed) Ecology and biogeography of *Pinus*. Cambridge Univ Press, Cambridge, pp 193–218

Agee JK, Finney M, Degouvenain R (1990) Forest fire history of Desolation Peak, Washington. Can J For Res 20:350–356

Alexeyev VN, Birdsey RA (1996) Carbon storage in forests and peatlands of Russia. USDA Forest Service, Northeastern Research Station, Delaware, p 136

Amiro BD (2001) Paired-tower measurements of carbon and energy fluxes following disturbance in the boreal forest. Global Change Biol 7:253–268

Amiro BD, Todd JB, Wotton BM, Logan KA, Flannigan MD, Stocks BJ, Mason JA, Martell DL, Hirsch KG (2001) Direct carbon emissions from Canadian forest fires, 1959–1999. Can J For Res 31:512–525

Anderson L, Carlson CE, Wakimoto RH (1987) Forest fire frequency and western spruce budworm outbreaks in Western Montana. For Ecol Manage 22:254–260

Anthoni PM, Unsworth MH, Law BE, Irvine J, Baldocchi DD, Tuyl SV, Moore D (2002) Seasonal differences in carbon and water vapor exchange in young and old-growth ponderosa pine ecosystems. Agric For Meteorol 111:203–222

Arbatskaya MK, Vaganov EA (1997) Long-term variation in fire frequency and radial increment in pine from the middle taiga subzone of central Siberia. Rus J Ecol 28:291–297

Archibold OW (1995) Ecology of world vegetation. Chapman and Hall, London

Arno SF (1976) The historical role of fire on the Bitterroot National Forest. Intermountain Forest and Range Experimental Station, Ogden, Utah 84401, Ogden

Arno SF (1980) Forest fire history in the northern Rockies. J For 78:460–465

Bergeron Y (1991) The influence of island and mainland lakeshore landscapes on boreal forest-fire regimes. Ecology 72:1980–1992

Bergeron Y, Brisson J (1990) Fire regime in red pine stands at the northern limit of the species' range. Ecology 71:1352–1364

Bergeron Y, Dubuc M (1989) Succession in the southern part of the Canadian boreal forest. Vegetatio 79:51–63

Bergeron Y, Gauthier S, Kafka V, Lefort P, Lesieur D (2001) Natural fire frequency for the eastern Canadian boreal forest: consequences for sustainable forestry. Can J For Res 31:384–391

Black RA, Bliss LC (1980) Reproductive ecology of *Picea mariana* (Mill) Bsp, at tree line near Inuvik, Northwest Territories, Canada. Ecol Monogr 50:331–354

Bond WJ, van Wilgen B (1996) Fire and plants. Chapman and Hall, London

Bourgeau-Chavez LL, Kasischke ES, Mudd JP, French NHF (2000) Characteristics of forest ecozones in the North American Boreal Region. In: Kasischke ES, Stocks BJ (eds) Fire, climate change, and carbon cycling in the boreal forest, vol 138. Springer, Berlin Heidelberg New York, pp 258–273

Bradshaw RHW, Hannon G (1992) Climatic change, human influence and disturbance regime in the control of vegetation dynamics within Fiby Forest, Sweden. J Ecol 80:625–632

Brown JK, Debyle NV (1987) Fire damage, mortality, and suckering in aspen. Can J For Res 17:1100–1109

Cahoon DR, Stocks BJ, Levine JS, Cofer WR, Pierson JM (1994) Satellite analysis of the severe 1987 forest-fires in northern China and southeastern Siberia. J Geophys Res Atmos 99:18627–18638

Cahoon DR, Stocks BJ, Levine JS, Cofer WR III, Barber JA (1996) Monitoring the 1992 forest fires in the boreal ecosystem using NOAA-AVHRR satellite imagery. In: Levine JS (ed) Biomass burning and global change, vol II. MIT Press, Cambridge, MA, pp 795–801

Campbell ID, Flannigan MD (2000) Long-term perspectives on fire-climate-vegetation relationships in the North American boreal forest. In: Kasischke ES, Stocks BJ (eds) Fire, climate change, and carbon cycling in the boreal forest, Ecological Studies, vol 138. Springer, Berlin Heidelberg New York, pp 151–172

Canadian National Forestry Database Program http://nfdp.ccmf.org/

Carrol SB, Bliss LC (1982) Jack pine – lichen woodland on sandy soils in northern Saskatchewan and northeastern Alberta. Can J Bot 60:2270–2282

Cogbill CV (1984) Dynamics of boreal forests of the Laurentian Highlands, Canada. Can J For Res 15:252–261

Conard SG, Ivanova GA (1997) Wildfire in Russian boreal forests – potential impacts of fire regime characteristics on emissions and global carbon balance estimates. Environ Pollut 98:305–313

Conard SG, Sukhinin AI, Stocks BJ, Cahoon DRJ, Davidenko EP, Ivanova GA (2002) Determining effects of area burned and fire severity on carbon cycling and emission in Siberia. Clim Change 55:197–211

Conell JH (1979) Tropical rainforests and coral reefs as open non-equilibrium systems. In: Anderson AM, Turner BD, Taylor LR (eds) Population dynamics: 20th symposium of the British Ecological Society. Blackwell, Oxford, pp 141–163

Cooke WF, Wilson JJN (1996) A global black carbon aerosol model. J Geophys Res Atmos 101:19395–19409

Cwynar LC (1987) Fire and the forest history of the North Cascade Range. Ecology 68:791–802

Day RJ (1972) Stand structure, succession, and use of southern Albertas Rocky Mountain Forest. Ecology 53:472–478

DeBano LF, Neary DG, Ffolliott PF (1998) Fire's effects on ecosystems. Wiley, New York

Dekort I (1993) Wood production and latewood percentage of Douglas-fir from different stands and vitality classes. Can J For Res 23:1480–1486

Dixon RK, Brown S, Houghton RA, Solomon AM, Trexler MC, Wisniewski J (1994) Carbon pools and flux of global forest ecosystems. Science 263:185–190

Donnegan JA, Veblen TT, Sibold JS (2001) Climatic and human influences on fire history in Pike National Forest, central Colorado. Can J For Res 31:1526–1539

Engelmark O (1987) Fire history correlations to forest type and topography in northern Sweden. Ann Bot Fenn 24:317–324

FIRESCAN science team (1996) Fire in ecosystems of boreal Eurasia: the Bor forest island fire experiment fire research campaign Asia-North (FIRESCAN). In: Levine JS (ed) Biomass burning and global change, vol II. MIT Press, Cambridge, MA, pp 848–873

Foster DR (1985) Vegetation development following fire in *Picea mariana* (Black Spruce)-*Pleurozium* forests of southeastern Labrador, Canada. J Ecol 73:517–534

Foster DR, King GA (1986) Vegetation pattern and diversity in Se Labrador, Canada – *Betula-Papyrifera* (birch) forest development in relation to fire history and physiography. J Ecol 74:465–483

French NHF, Kasischke ES, Stocks BJ, Mudd JP, Martell DL, Lee BS (2000) Carbon Release from Fires in the North American Boreal Forest. In: Kasischke ES, Stocks BJ (eds) Fire, climate change, and carbon cycling in the boreal forest, vol 138. Springer, New York, pp 377–388

Frissell SSJ (1973) The Importance of fire as a natural ecological factor in Itasca State Park, Minnesota. Quat Res 3:397–407

Furyaev VV, Wein RW, MacLean DA (1983) Fire influences in *Abies*-dominated forests. In: Wein RW, MacLean DA (eds) The role of fire in northern circumpolar ecosystems, 18th edn. Wiley, Chichester, pp 221–232

Furyaev VV, Vaganov EA, Tchebakova NM, Valendik EN (2001) Effects of fire and climate on successions and structural changes in the Siberian boreal forest. Euras J For Res 2:1–15

Gabriel HW (1976) Wilderness ecology: the Danahar Creek drainage, Bob Marshall Wilderness, Montana. School of forestry. University of Montana, Missoula, p 224

Gill AM (1981) Fire adaptive traits of vascular plants. Fire regimes and ecosystem properties: proceedings of the conference, vol WO-26. USDA Forest Service, Honolulu, pp 208–230

Gorbachev VN, Popova EP (1996) Fires and soil formation. In: Goldammer JG, Furyaev VV (eds) Fire in ecosystems of boreal Eurasia, Forestry Sciences, vol 48. Kluver, Dordrecht, pp 331–336

Gower ST, McMurtrie RE, Murty D (1996) Aboveground net primary production decline with stand age: Potential causes. Trends Ecol Evol 11:378–382

Green DF, Zasada JC, Sirois L, Kneeshaw D, Morin H, Charron I, Simard MJ (1999) A review of the regeneration dynamics of North American boreal forest tree species. Can J For Res 29:824–839

Greene DF, Johnson EA (2000) Tree recruitment from burn edges. Can J For Res 30:1264–1274

Gutsell SL, Johnson EA (1996) How fire scars are formed: coupling a disturbance process to its ecological effect. Can J For Res 26:166–174

Hakkila P (1989) Utilization of residual forest biomass. Springer, Berlin Heidelberg New York

Harden JW, Trumbore SE, Stocks BJ, Hirsch A, Gower ST, O'Neill KP, Kasischke ES (2000) The role of fire in the boreal carbon budget. Global Change Biol 6 [Suppl 1]:174–184

Hare RC (1965) Contribution of bark to fire resistance. J For 63:248–251

Hawkes BC (1979) Fire history and appraisal study of Kananskis Provincial Park, Alberta Contract Rep for Alberta Provincial Parks. Department of Forest Science, University of Alberta, Edmonton, p 172

Heinselman ML (1981) Fire intensity and frequency as factors in the distribution and structure of northern ecosystems. Fire regimes and ecosystem properties: proceedings of the conference, vol WO-26. USDA Forest Service, Honolulu, pp 7–57

Hengst GE, Dawson JO (1994) Bark properties and fire resistance of selected tree species from the central hardwood region of North America. Can J For Res 24:688–696

Henry JD, Swan JMA (1974) Reconstruction forest history from live and dead plant material – an approach to the study of forest succession in southwest New Hampshire. Ecology 55:772–783

Horn HS (1974) The ecology of secondary succession. Annu Rev Ecol Syst 5:25–37

Houston DB (1973) Wildfires in northern Yellowstone National Park. Ecology 54:1111–1117

Ivanova GA (1998/1999) The history of forest fire in Russia. Dendrochronologia 16/17:147–161

Jackson JF, Adams DC, Jackson BU (1999) Allometry of constitutive defense: a model and a comparative test with tree bark and fire regime. Am Nat 153:614–632

Johnson EA (1979) Fire recurrence in the subarctic and its implications for vegetation composition. Can J Bot 57:1374–1379

Johnson EA, Larsen CPS (1991) Climatically induced change in fire frequency in the southern Canadian Rockies. Ecology 72:194–201

Johnson EA, Miyanishi K (2001) Forest fires – behaviour and ecological effects. Academic Press, San Diego

Johnson EA, Rowe JS (1975) Fire in the subarctic wintering ground of the Beverly caribou herd. Am Midl Nat 94:1–14

Johnson EA, Fryer GI, Heathcott MJ (1990) The influence of man and climate on frequency of fire in the interior wet belt forest, British-Columbia. J Ecol 78:403–412

Kasischke ES (2000) Boreal ecosystems in the global carbon cycle. In: Kasischke ES, Stocks BJ (eds) Fire, climate change, and carbon cycling in the boreal forest, Ecological Studies, vol 138. Springer, Berlin Heidelberg New York, pp 19–30

Kasischke ES, O'Neill KP, French NHF, Bourgeau-Chavez LL (2000) Controls on patterns of biomass burning in Alaskan boreal forests. In: Kasischke ES, Stocks BJ (eds) Fire, climate change, and carbon cycling in the boreal forest, Ecological Studies, vol 138. Springer, Berlin Heidelberg New York, pp 173–196

Kaufmann YJ, Remer LA, Ottmar RD, Ward DE, Li RR, Kleidman R, Fraser RS, Flynn L, McDougal D, Shelton G (1996) Relationship between remotely sensed fire intensity and rate of emission of smoke: SCAR-C Experiment. In: Levine JS (ed) Biomass burning and global change, vol II. MIT Press, Cambridge, MA, pp 685–696

Keeley JE, Bond WJ (2001) On incorporating fire into our thinking about natural ecosystems: a response to Saha and Howe. Am Nat 158:664–670

Keeley JE, Zedler P (1998) Evolution and life histories of *Pinus*. In: Richardson DM (ed) Ecology and biogeography of *Pinus*. Cambridge Univ Press, Cambridge

Kilgore BM (1981) Fire in ecosystem distribution and structure: Western forests and shrublands. Fire regimes and ecosystem properties: Proceedings of the Conference, vol WO-26. USDA Forest Service, Honolulu, pp 58–89

Kolchugina TP, Vinson TS (1995) Role of Russian forests in the global carbon balance. Ambio 24:258–264

Krebs CJ (1999) Ecological methodology. Benjamin/Cummings, Menlo Park

Larsen CPS (1997) Spatial and temporal variations in boreal forest fire frequency in northern Alberta. J Biogeogr 24:663–673

Kurz WA, Apps MJ (1999) A 70-year retrospective analysis of carbon fluxes in the Canadian forest sector. Ecol Appl 9:526–547

Laursen KK, Radke LF (1996) Biomass burning smoke in the tropics: from sources to sinks. In: Levine JS (ed) Biomass burning and global change, vol II. MIT Press, Cambridge, MA, pp 193–201

Lehtonen H, Huttunen P, Zetterberg P (1996) Influence of man on forest fire frequency in North Karelia, Finland, as evidenced by fire scars on Scots pines. Ann Bot Fenn 33:257–263

Liousse C, Penner JE, Chuang C, Walton JJ, Eddleman H, Cachier H (1996) A global three-dimensional model study of carbonaceous aerosols. J Geophys Res Atmos 101:19411–19432

Loope LL, Gruell GE (1973) The ecological role of fire in the Jackson Hole Area, Northwestern Wyoming. Quat Res 3:425–443

Lorbert JM, Warnatz J (1993) Emission from the Combustion Process in Vegetation. In: Crutzen PJ, Goldammer JG (eds) Fire in the environment – the ecological, atmospheric, and climatic importance of vegetation fires, 13th edn. Wiley, Chichester, pp 15–38

Lynham TJ, Stocks BJ (1991) The natural fire regime of an unprotected section of the boreal forest in Canada. 17th Tall Timbers fire ecology conference. Tall Timbers Research Station, Tallahassee, FL, pp 99–109

MacDonald GM, Cwynar LC, Whitlock C (1998) The late Quaternary dynamics of pines in northern America. In: Richardson DM (ed) Ecology and biogeography of pinus. Cambridge Univ Press, Cambridge, pp 122–152

Maikawa E, Kershaw KA (1976) Studies on lichen-dominated systems. XIX. The postfire recovery sequence of black spruce-lichen woodland in the Abitau Region, N. W. T. Can J Bot 54:2679–2687

Masters AM (1990) Changes in forest fire frequency in Kootenay-National-Park, Canadian Rockies. Can J Bot 68:1763–1767

McGuire AD, Wirth C, Apps M, Beringer J, Clein J, Epstein H, Kicklighter DW, Bhatti J, Chapin FS, de Groot B, Efremov D, Eugster W, Fukuda M, Gower T, Hinzman L, Huntley B, Jia GJ, Kasischke E, Melillo J, Romanovsky V, Shvidenko A, Vaganov E, Walker D (2002) Environmental variation, vegetation distribution, carbon dynamics and water/energy exchange at high latitudes. J Veg Sci 13:301–314

McLean DA, Woodley SJ, Weber MG, Wein RW (1983) Fire and nutrient cycling. In: Wein RW, McLean DA (eds) The role of fire in northern circumpolar ecosystems, 18th edn. Wiley, Chichester, pp 111–134

McRea DJ, Conard SG, Ivanova GA, Sukhinin AI, Hao WM, Koutzenougij KP, Baker SP, Ivanov VA, Samosonov YN, Churkina TV, Ivanov AV, Blake TW (2004) Fire regimes, variability in fire behaviour, and fire effects on combustion and chemical and carbon emission in Scots pine forests of Central Siberia. International Journal for Mitigation and Adaptation Strategies for Global Change (MITI) (in press)

Michalek JL, French NHF, Kasischke ES, Johnson RD, Colwell JE (2000) Using Landsat TM data to estimate carbon release from burned biomass in an Alaskan spruce forest complex. Int J Remote Sensing 21:323–338

Millar CI (2000) Early evolution of pines. In: DM Richardson (ed) Ecology and biogeography of *Pinus*, Cambridge University Press, Cambridge

Mollicone D, Achard F, Marchesini LB, Federici S, Wirth C, Leipold M, Rosellini S, Schulze ED, Valentini R (2002) A remote sensing based approach to determine forest fire cycle: case study of the Yenisei Ridge dark taiga. Tellus B54:688–695

Murphy PJ, Mudd JP, Stocks BJ, Kasischke ES, Barry D, Alexander ME, French NHF (2000) Historical fire records in the North American boreal forest. In: Kasischke ES, Stocks BJ (eds) Fire, climate change, and carbon cycling in the boreal forest, vol 138. Springer, Berlin Heidelberg New York, pp 274–288

Neary DG, Klopatek CC, DeBano LF, Ffolliott PF (1999) Fire effects on belowground sustainability: a review and synthesis. For Ecol Manage 122:51–71

Nikolov N, Helmisaari H (1992) Silvics of the circumpolar boreal forest species. In: Shugart HH, Leemans R, Bonan GB (eds) A systems analysis of the global boreal forest. Cambridge Univ Press, Cambridge, p 565

Nilsson S, Shvidenko A, Stolbovoi V, Gluck M, Jonas M, Obersteiner M (2000) Full carbon account of Russia. interim report 00-021, International Institute for Applied Systems Analysis, Laxenburg, Austria

Olsen JS (1981) Carbon balance in relation to fire regimes. Fire regimes and ecosystem properties: proceedings of the conference, vol WO-26. USDA Forest Service, Honolulu, pp 327–378

Payette S, Morneau C, Sirois L, Desponts M (1989) Recent fire history of the northern Quebec biomes. Ecology 70:656–673

Pitkäninen, A, Grönlund E (2001) A 600-year forest fire record in a varved lake sediment (Ristijärvi, northern Karelia, eastern Finland). Ann Bot Fenn 38: 63–73

Poorter H, Villar R (1997) The fate of acquired carbon in plants: chemical composition and construction costs. In: Bazzaz FA, Grace J (eds) Plant resource allocation. Academic Press, San Diego, pp 39–72

Prentice IC, Cramer W, Harrison SP, Leemans R, Monserud RA, Solomon AM (1992) A global biome model based on plant physiology and dominance, soil properties and climate. J Biogeogr 19:117–134

Raison RJ (1979) Modification of the soil environment by vegetation fires, with particular reference to nitrogen transformations – review. Plant Soil 51:73–108

Roberts MR, Gilliam FS (1995) Patterns and mechanisms of plant diversity in forested ecosystems – implications for forest management. Ecol Appl 5:969–977

Romme WH (1982) Fire and landscape diversity in subalpine forests of Yellowstone-National-Park. Ecol Monogr 52:199–221

Romme WH, Knight DH (1981) Fire frequency and subalpine forest succession along a topographic gradient in Wyoming. Ecology 62:319–326

Rowe JS (1983) Concepts of fire effects on plant individuals and species. In: Wein RW, McLean DA (eds) The role of fire in northern circumpolar ecosystems, 18th edn. Wiley, Chichester

Rowe JS, Spittelhouse D, Johnson E, Jasenuik M (1975) Fire studies in the upper Mackenzie Valley and adjacent Precambrian uplands. Canadian Department of Indian and Northern Affairs (DINA), Arctic Land Use Research. Information, Ottawa, Canada

Rundel PW (1981) Structural and chemical components of flammability. Fire regimes and ecosystem properties: proceedings of the conference, vol WO-26. USDA Forest Service, Honolulu, pp 183–207

Ryan KC, Reinhardt ED (1988) Predicting postfire mortality of 7 western conifers. Can J For Res 18:1291–1297

Sannikov SN, Goldammer JG (1996) Fire ecology of pine forests of northern Eurasia. In: Goldammer JG, Furyaev VV (eds) Fire in ecosystems of boreal Eurasia, Forestry Sciences, vol 48. Kluver, Dordrecht, pp 151–167

Schulze ED, Schulze W, Kelliher FM, Vygodskaya NN, Ziegler W, Kobak KI, Koch H, Arneth A, Kusnetsova WA, Sogachev A, Issajev A, Bauer G, Hollinger DY (1995) Aboveground biomass and nitrogen nutrition in a chronosequence of pristine dahurian larix stands in eastern Siberia. Can J For Res 25:943–960

Schwilk DW, Ackerly DD (2001) Flammability and serotiny as strategies: correlated evolution in pines. Oikos 94:326–336

Shepashenko D, Shvidenko A, Nilsson S (1998) Phytomass (live biomass) and carbon of Siberian forests. Biomass Bioenerg 14:21–31

Sheppard PR, Lassoie JP (1998) Fire regime of the Lodgepole pine forest of Mt. San Jacinto, California. Madrono 45:47–56

Shvidenko A, Nilsson S (1994) What do we know about the Siberian forests. Ambio 23:396–404

Shvidenko A, Nilsson S (2000a) Extent, distribution, and ecological role of fire in Russian forests. In: Kasischke ES, Stocks BJ (eds) Fire, climate change, and carbon cycling in the boreal forest, Ecological Studies, vol 138. Springer, Berlin Heidelberg New York, pp 132–150

Shvidenko A, Nilsson S (2000b) Fire and the carbon budget of Russian forests. In: Kasischke ES, Stocks BJ (eds) Fire, climate change, and carbon cycling in the boreal forest, Ecological Studies, vol 138, Springer, Berlin Heidelberg New York, pp 289–311

Shvidenko A, Nilsson S (2002) Dynamics of Russian forests and the carbon budget in 1961–1998: an assessment based on long-term forest inventory data. Clim Change 55:5–37

Sneck KM (1970) The fire history Coram experimental forest. University of Montana, Missoula, p 134

Snytkin GV (1996) Fires in ecosystem of the far northeast of Siberia. In: Goldammer JG, Furyaev VV (eds) Fire in ecosystems of boreal Eurasia, Forestry Sciences, vol 48. Kluver, Dordrecht, pp 197–210

Stocks BJ (1989) Fire behavior in mature jack pine. Can J For Res 19:783–790

Stocks BJ, Cahoon DRJ, Cofer WR III, Levine JS (1996) Monitoring large-scale forest fire behavior in Northeastern Siberia using NOAA-AVHRR satellite imagery. In: Levine JS (ed) Biomass burning and global change, vol II. MIT Press, Cambridge, MA, pp 802–807

Stuart JD, Salazar LA (2000) Fire history of white fir forests in the coastal mountains of northwestern California. Northwest Science 74:280–285

Swetnam TW (1993) Fire history and climate-change in giant sequoia groves. Science 262:885–889

Sykes MT (2001) Modelling the potential distribution and community dynamics of lodgepole pine (*Pinus contorta* Dougl. ex. Loud.) in Scandinavia. For Ecol Manage 141:69–84

Tande GF (1977) Forest fire history around Jasper Townsite, Jasper National Park, Alberta. Department of Botany. University of Alberta, Edmonton, p 169

Tande GF (1979) Fire history and vegetation pattern of coniferous forests in Jasper National Park, Alberta. Can J Bot 57:1912–1931

Taylor KL, Fonda RW (1990) Woody fuel structure and fire in sub-alpine fir forests, Olympic-National-Park, Washington. Can J For Res 20:193–199

Tilman D, Lehman C (2001) Biodiversity, composition, and ecosystem processes: theory and concepts. In: Kinzig AP, Pacala SW, Tilman D (eds) The functional consequences of biodiversity – empirical progress and theoretical extensions. Princeton Univ Press, Princeton, pp 9–41

Tsvetkov PA (1996) Fire effects on larch forests. In: Goldammer JG, Furyaev VV (eds) Fire in ecosystems of Boreal Eurasia, Forestry Sciences, vol 48. Kluver, Dordrecht, pp 387–392

United Nations Economic Commission for Europe, Food and Agriculture Organisation UN-ECE/FAO (2000) Forest Resources of Europe, CIS, North America, Australia, Japan, New Zealand. Geneva Timber and Forest Study Papers, No 17 United Nations Publications, New York, Geneva

Utkin A (1965) Forests of Central Yakutia. Nauka Publ, Moscow

Vaganov EA, Arbatskaya MK, Shashkin AV (1996) Climate history and fire incidence in the central part of Krasnoyarsk Krai. II. Dendrochronological analysis of relationship between variability of tree increment, climate, and fire incidence. Sib J Ecol 1:19–27

Valendik EN (1996) Ecological aspects of forest fires in Siberia. Sib J Ecol 1:1–8

Van Wagner CE (1972) Height of crown scorch in forest fires. Can J For Res 3:373–378

Van Wagner CE (1978) Age-class distribution and forest fire cycle. Can J For Res 8:220–227

Van Wagner CE (1983) Fire behaviour in northern conifer forests and shrublands. In: Wein RW, McLean DA (eds) The role of fire in northern circumpolar ecosystems, 18th edn. Wiley, Chichester, pp 65–80

Viereck LA (1973) Wildfire in the Taiga of Alaska. Quat Res 3:465–495

Vines RG (1968) Heat transfer through bark, and the resistance of trees to fire. Aust J Bot 16:499–514

Vitousek PM, Howarth RW (1991) Nitrogen limitation on land and in the sea – how can it occur. Biogeochemistry 13:87–115

Vygodskaya NN, Schulze ED, Tchebakova NM, Karpachevskii LO, Kozlov, Sidorov KN, Panfyorov MI, Abrazko MA, Shaposhnikov ES, Solnzeva ON, Minaeva TY, Jeltuchin AS, Wirth C, Pugachevskii AV (2002) Climatic control of stand thinning in unmanaged spruce forests of the southern taiga in European Russia. Tellus B 54:443–461

Waldrop TA, Brose PH (1999) A comparison of fire intensity levels for stand replacement of table mountain pine (*Pinus pungens* Lamb.). For Ecol Manage 113:155–166

Wang CK, Gower ST, Wang YH, Zhao HX, Yan P, Bond-Lamberty BP (2001) The influence of fire on carbon distribution and net primary production of boreal *Larix gmelinii* forests in north-eastern China. Global Change Biol 7:719–730

Weaver H (1959) Ecological changes in the ponderosa pine forest of the warm springs Indian reservation in Oregon. J For 57:15–20

Wirth C, Schulze ED, Schulze W, von Stünzner-Karbe D, Ziegler W, Miljukova IM, Sogatchev A, Varlagin AB, Panvyorov M, Grigoriev S, Kusnetzova W, Siry M, Hardes G, Zimmermann R, Vygodskaya NN (1999) Above-ground biomass and structure of pristine Siberian Scots pine forests as controlled by competition and fire. Oecologia 121:66–80

Wirth C, Czimczik CI, Schulze E-D (2002a) Beyond annual budgets: carbon flux at different temporal scales in fire-prone Siberian Scots pine forests. Tellus B 54:611–630

Wirth C, Schulze ED, Lühker B, Grigoriev S, Siry M, Hardes G, Ziegler W, Backor M, Bauer G, Vygodskaya NN (2002b) Fire and site type effects on the long-term carbon and nitrogen balance in pristine Siberian Scots pine forests. Plant Soil 242:41–63

Wirth C, Schulze E-D, Kusznetova V, Milyukova I, Hardes G, Siry M, Schulze B, Vygodskaya NN (2002 c) Comparing the influence of site quality, stand age, fire and climate on aboveground tree production in Siberian Scots pine forests. Tree Physiol 22:537–552

Wright HA, Bailey AW (1982) Fire ecology – United States and southern Canada. Wiley, New York

Yarie J (1980) Forest fire cycles and life tables: a case study from interior Alaska. Can J For Res 11:554–562

Yarie J, Billings S (2002) Carbon balance of the taiga forest within Alaska: present and future. Can J For Res 32:757–767

Yevdokimenko MD (1996) Fire-induced transformations in the productivity of light coniferous stands of the Trans-Baikal region and Mongolia. In: Goldammer JG, Furyaev VV (eds) Fire in ecosystems of Boreal Eurasia, Forestry Sciences, vol 48. Kluver, Dordrecht, pp 211–218

Zackrisson O (1977) Influence of forest fires on the North Swedish boreal forest. Oikos 29:22–32

Zasada JC, Sharik TL, Nygren M (1992) The reproductive process in boreal forest trees. In: Shugart HH, Leemans R, Bonan GB (eds) A systems analysis of the global boreal forest. Cambridge Univ Press, Cambridge, pp 85–125

Part E Perspectives

16 The Design of Experimental Tree Plantations for Functional Biodiversity Research

M. Scherer-Lorenzen, C. Potvin, J. Koricheva, B. Schmid,
A. Hector, Z. Bornik, G. Reynolds, and E.-D. Schulze

16.1 Introduction, or "Why Do We Need Diversity Experiments with Trees?"

One way to assess the functional significance of biodiversity in forests would be to compare existing stands of contrasting diversity and to study ecosystem functioning and/or the provision of ecosystem goods and services in these stands (see contributions in this book). However, unless site conditions are extremely similar, across-habitat or across-locality comparisons can be misleading, because environmental differences between stands may hide potential within-habitat effects of differences in biodiversity on ecosystem processes in such sample surveys or comparative studies (Lawton et al. 1998; Schmid 2002; Vilá, Chap. 4, this Vol.). Almost all forests, particularly in Europe, have been managed for long time periods, so land-use history will certainly have long-term influences on both biodiversity and ecological processes (Mund and Schulze, Chap. 10, this Vol.). Thus, sample surveys or comparative studies can be used to document correlations between diversity and ecosystem processes, but they cannot be used to establish causality or underlying mechanisms of this relationship (Caspersen and Pacala 2001). Therefore, manipulative experiments, with random allocation of biodiversity treatments to plots while keeping environmental conditions as constant as possible, are needed to complement observational diversity–functioning studies. Such experimental approaches have successfully been applied during the last decade in the study of the interlinkages between biodiversity and ecosystem functioning of grasslands, and of terrestrial and aquatic microcosms (Kinzig et al. 2002; Loreau et al. 2002). In our view, an intriguing challenge will be the adoption of the basic principles and lessons learned from the design and interpretation of these small-scale and short-term experiments to long-lived forest communities.

As life cycles of tree species are in the order of decades or centuries, such experiments have to be planned with a long-term perspective. Such long-last-

Ecological Studies, Vol. 176
M. Scherer-Lorenzen, Ch. Körner, and E.-D. Schulze (Eds.)
Forest Diversity and Function: Temperate and Boreal Systems

ing studies exceed normal funding periods, but are regularly demanded by scientific advisory bodies (e.g., WBGU 2000). However, experiences from silvicultural investigations show that the maintenance of such long-term experiments is generally possible.

In this chapter, we discuss the setup of long-term biodiversity experiments manipulating tree species diversity. We do not discuss general aspects of the design and interpretation of manipulative biodiversity–ecosystem functioning experiments; these have been discussed in depth elsewhere (e.g., Huston and McBride 2002; Schmid et al. 2002).

16.2 Experimental Approaches

In order to study the relation between tree diversity and ecosystem functioning, a gradient of species number ranging from single-species stands to multi-species mixtures is necessary. Alternatively, other aspects of diversity such as functional diversity or genetic diversity may be manipulated. In principle, such a gradient may be created by two different approaches. First, certain species may be removed from an already established natural or semi-natural multi-species forest stand (so-called "removal experiments"). The inverse case, i.e., "addition experiments", where species are added to an existing monoculture or low diversity stand, may not be feasible in mature stands, but would merit consideration in recently planted afforestations or natural regrowth in gaps or after clear-cut. Removal or addition experiments have some drawbacks (e.g., large disturbance effects, change in density, spatial segregation of species), but can be useful under certain circumstances (Freckleton and Watkinson 2000; Díaz et al. 2003). In the second approach, forest stands differing in tree diversity may be created by new planting, similar to the experiments with herbaceous species (e.g., Hector et al. 1999; Tilman et al. 2001). It may thus be called the "synthetic-community approach" – on which we will focus in this chapter.

16.2.1 Basics: Some Population Biological Rules to Be Considered in Plantation Experiments

There are several aspects specific to plants that should be considered in biodiversity experiments. In particular, plants as modular organisms grow continuously by adding new parts to their "body" and, therefore, individuals can vary greatly in size (Harper 1977; Hallé 1986). Indeed, an individual woody plant can itself be viewed as a population of twigs and branches. In most plantation experiments, stand or community dynamics are entirely due to processes at this within-plant population level: individuals increase or decrease in size, they may die, but usually there is no new generation establishing itself from seeds. Two basic rules have been observed in such one-generation plantation experiments using single species. The constant-yield law (Kira et al. 1953) states that over a large range of planting densities the size of individuals is adjusted plastically in such a way that the total amount of biomass in the stand remains constant. If planting densities are very high, however, size plasticity cannot absorb all competition pressure and therefore some individuals die. This leads to an ordered process of size-dependent plant death described as the self-thinning rule (Yoda et al. 1963), in which, for each mortality-related reduction in density, the final yield of a stand increases by a greater amount than accounted for by the losses.

The constant-yield law and the self-thinning rule suggest that similar processes may occur in plant stands consisting of a mixture of species. It is therefore important that biodiversity experiments control for potentially confounding effects of density. This is most often done by holding total density constant and, instead of adding individuals of different species to a mixture, substituting a number of individuals of one species with the same number of individuals of another species (so-called substitutive experiments or replacement series, see e.g., de Wit 1960; Harper 1977). This approach assumes a null hypothesis of equivalence of individuals between species. If this assumption is not justified, a group of individuals or a patch of unit size may be used instead, although the term "total density" then loses some of its meaning for mixed stands. Similarly, the term "abundance" in a plant stand may sometimes be better defined in terms of cover or biomass than in terms of number of individuals, especially if these vary in size both between and within species.

16.2.2 The Use of Existing Experiments in Forestry

For a long time, foresters tried to understand the factors influencing the performance of forest stands (in terms of growth, timber yield, nutrient-use efficiency, or stability) using experimental plantations. However, different species have mostly been grown only in monocultures or two-species mixtures. Nevertheless, these experiments have yielded a wealth of information concerning the effects of species mixtures on ecosystem processes (Jones et al., Chap. 6, this Vol.). Because establishing a new forest diversity experiment is an expensive and time-consuming task, and is unlikely to provide many results during the first few years, and because environmental conditions are very variable, it may be appropriate to identify already existing experiments from forestry with similar goals and designs. Such existing experiments may complement the new ones, for example, by allowing comparisons of different stages of stand development at the same time, or by elucidating differences related to soil chemistry. Although forestry plantations were not planned from a biodiversity perspective, and only cover the very low end of the diversity gradient (usually one or two species), they may save some tests and treatments in new experiments, yield extra data, and promote cooperation between forestry and ecosystem science.

Suitable tools to find such experiments are databases on the Internet, which can be used to search for different tree species, mixtures, ages, objectives and geographical locations. Examples from Europe include the European Forest Ecosystem Research Network EFERN (http://ifff.boku.ac.at/efern/) and the Nordic Database for Long-Term Forest Experiments NOLTFOX (http://noltfox.metla.fi/).

16.2.3 A New Generation of Biodiversity Experiments with Trees

Very recently, ecologists have begun to perform manipulative biodiversity experiments with multiple species of trees, similar to those in grasslands or aquatic ecosystems that exceed the range of species grown together in traditional silvicultural experiments. To our knowledge, there are only seven experiments of this type: two in Finland, two in Germany, two in Panama, and one in Borneo (Table 16.1). We do not consider here the experiment by Ewel and colleagues (Berish and Ewel 1988; Ewel et al. 1991), which was not designed to test diversity effects on ecosystem processes, but to explore the possibility of using natural succession as a model for sustainable, low-input agroecosystems for the humid tropics. However, the experiment was analyzed for its diversity effects by Vitousek and Hooper (1993), who pointed out that the results were consistent with an effect of biodiversity on biogeochemistry, but that they did not prove a relationship, mainly because all of the observed dynamics occurred in the transition between maize (planted as monoculture in the first two years) and highly diverse treatments consisting of more than 100 plant species.

16.2.3.1 The Boreal Tree Diversity Experiments in Finland

The Finnish tree species diversity experiment (coordination: Julia Koricheva) was established in spring 1999 on three clear-cut areas (about 1.5–2 ha each) located 20–30 km from each other in the Satakunta area, western Finland (61° N, 22° E). The sites are within the boreal coniferous forest belt, dominated by spruce, Scots pine, and birch. Each experimental area contains 38 plots randomly allocated to 19 treatments which represent monocultures and two-, three- and five-species combinations of five tree species (see Fig. 16.1, Table 16.1). Tree species used in the experiment include the locally dominating and economically important species for Finland (*Pinus sylvestris*, *Picea abies*, and *Betula pendula*), one nitrogen-fixing species (*Alnus glutinosa*), and one exotic conifer (*Larix sibirica*). Species mixtures are composed in such a

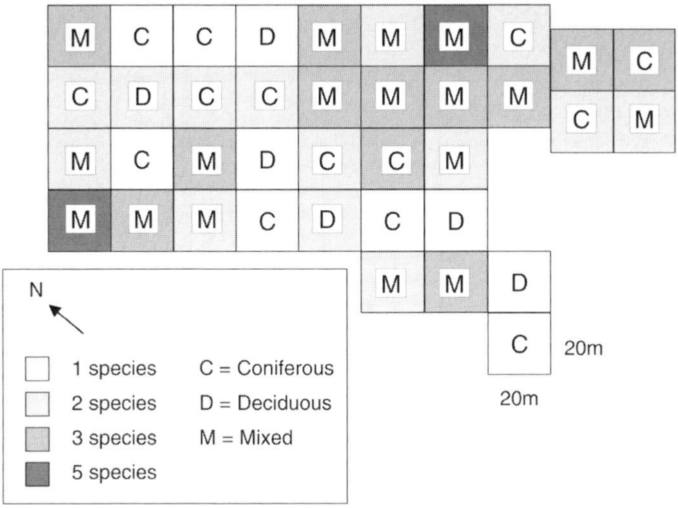

Fig. 16.1. Experimental layout at one of the three sites (Pomarkku) of the Finnish diversity experiment with boreal tree species

way that they represent a gradient from completely coniferous forest (pine, spruce, and larch) through mixed conifer/deciduous stands to deciduous ones (birch and alder). There are two replicates of each species mixture per area (six replicates altogether). Mixed plots contain the same number of saplings of each species; the positions of saplings of each species are randomized within plots.

Another experiment established in the Satakunta area in summer 2000 is aimed at studying the effects of within-species genetic diversity. It consists of an approximately 2-ha clear-cut area which contains 49 plots planted with micropropagated plantlets of eight clones of silver birch (*Betula pendula*). Plots are randomly allocated to the following treatments: single-clone stands, five different two-clone mixtures, five different four-clone mixtures, and eight-clone mixtures (Table 16.1). Each particular clone combination is replicated two to three times within the experimental area to allow the separation of effects of the number of clones, particular clone identity or mixture, and residual variation among plots of identical clone composition. Birch clones chosen for this experiment are of southern Finnish origin and are known to display different degrees of resistance to herbivores and pathogens. Clone combinations were selected on the basis of clone resistance, e.g., two-clone mixtures include either two susceptible clones, two resistant clones or one susceptible and one resistant clone. This design allows testing of the "associational resistance hypothesis," which predicts that susceptible plants growing in association with genetically or taxonomically diverse plants may gain benefits in terms of reduced herbivore or pathogen attacks (Andow 1991).

Table 16.1. Main characteristics of existing biodiversity experiments with trees

Biome	Country	Planting	No. of sites	Plot size	Total no. of plots	Diversity variables	Diversity gradients	Replication[a]	Size of species pool	Tree-to-tree planting distance
Boreal	Finland	1999	3	0.04 ha	114	Species richness Number of functional groups	1, 2, 3, 5 Species Deciduous/coniferous vs. mixed	s, f, m	5	1.5 m
Boreal	Finland	2000	1	0.04 ha	49	Within-species genetic diversity of *Betula pendula*	1, 2, 4, 8 Clones	c, cm	8 Clones	2 m
Temperate	Germany	2003/2004	2	1.2 ha 0.6 ha	40 16	Species richness Species richness	1, 2, 4, 6 Species 1, 2, 3, 4 Species	s, m[b] s, m[b]	6 4	2 m[c] 2 m[c]
Temperate	Germany	2003	1	0.2 ha	25	Functional diversity (FD[d])	Very low, low, high, very high FD	FD	16	2 m, 1 m[e]
Tropical	Panama	2001	1	0.2 ha	24	Species richness Identity of functional groups	1, 3, 6 Species Pioneer, shade-tolerant, intermediate	s, m[b]	6	3 m
Tropical	Panama	2003	1	0.03 ha	24	Species richness Assemblage identity	6, 9, 18 Species 4 Different assemblages	s, m	28	3 m
Tropical	Borneo	2002/2003	1	4.0 ha	124	Species richness Generic diversity (no. of genera) Tree height (short, medium, tall)	1, 4, 16 Species 2 vs. 4 Genera 2 vs. 3 Height classes	s, g, h s, g, h	16 16	10 m, 3 m[f] 10 m, 3 m[f]

[a] Given are the experimental factors that are replicated. *s* each level of species richness is replicated with different species mixtures, except the highest level. *f* each level of functional group richness is replicated with different species mixtures. *m* each particular species mixture (= individual species composition or assemblage) is replicated identically. *c* each level of clone diversity is replicated with different clone mixtures. *cm* each particular clone mixture (= individual clone composition) is replicated identically. *FD* each level of FD is replicated with different species mixtures. *g* each level of generic diversity is replicated with different species mixtures. *h* each level of height diversity is replicated with different species mixtures

[b] Individual species mixtures are replicated only at the highest diversity level, but not at lower diversity levels

[c] Distance between rows: 2 m; distance within row, depending on species (1.0, 1.5, or 2.0 m)

[d] Based on FD approach by Petchey and Gaston (2002)

[e] Distance between rows: 2 m; distance within rows: 1 m

[f] Distance between rows: 10 m; distance within rows: 3 m

16.2.3.2 BIOTREE: BIOdiversity and Ecosystem Processes in Experimental TREE Stands

In Germany, two experiments with temperate tree species have been planned since 2000, and have now been established in close cooperation between the Max Planck Institute for Biogeochemistry and the State Forest Research Institution of Thuringia (coordination: Michael Scherer-Lorenzen and Ernst-Detlef Schulze). The planting of a total of 250,000 trees was started in spring 2003, covering a total area of approximately 80 ha, located at three sites (51° N, 11° E) with different geology (one acidic and two calcareous sites, see Table 16.1, Fig. 16.2). The natural vegetation would be beech-dominated forests with oak, the latter being in higher abundances at the drier calcareous site. Prior to planting, the sites were in agricultural use until the mid-twentieth century and then converted into mown and grazed grasslands.

Two different approaches have been used to establish a gradient in tree diversity. First, varying the number of tree species (BIOTREE-SPECIES), and

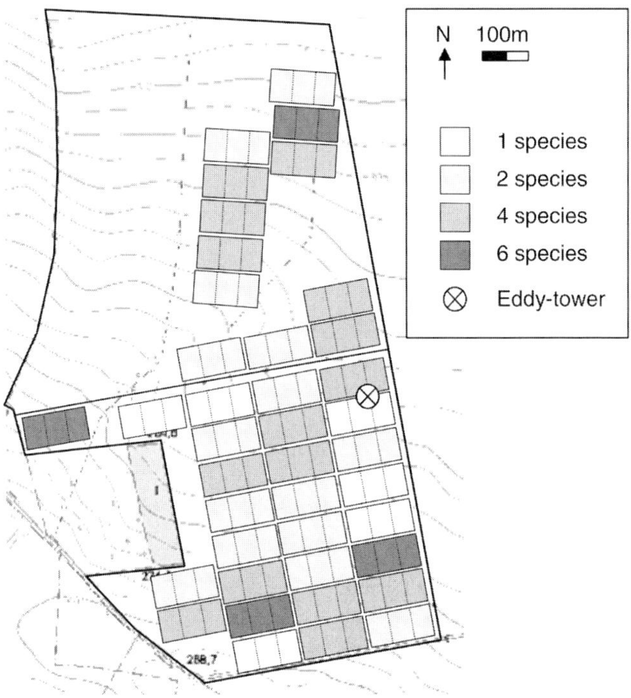

Fig. 16.2. Experimental layout at one of the three sites (Mehrstedt, calcareous) of the German diversity experiment BIOTREE with temperate tree species. Ecosystem C fluxes will be measured by eddy covariance over the plantation

second, varying the functional diversity within four-species mixtures (BIOTREE-FD). Within BIOTREE-SPECIES, all possible species combinations at four diversity levels (one, two, three, four species at the acidic site, and one, two, four, six species at the calcareous site, respectively) are grown, which enables the following hypotheses to be tested: (1) that random species loss matters for ecosystem functioning, (2) that particular species affect ecosystem functioning, and (3) that there is a minimum set of complementary species that is sufficient to explain diversity effects (Spaèkova and Lepš 2001; Loreau et al. 2002; Schmid et al. 2002). Similar to the Panama experiment, the diversity levels are replicated with different species mixtures, while replications of the specific mixture are only done at the highest diversity level. Besides the manipulation of tree species richness, the impact of silvicultural management and the addition of rare species are considered, using a split-plot design with three treatments: unmanaged ("U"), managed according to prevalent silvicultural practices ("M"), and managed with additional species ("M+"; Fig. 16.3). The comparison between unmanaged and managed subplots will help to solve the debate about whether diversity effects might be obscured by silvicultural practice (Mund and Schulze, Chap. 10, this Vol.). The inclusion of subplot "M+" allows the testing of the additional hypothesis that the deletion of subdominant or rare species is irrelevant for ecosystem functioning (Walker et al. 1999). Species have been planted in a random checkerboard pattern of 64 m^2 patches for each species (Fig. 16.3), thus minimizing

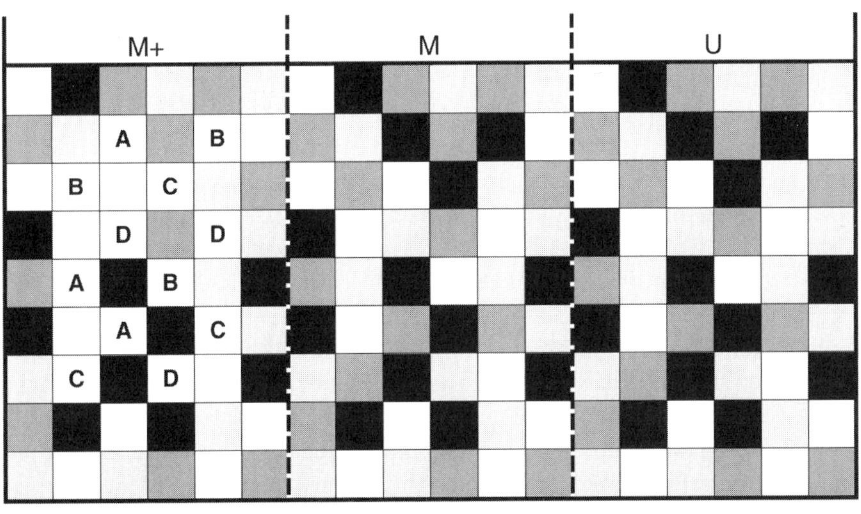

Fig. 16.3. Within-plot design of the BIOTREE-SPECIES experiment in Germany, shown for a four-species mixture as an example. Different *gray scales* represent different species, different *letters* represent different subdominant species. Each single patch has an area of 8×8 m and was planted in rows with 2 m row-to-row distance. *M+* Managed with addition of subdominant species, *M* managed, *U* unmanaged

species loss due to interspecific competition at an early stage of establishment. This should also prevent dominance of species with particular traits (at least during the first decades until species start to regenerate), which reduces the possibility that "selection effects" (sensu Loreau and Hector 2001) may occur.

In contrast to BIOTREE-SPECIES, the diversity gradient in the BIOTREE-FD experiment has not been established by manipulating the number of species, but by mixing species with different functional attributes. Based on the "FD"-approach by Petchey and Gaston (2002), the functional diversity (FD) of all 1,820 possible four-species mixtures from a pool of 16 tree species was calculated. The trait matrix contained nine criteria for which data for all species were available. The criteria selected represent attributes which are indicative for complementary resource use and nutrient cycling, the two main functions of interest of this experiment. Six mixtures were randomly selected out of four groups, representing "very low" (i.e., a mixture of functionally very similar species), "low", "high", and "very high" (i.e., a mixture of functionally very different species) functional diversity (Table 16.1).

16.2.3.3 The Forest Biodiversity Experiments of Panama

The Forest Biodiversity Experiment of Panama (coordination: Catherine Potvin) was specifically designed to test how increases in tree species richness might affect the cycling and storing of carbon (C). The native tropical semi-deciduous lowland forest at the experimental site 55 km north of Panama City (9° N, 79° W) was logged in 1952/1953. The area was used for agriculture for 2 years and then converted into pasture by seeding grasses.

An area of 9 ha was planted with six native tree species: two pioneer (*Luehea seemanii* and *Cordia alliodora*), two light-intermediate (*Anacardium excelsum* and *Hura crepitans*), and two shade-tolerant species (*Cedrela odorata* and *Tabebuia rosea*). Functional groups were classified based on relative growth rates (9.1 and 7.0 %; 5.9 and 4.9 %; 2.3 and 3.4 %, respectively) and on frequency of encounter in gaps or closed forests of the 50-ha permanent plot of Barro Colorado Island (BCI). Twenty-four diversity plots were established with either one, three, or six species (Fig. 16.4, Table 16.1). The design allows testing for the effect of species richness without confounding it with species identity. It also controls the effect of certain functional groups, but the number of functional groups is not varied. Thus, using the monoculture plots, one can test for a functional-group effect but cannot test the relation between functional diversity and ecosystem functioning. The basic analysis of the variance (ANOVA) model compares six monocultures, six triplets and six six-species plots. In this model, the type of replication differs among diversity levels, i.e., the triplets differ in species composition while the six-species plots do not (Table 16.1). This might potentially lead to violation of the homogeneity-

Fig, 16.4. Experimental layout of the main plantation in the Forest Biodiversity Experiment of Panama with tropical tree species. Ecosystem C fluxes will be measured by eddy covariance over the plantation and over an adjacent pasture. The *letters* represents different species grown in monoculture

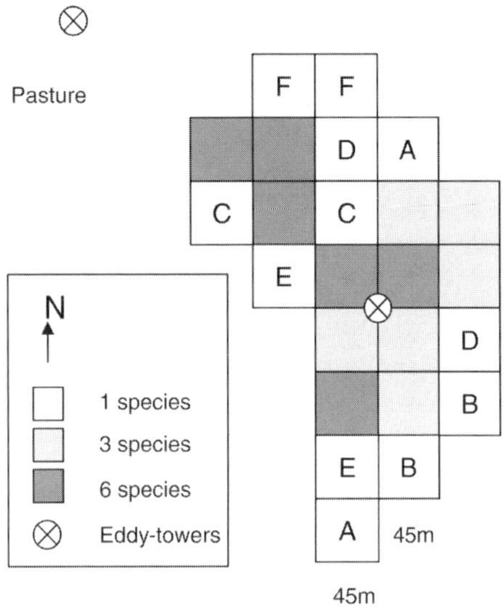

of-variance assumption. If needed, corrections, e.g., those developed by Dutilleul and Potvin (1995) in the context of genotype-by-environment analysis, will be applied.

The results of BIODEPTH and other grassland experiments suggested that the diversity-by-function relationship might be asymptotic (e.g., Hector et al. 1999; Tilman et al. 2001). Thus, the main experiment in Panama was set up with plots of one, three and six species. We hypothesized that these three levels of species richness would be within the linear portion of the diversity–ecosystem functioning relationship. This design however leaves two questions unanswered: (1) is the high diversity treatment relevant to the natural level of tree diversity in the forest? (2) What is the importance of community composition at high diversity? Another experiment was therefore set up in July 2003 in which high-diversity plots, comparable to the diversity of natural forests, were established (Table 16.1). The aim of this second experiment was to study the asymptotic portion of the diversity vs. functioning relationship. In addition, the plantation specifically accounts for environmental heterogeneity by embedding replicated blocks in the landscape. Eight different blocks, containing three plots of 36 saplings, were planted on either steep slopes or flatter landscape. The blocks correspond to four different species assemblages, each replicated on steep and flat landscapes. The idea of multiple experiments at one site could be a practical way to address more than one question while remaining in budget.

The Panama Forest Biodiversity Experiment provides preliminary data allowing examination of the adequacy of the experimental approach. Six months after planting, basal diameter, representing secondary growth

(Kohyama and Hotta 1990), was measured for each individual planted seedling. Two different nested ANOVAs were used to test for the effect of (1) species richness and (2) species identity on seedlings' diameter. In both ANOVAs, plots were nested under the main effect of interest (six plots per level for species richness and two plots per level for species identity). Accordingly, the mean square (MS) for plots was used as the MS error for either species richness or species identity. The ANOVAs indicated that species identity exerted a significant effect on mean diameter ($F_{4,5}$=7.05, P<0.05), and that the effect of species richness was nearly, but not quite, statistically significant ($F_{2,15}$=2.53, P<0.08). The six species differed significantly from each other, the largest one being *Tabebuia rosea* (13.7±5.2 cm), while the smallest was *Luehea seemanii* (6.9±3.8 cm). Although the difference was not statistically significant, seedlings tended to be smallest in the monoculture plots and largest in the triplets (9.6±5.1 cm for monocultures and 13.5±8.5 cm for triplets).

Several authors suggested that productivity may increase with diversity because of trait complementarity among species (e.g., Hector et al. 1999; Tilman et al. 2001). The observation that seedlings planted in the monoculture tended to be the smallest supports that hypothesis. We anticipate that, with time, the effect of species richness on seedling growth and plot productivity will become stronger and clearer as saplings begin to compete for light and nutrients.

16.2.3.4 The Sabah Tropical Forest Biodiversity Experiment, Malaysian Borneo

The southeast Asian dipterocarp forests are some of the most diverse on earth, and are also the most productive tropical forests in terms of timber yield. Because large areas have been lost through logging, replanting schemes have been developed that aim to provide future logging timber, and to offset industrial C emissions. These plantations generally consist of monocultures or mixtures much lower in species diversity than natural forests. However, the availability of seedlings from a wide variety of species means that replanting could also be conducted at levels closer to natural diversity. The aim of the Sabah Forest Biodiversity Experiment (Sabah is the eastern state of Malaysian Borneo) is to compare community and ecosystem processes in replanted plots of low and high tree diversity (Holden 2003). The experiment (coordination: Andy Hector and Charles Godfray) is a collaboration between the NERC Centre for Population Biology at Imperial College, The Royal Society's research station in Danum Valley, Sabah, and the Innoprise Corporation (carbon offset, timber replanting, and forestry).

The native dipterocarp forest at the site (5° N, 118° W) had been logged in the early 1990s and then left unmanaged until replanting started in

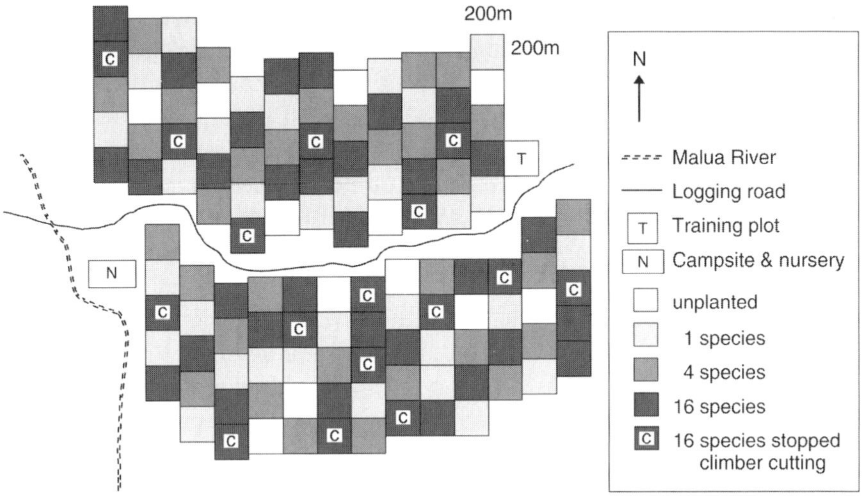

Fig. 16.5. Experimental layout of the Sabah Forest Biodiversity Experiment with tropical tree species

2002/2003. The total area of the experiment is approximately 500 ha (Fig. 16.5). Sixteen species were available for the experiment and these are being grown alone and all together to provide the two possible extremes of diversity along with a number of intermediate four-species mixtures (total of 96 plots, see Table 16.1). The 16 four-species treatments are a factorial design that separates the effects of generic diversity (2 vs. 4) from canopy thickness (combinations of short, medium, plus tall species versus less varied combinations of only tall plus medium or short plus medium). In addition, 12 plots were left as uncut and unplanted controls. A further 16 plots were planted with the 16-species mixture, but will receive reduced climber cutting once the trees are established – to address this usual management option. Finally, in the center of each plot, two replicate seedlings of each of the species were planted to compare levels of herbivory and mortality of the 16 dipterocarps in the different background communities provided by the experimental plots.

16.3 Methodological and Design Considerations

There are numerous difficulties in designing, executing, and interpreting manipulative diversity experiments (Lamont 1995; Huston 1997; Allison 1999; Huston and McBride 2002; Mikola et al. 2002; Schmid et al. 2002), and different designs may yield different answers to the same research question. Obviously, there cannot be one single optimal design for an experiment to analyze the relation between tree species diversity and ecosystem functioning, and/or the provision of goods and services. Even if the aim of the study is defined in such a broad way, the functions or response variables of interest may be quite distinct, requiring different designs. For instance, if the aim of the project is more applied (e.g., to provide recommendations for forest management strategies), the design should be based more closely on current forestry practices. In contrast, if the main interest is on the effects of species numbers the inclusion of certain mixtures that would never be planted by foresters because they are of no economic value becomes more important, as done in the BIOTREE experiment, for example. Additionally, logistic constraints and limitations of resources very often prevent the adoption of an appropriate design so that pragmatic compromises have to be chosen. Below we discuss some factors which may affect the experimental design of forest diversity experiments.

16.3.1 Environmental Heterogeneity

In order to exclude confounding influences of environmental variables, abiotic factors within the field site should be as homogeneous as possible. For example, small variations in soil conditions can have large effects on growth of trees (Oliver and Larson 1996), which introduces the possibility that soil heterogeneity can act as a "hidden treatment" (Huston and McBride 2002). On the other hand, one important effect of biodiversity may be to allow efficient exploitation of a heterogeneous habitat. Because environmental heterogeneity is presumably the rule rather than the exception in forest ecosystems, and because it will frequently be in large part due to the big areas required for replicated experiments, tree diversity experiments may be well suited for testing this potential biodiversity effect.

Obviously, environmental heterogeneity can be dealt with *a posteriori* and measured environmental properties can be used as covariates in the analyses. Another possibility is to account for it at the planning stage by blocking. Randomized block and Latin square designs have been shown to be an efficient means of controlling and accounting for environmental heterogeneity or gradients (Potvin 2001). However, because simple randomization of treatments in space may not be sufficient to "equal out" this heterogeneity, without the number of replicates becoming prohibitively large, the spatial arrangement of plots in forest diversity experiments should be considered carefully. Furthermore, any analysis of data obtained from these experiments should include a "geographical" model, in which spatial information and neighbor-relationships between plots are included as additional explanatory variables (see, e.g., Ford and Renshaw 1984; Kempton and Lockwood 1984; Schmid et al. 2002).

The Panama forest diversity experiment provides an example of the possible importance of the environment on seedling growth. Over the whole landscape, topography, drainage, facing, height and density, and diversity of herbaceous vegetation were scored on an ordinal scale. Additionally, soil color was rated, based on Munsell soil color chart classifications (Munsell Colour 1990). To provide an adequate scale for measuring these micro-environmental characteristics, each diversity plot was divided into four sub-plots. Redundancy analysis (RDA) indicates that 86.3 % of the species–environment variance in height can be explained by the first canonical axis (F=17.154, P<0.001). RDA for diameter similarly shows that 82.4 % of the variance is explained by the first canonical axis (F=15.499, P<0.001). The environmental characteristics that correlate most strongly with the first canonical axis for both variables are topography and density of herbaceous vegetation. The inter-set correlations (topography, density) are –0.4241 and +0.3385 for height and –0.3811 and +0.3292 for diameter. The biplots for diameter (Fig. 16.6) and height reveal an almost identical distribution of species, suggesting that in the first year, primary and secondary growth responded in the

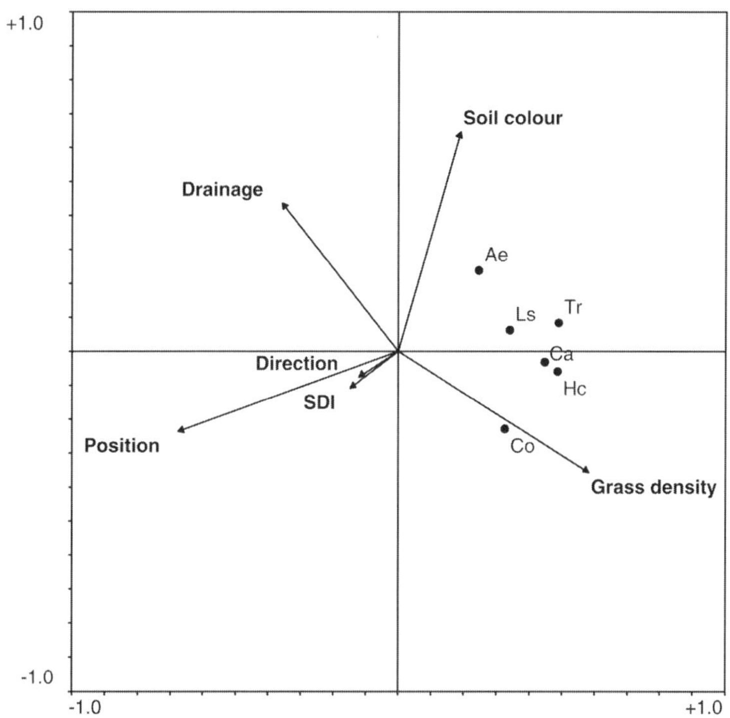

Fig. 16.6. Redundancy analysis (RDA) biplots for the species–environment variance in diameter with species scores positioned with regard to the first two canonical axes. Species abbreviations are as follows: (*Ls*) *Luehea seemanii*, (*Ca*) *Cordia alliodora*, (*Ae*) *Anacardium excelsum*, (*Hc*) *Hura crepitans*, (*Co*) *Cedrela odorata*, and (*Tr*) *Tabebuia rosea*

same way to environmental characteristics. They also reveal that all six species grow taller on the top of a hill than in depressions. Growth also has a positive correlation with areas of dense herbaceous vegetation. Herbaceous vegetation in contrast appeared to correlate mainly to soil color (inter-set correlation with first axis –0.4694) and litter accumulation (inter-set correlation with second canonical axis 0.4685). The cumulative percentage variance of species–environment relation explained by the two first canonical axes was 89.6 %.

These results suggest that in theory plots could have been established as blocks positioned differently in the landscape (hilltop, slope, depression). In practice, however, this was not possible because a sloped terrain offers unequal surface area of the three topographies, preventing the establishment of an equal number of plots for each topography.

Within-site heterogeneity is not the only factor to be considered. Variability of environmental conditions across larger areas must be included too. For instance, the outcome of species interactions critically depends on soil fertility. A nitrogen-fixing species in a mixture usually improves growth of other tree species on poor soils but may result in increased competition on rich ones (Binkley 1992). Therefore, it is important for the experiment to be replicated at several sites differing in geology, soil type or climate, as done in the Finnish and German experiments. Interestingly, within the BIODEPTH project it was shown for grasslands that, besides strong differences in environmental conditions – a general diversity – productivity relationship could be detected across eight sites (Hector et al. 1999).

16.3.2 Unit of Diversity

The term biodiversity encompasses several levels of biological variability, from genes to species and ecosystems (Heywood and Watson 1995). It is therefore necessary to define the term for the purpose of the study. For manipulative experiments with trees, one has to decide whether genotypes, species or some sort of functional type (groups of species with contrasting effects on ecosystem functioning) should be the basic unit of diversity. Most of the previous and existing forest experiments manipulated tree species richness. There is growing consensus, however, that any biodiversity effects on ecosystem functioning will arise from phenotypic variation between species, i.e., from their functional traits or from species' interactions (Loreau 2000), and even within species, i.e., from their age and stage classes or ecotypes (Schmid et al. 2002). Effects of species (or phenotypic variants within species) will thus be related to the size of their functional differences (Petchey and Gaston 2002), and the use of functional types instead of, or in addition to, species might be appropriate for many studies (Díaz and Cabido 2001; Hooper et al. 2002). However, grouping always reflects functional differences of a particular kind, and if several functional dimensions are combined, one often arrives at as many groups as there are species or phenotypic variants (Körner 1993; Smith et al. 1997; Wirth, Chap. 15, this Vol.).

In addition, because of the large plot sizes required for forest experiments, manipulations of both species number and functional diversity may not be feasible in a single experiment (but see examples in Table 16.1). Nevertheless, functional characteristics of individual species have to be considered when choosing the species for an experiment because they may affect the outcome of species interactions in mixtures. For instance, complementarity in resource use is more likely to be observed in mixtures of species that are using resources in different ways (e.g., between deciduous and evergreen species or between deep- and shallow-rooted species) than among species with similar requirements. Certain functional characteristics of tree species may even put constraints on the experimental design. For instance, it might be impossible to grow monocultures of late successional species on a clear-cut area because the species may only establish under a canopy of pioneers. Similarly, some species mixtures are unlikely to be maintained without human interference in the form of thinning, trimming, etc.

While functional characteristics of most native boreal and temperate species are well known, and the outcomes of interactions between these species may be to some extent predictable, ecological consequences of an introduction of an exotic tree species are more difficult to foresee (Engelmark et al. 2001; Peterken 2001). Therefore, some of the new forest diversity experiments include exotic tree species such as Siberian larch in Finland and Douglas fir in Germany, making it possible to assess interactions between native and introduced tree species.

Experiments manipulating genetic diversity within tree species have been rare, probably because within-species variation in functional characteristics, which may affect ecosystem functioning, is considered to be less than between-species differences. Yet, genetic diversity is the basis of all biodiversity, because it provides raw material for the adaptation, evolution and survival of species and individuals, especially under changing environment and disease conditions. Forest management practices, including tree improvement, can significantly affect the genetic variability of forest plantations (Gomory 1992; Rajora 1999). Several recent studies in Europe suggest that reduction in genetic diversity predisposes forests to an environmentally-related decline in health and productivity (Bergmann et al. 1990; Oleksyn et al. 1994; Raddi et al. 1994; Müller-Starck, Chap. 5, this Vol.) and to attacks by pests and pathogens (McCracken and Dawson 1998). Thus, genetic diversity is the foundation for forest sustainability and ecosystem stability. An experiment recently started in Finland (see Sect. 16.2.3, Table 16.1) specifically addresses the importance of within-species genetic variation in stand susceptibility to herbivores and pathogens by manipulating the number of silver birch clones per plot. In the BIOTREE experiment in Germany, the effects of high versus reduced genetic variation will be addressed by adding subplots at the one-species level planted with single clones or individuals derived from the same parent tree.

16.3.3 Diversity Gradient and Creation of Mixtures

The decision on the appropriate diversity gradient depends very much on the general aim of the study. If feasible, the gradient should range from single species stands to a certain maximum diversity mixture that could mimic either the diversity of natural or managed forest, or multi-species plantations of practical relevance.

Including monocultures of all species allows the comparison of the performance of mixtures vs. pure stands and the separation of selection versus complementarity effects. However, as discussed by Schmid et al. (2002), there are alternatives available to test for selection effects. In addition, having all monocultures or low-diversity mixtures of a restricted species pool also has some drawbacks, such as the restriction of random sampling of different species combinations at high diversity levels or the variance reduction effect (Huston 1997; Schmid et al. 2002). This implies the use of different communities at high diversity levels or repeating an entire design with several species pools.

From a practical point of view, one should ask what the "appropriate" background diversity level is in highly diverse systems. In the grassland BIODEPTH experiment, for example, the various sites used best estimates of natural "background" diversity as their reference point for determining the highest diversity level (Hector et al. 1999, 2002). In Panama, the 50-ha permanent forest plot on the island of Barro Colorado (BCI), near the tree diversity experiment, contains 177 tree species with diameter at breast height (dbh) larger than 20 cm. At a smaller scale in the forest, however, diversity is reduced (Palmer et al. 2000) because of the clustered distribution of species (Thorington et al. 1991). Thus we suggest that the appropriate background level of diversity to establish high diversity plots should be obtained from forest plots of equal size as the reforestation ones. In Panama, for example, preliminary work on BCI shows that, on average, a forest plot of 45×45 m has 19 different tree species of dbh >20 cm.

However, practical difficulties will complicate the design of tree diversity experiments in species-rich regions. First, most of the species are rare and therefore obtaining seeds is seldom possible except for those of the most common species. Once seeds have been obtained, and in contrast with the well-established forestry practices of northern countries such as Finland and Germany, germination protocols have to be developed. Even if seedlings are successfully germinated, establishment in the field is far from guaranteed. In Panama, the PRORENA project, which develops reforestation protocols for native trees, has been able to establish reforestation plots successfully with 40 species (www.prorena.org). We therefore believe that the maximum species number available for planting high tree-diversity plots will be limited by the existing knowledge in tropical forestry practices.

If the species pool is small (as in the experiments in Finland and Germany), a complete design with the planting of all possible species mixtures may be feasible (e.g., a pool of six species and one-, two-, four- and six-species mixtures: 37 possible combinations), which minimizes problems associated with the use of random-selection experiments (Huston and McBride 2002; Schmid et al. 2002; Schmid and Pfisterer 2003). With large species pools, as in the tropical systems, either a random selection of mixtures or a specific extinction or planting scenario may be used to create the gradient of diversity.

As an alternative, one could combine both approaches as in the Panama-experiment, planting a complete design with a small number of species plus some high-species-diversity plots (see Sect. 16.2.3.3; Table 16.1).

16.3.4 Plot Size

The appropriate size of the plots certainly will very much depend on the aim of the study. For example, if one is mainly interested in decomposition rates and effects on soil quality, plots smaller than 1 ha would probably be sufficient. However, if one wants to know how forest diversity affects diversity of birds, mammals, insects, etc., or if silvicultural management is to be performed, larger plots are required. In most cases, one will face a trade-off between size and number of plots due to the available field size. In attempting to solve this dilemma, the main factors influencing the ecosystem processes and characteristics of interest should be kept in mind. For instance, the influence of neighboring stands on soil chemistry of a target plot due to litter input can easily reach distances of several tens of meters (Rothe and Binkley 2001). This is dependent on the neighboring species (e.g., litter of broad-leaved species may be blown farther than needle litter), the main wind direction, and the presence of a closed forest edge that acts as a windbreaker. Thus, a checkerboard design with plots of say 20×20 m might even be too small to study effects on decomposition and soil chemistry.

The optimal plot would be large enough for a typical interior forest microclimate to develop, which for most forest types will certainly be larger than 1 ha, as edge zones with altered physical conditions may be up to two to three tree heights (e.g., Chen et al. 1995). Principles of conservation biology suggest that patches of 10 or even 100 ha may contain no true interior forest habitat. Because such large areas are never likely to be available for a replicated experiment, and would also exceed any reasonable level of financial and labor resources, one has to go for a pragmatic selection of plot sizes. As a rule of thumb we would recommend the use of double the height of the final tree height as side length of a plot, corresponding to a size of 0.5–1.0 ha, while recognizing that this size is still difficult to obtain and to manage. Given a replication of 20–40 plots, the whole experimental area would then be a size that also allows for the development of a typical forest microclimate.

16.3.5 Within-Plot Design: Spatial Arrangements

In most grassland biodiversity–functioning experiments that adopted the "synthetic community approach", plant mixtures were sown in bare soil. As a result of sowing, the spatial distribution of species within a plot initially is more or less homogeneous (random at the smallest scales – at least at initial sowing), and the identity of neighbors cannot be controlled. Nevertheless, the spatial pattern of mixed-species stands can influence community variables such as biomass in experiments with herbaceous species (Harper 1977; Schmid and Harper 1985; Stoll and Prati 2001). In particular, regular arrangements may lead to more rapid suppression of competitively inferior species by superior ones than random arrangements, and even more so than clumped ones. This is because regular arrangements maximize the frequency of interspecific neighbor relationships. Because trees are usually planted, it is possible to choose a spatial distribution of species and of individuals within a plot that yields a particular frequency (number of neighbors) and intensity (distance to neighbors) of neighbor relationships. This enables a plant-to-plant view in adopting techniques for neighborhood analyses (see e.g. Stoll et al. 1994).

In principle, within a plot, individuals could be arranged at random, in a regular manner, or in clumped patches (Fig. 16.7, upper part). For practical reasons, one would mostly adopt a regular planting scheme, which also mimics silvicultural practice for even-aged afforestations. Within such a pattern, the spatial distribution of different species may also be at random (as in the experiments in Finland and Borneo), in regular intervals (as in the Panama experiment), or in clumps (as in the BIOTREE experiment, Fig. 16.7, lower part). However, random placement of large numbers of seedlings may generate a wide range of spatial patterns within the plots. Finally, within a planting scheme in rows, the tree-to-tree distance between all neighbors could either be the same (hexagonal planting), or it could differ (quadratic planting, as shown in Fig. 16.7). The former is preferable for studying species interactions (Kelty and Cameron 1995; Gibson et al. 1999), whereas the latter is commonly used in silviculture.

The decision as to whether to plant in an aggregated pattern or not mainly depends on the expected duration of the experiment. Clumping is an effective way to prevent species loss due to interspecific competition at an early stage of the experiment, and to ensure the maintenance of the initial diversity gradient over long time intervals in terms both of species numbers and area occu-

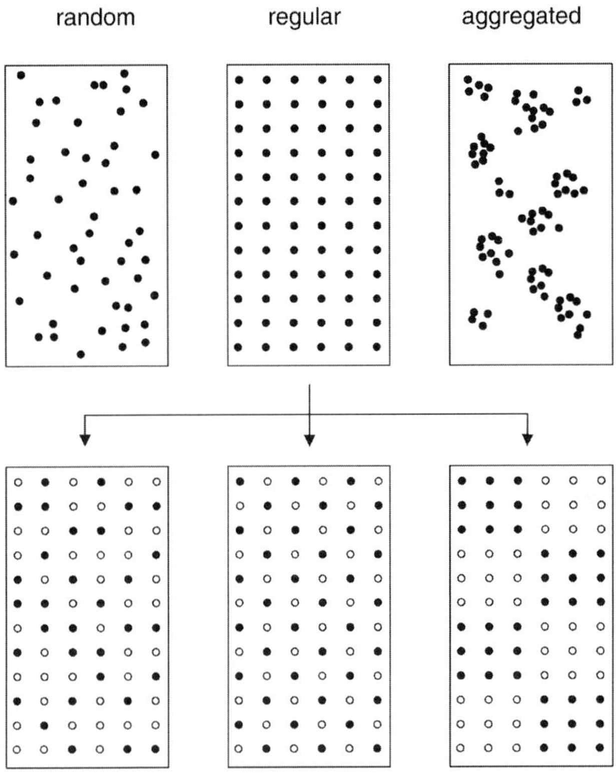

Fig. 16.7: Spatial distribution of individual trees within a plot. For manipulative diversity experiments adopting the "synthetic community approach," individuals will mainly be planted in a regular pattern (*upper part, center*). Within such a regular pattern, the distribution of different species can follow random, regular, or aggregated patterns (*lower part*), as shown for a two-species mixture here

pied by species. In contrast, if the expected duration of the experiment is short, a random or even a regular planting of each species might be preferable because this allow early interspecific interactions. However, strong competitors may outcompete inferior species, thus leading to changes in species composition. Such "species takeover" is surely less probable in tree than in grassland experiments, although fast-growing, clonal species like aspen or alder in boreal sites, for instance, may quickly outcompete neighboring species unless management interventions such as weeding or thinning are adopted (see Sect. 16.3.6).

In summary, we expect that the spatial distribution of species within a plot strongly affects ecosystem processes through alteration of competitive interactions (Pacala and Deutschman 1995; Stoll and Prati 2001). It would thus also be of interest for the outcome of the diversity–functioning relationship to test this assumption experimentally by comparing different spatial planting schemes.

16.3.6 Management Options

It is important to consider the maintenance strategy when designing large-scale experiments with trees. Usually, some sort of management has to be applied to ensure the establishment and maintenance of the communities. However, should the experiment mimic current silvicultural practices, or should management interventions be restricted to a minimal level? In most cases, site-specific trade-offs between requirements of science and of practice will have to be solved in pragmatic ways. Some examples should illustrate this:

- Weeding: If the experiment is to mimic natural forests, one should not weed. On the other hand, if the aim is to examine effects of a particular species loss on ecosystem functioning, the presence of this species as a weed on plots where it should not occur may spoil the whole experiment.
- Mowing: If the aim is to study successional dynamics after planting, it would be inappropriate to mow the understory. However, initial mowing will usually be necessary to reduce competition by grasses or herbs and to ensure successful tree establishment. Decisions concerning mowing could also have unforeseen side effects: in the Panama experiment, for instance, unmown plots were quickly infested by venomous snakes.
- Fencing: The need for this will relate to the density of mammalian herbivores in the experimental area. If it is so high that without fencing the experiment (or some parts of it) would be eliminated within a short time, one certainly should fence. On the other hand, if a very promising tree mixture (in terms of productivity, ecosystem processes, biodiversity maintenance, etc.) is identified in a fenced experiment, it may turn out that this mixture would not exist without a fence because it is much more susceptible to herbivores than other mixtures.
- Fertilization, fumigation: Given the financial investment needed to establish such experiments, tree survivorship should be maximized, at least in the first years. Thus, initial fertilization after planting must be carefully considered, as well as spot treatments with insecticides (in the tropics) or rodenticides (in temperate or boreal zones) if needed. However, since such treatments may also affect ecological processes at a later stage, they must be applied at an absolute minimum.
- Replacement: Judging whether tree mortality at an early stage should be regarded as a "valid" biodiversity effect (e.g., herbivores could be more attracted to high-diversity mixtures, see Jactel et al., Chap. 12, this Vol.), or as a danger for the success of the whole experiment might sometimes be difficult. Again, pragmatic choices depending on the site-specific situation will be necessary.

- Thinning: Requirement for this will depend on the duration of the experiment and on the initial planting density. If the planting density is high, seedlings will start to interact sooner, which might provide some results earlier. However, one probably needs to thin at some stage, otherwise alien thinning might eliminate some tree species in the mixtures. If thinning is planned, then the spatial arrangement within a plot should allow for that. On the other hand, natural thinning processes may reveal important species interactions, and only unmanaged plots can be used to determine the carrying capacity (in terms of total biomass for example) of the communities.
- Trimming: Plantation managers would usually trim to favor height growth or stemwood quality. However, if the experiment is considered as a model system for C-sink plantations, for instance, then any losses of C by management interventions should be avoided.

Because tree diversity experiments are more likely to mimic situations of managed forests or even plantations than natural, undisturbed forests, some sort of management will certainly have to be applied. The specific management options will first and foremost be dictated by the aim of each experiment. We believe that management strategies will often have to take into account both the risk of losing the experiment and the cost of the various options. In the context of dichotomous management strategies, we suggest the possibility of establishing split-plot experiments with managed and unmanaged subplots. Split-plot designs are indeed amenable to all the above questions (e.g., to fence or not, to fertilize or not, to thin or not – the latter is done in BIOTREE, for instance) and their analysis is well known (Winer 1971; Potvin 2001). The advantage of using them is to allow testing of additional experimental factors, while the disadvantage is a reduction in plot size as each main plot would be subdivided into subplots.

16.4 Response Variables

The general aim of forest diversity experiments is to examine the relationship between forest diversity and ecosystem functioning and/or the provision of goods and services. Optimally, the experimental design should enable responses in terms of different ecosystem processes and properties to be examined. The following list of key response variables is quite ambitious but would allow comparisons across different experiments:

- Tree mortality
- Tree growth and stand productivity (timber production)
- Canopy architecture (leaf area index, crown stratification)
- Phenology

- Carbon sequestration (in plants and soil)
- Nutrient retention in plants and soil (N, P, cations)
- Nutrient loss to groundwater (leaching)
- Decomposition, mineralization
- Soil respiration
- Microclimate, including soil temperature and water content
- Associated diversity (plants, above- and belowground arthropods, nematodes, earthworms, birds)
- Stand stability (resistance, resilience, and invariability)

Yet, depending on the specific interests and fields of expertise of the research team establishing the experiment, some response variables may have priority over others. Different research aims may put different constraints on the experimental design; and the design, in turn, has important consequences for the sampling procedures. For instance, plots with a clumped arrangement of species require a sampling that allows merging the properties of the many small, single-species patches at the whole-plot level. In Table 16.2, we attempt to analyze which features of experimental design are likely to be influenced by different response variables. Several generalizations emerge:

1. Variables that involve responses of individual trees (e.g., tree survival, growth, and canopy architecture) or a sum of individual tree responses (e.g., stand productivity) seem to be affected mainly by design features which concern tree distribution within plots (spatial arrangement, density, and proportion of each species in a mixture).
2. Response variables that are measured at the stand level and that are the products of stand functioning as a whole (e.g., nutrient retention, decomposition, hydrology- and microclimate-associated diversity) may be largely dependent on plot size and environmental heterogeneity. It is better, therefore, if experiments aimed at measuring the above variables are planted on the largest possible plots, and replicated on different soil types.
3. If response variables of interest involve dynamic properties of ecosystems (Schläpfer and Schmid 1999), such as resistance to invasions and frequency and severity of attacks by forest pests and pathogens, the duration of the experiment is crucial because it has to allow the above events to take place.

16.5 Major Caveats

Experiments using the "synthetic community approach" to study diversity effects on ecosystem functioning clearly have several general limitations (Huston and McBride 2002). In those with trees, additional restrictions have to be kept in mind. The most obvious ones are:

Table 16.2. Experimental design factors that can influence measured processes and properties of tree experiments in addition to the effects of abiotic factors

Key response variables	Design features which are likely to influence the outcome of the experiment
Tree mortality/survival	Spatial arrangement of trees within plots
	Planting density and proportion of each species in a mixture
	Presence/absence of herbivores (e.g., fencing)
Tree growth and stand productivity	Spatial arrangement of trees within plots
	Planting density and proportion of each species in a mixture
Canopy architecture	Spatial arrangement of trees within plots
	Planting density and proportion of each species in a mixture
Carbon sequestration	Stand management (trimming, thinning)
Nutrient retention and loss to groundwater	Plot size (microclimate)
	Proportion of each species in a mixture
Decomposition and mineralization rates	Plot size (litter input from adjacent plots, microclimate)
	Proportion of each species in a mixture
Soil respiration	Plot size (microclimate)
	Proportion of each species in a mixture
Microclimate	Plot size
Associated biodiversity	plot size (depending on size and mobility of the organisms)
Stability	Duration of the experiment

- It is an even-aged plantation and not a naturally established forest. Many multi-species forests established from natural seedling invasion and regeneration contain trees of several age classes. Differences in age also influence stand structure, which is another important factor of forest biodiversity (Franklin et al. 2002). However, there are also numerous examples for natural even-aged and homogeneous forests, e.g., after stand-replacing fires in the boreal zone.
- Large field sites are needed. Depending on the number of replicate communities and the chosen plot size, the field sites may extend over large areas. For example, the experiments described here use single field sites

ranging from <2 ha (Finland), 9 ha (Panama), 50 ha (Germany), up to 500 ha (Borneo). This also restricts the options for an "optimal" location (in terms of environmental heterogeneity, infrastructure, accessibility, etc.). Communications with the state forestry administration or commercial forestry companies may help to find and get access to such large areas.

- High costs for establishment and maintenance. Due to the large field size needed, leasing costs for land may be substantial and regular research-funding institutions will presumably not be in a position to pay them. Unorthodox alliances may help to solve this problem. In the Borneo experiment, the planting of such a large area was possible, for example, by linking the infrastructure of a commercial carbon offset project with the biodiversity experiment. In Germany, loss of forest area due to the recent construction of a new highway has to be offset by afforestation, which will be achieved by the proposed experiment. Researchers could take advantage of similar legislation that may exist in other countries. Together with the increasing demand for afforestation in implementing the Kyoto Protocol (Schulze et al. 2002), such large-scale experiments might then be feasible.

16.6 Outlook

Existing manipulative tree biodiversity experiments demonstrate that such large-scale and long-term projects are feasible and furthermore indicate biodiversity effects even at a very early stage, contrary to most expectations. Of course, experiments using the "synthetic community approach", though promising, represent only one way to study biodiversity effects on ecosystem functioning and/or the provision of ecosystem goods and services. There are many possibilities for integration and comparison with complementary approaches, such as removal experiments and observational studies. Indeed, such integrated perspectives are needed in order to assemble the general picture about diversity–functioning relationships in forests.

We would thus like to encourage researchers to explore the possibilities to set up similar experiments that could be complementary to those presented here with respect to experimental design, size, species pool, or biome. The inclusion of manipulations of genetic diversity within species would be particularly important. A global network of tree diversity experiments would indeed be a fascinating opportunity to improve our understanding of ecosystem functioning, closing the gap between ecosystem process studies and population and biodiversity studies.

Acknowledgments: We thank Helen Jones for helpful comments on the manuscript. Parts of the study were supported by the German Science Foundation (DFG; FOR 456), by the Max-Planck-Society (Germany), by the Ministère de la Recherche, de la Science et de la Technologie (Quebec, Canada), through a grant from the Conseil de Recherche en Science Naturelle et en Génie du Canada to Catherine Potvin, by The Royal Society (UK) and by the Academy of Finland. Catherine Potvin and Zosia Bornik want to thank Jose Monteza for constant support in the field. Andy Hector and Glen Reynolds are grateful to Charles Godfray. Michael Scherer-Lorenzen and Ernst-Detlef Schulze acknowledge the Thüringer Landesanstalt für Wald, Jagd und Fischerei, and the Bundesforstamt Thüringer Wald for their fruitful cooperation; special thanks are extended to Eberhard Weller.

References

Allison GW (1999) The implications of experimental design for biodiversity manipulations. Am Nat 153:27–45

Andow DA (1991) Vegetation diversity and arthropod population response. Annu Rev Entomol 36:561–586

Bergmann F, Gregorius HR, Larsen JB (1990) Levels of genetic variation in European silver fir (*Abies alba*) – are they related to the species' decline? Genetica 82:1–10

Berish CW, Ewel JJ (1988) Root development in simple and complex tropical successional ecosystems. Plant Soil 106:73–84

Binkley D (1992) Functioning of nitrogen$_2$-fixing and non-nitrogen$_2$-fixing tree species. In: Cannell MGR, Malcolm DC, Robertson PA (eds) The ecology of mixed-species stands of trees. Blackwell, Oxford, pp 99–124

Caspersen JP, Pacala SW (2001) Successional diversity and forest ecosystem function. Ecol Res 16:895–903

Chen JQ, Franklin JF, Spies TA (1995) Growing-season microclimatic gradients from clear-cut edges into old-growth Douglas-fir forests. Ecol Appl 5:74–86

de Wit CT (1960) On competition. Verslagen Landbouwkundige Onderzoekingen 66:1–82

Díaz S, Cabido M (2001) Vive la différence: plant functional diversity matters to ecosystem processes. Trends Ecol Evol 16:646–655

Díaz S, Symstad AJ, Stuart Chapin IF, Wardle DA, Huenneke LF (2003) Functional diversity revealed by removal experiments. Trends Ecol Evol 18:140–146

Dutilleul P, Potvin C (1995) Among-environment heteroscedasticity and genetic autocorrelation: implications for the study of phenotypic plasticity. Genetics 139:1815–1829

Engelmark O, Sjoberg K, Andersson B, Rosvall O, Agren GI, Baker WL, Barklund P, Bjorkman C, Despain DG, Elfving B, Ennos RA, Karlman M, Knecht MF, Knight DH, Ledgard NJ, Lindelöw Å, Nilsson C, Peterken GF, Sörlin S, Sykes MT (2001) Ecological effects and management aspects of an exotic tree species: the case of lodgepole pine in Sweden. For Ecol Manage 141:3–13

Ewel JJ, Mazzarino MJ, Berish CW (1991) Tropical soil fertility changes under monocultures and successional communities of different structure. Ecol Appl 1:289–302

Ford ED, Renshaw E (1984) The interpretation of processes from pattern using two-dimensional spectral analysis: modelling single species pattern in vegetation. Vegetatio 56:113–123

Franklin JF, Spies TA, van Pelt R, Carey AB, Thornburgh DA, Berg DR, Lindenmayer DB, Harmon ME, Keeton WS, Shaw DC, Bible K, Chen JQ (2002) Disturbances and structural development of natural forest ecosystems with silvicultural implications, using Douglas-fir forests as an example. For Ecol Manage 155:399–423

Freckleton RP, Watkinson AR (2000) On detecting and measuring competition in spatially structured plant communities. Ecol Lett 3:423–432

Gibson DJ, Connolly J, Hartnett DC, Weidenhammer JD (1999) Designs for greenhouse studies of interactions between plants. J Ecol 87:1–16

Gomory D (1992) Effect of stand origin on the genetic diversity of Norway spruce (*Picea abies* Karst.) populations. For Ecol Manage 54:215–223

Hallé F (1986) Modular growth in seed plants. Philos Trans R Soc Lond B 313:77–87

Harper JL (1977) Population biology of plants. Academic Press, London

Hector A, Schmid B, Beierkuhnlein C, Caldeira MC, Diemer M, Dimitrakopoulos PG, Finn J, Freitas H, Giller PS, Good J, Harris R, Högberg P, Huss-Danell K, Joshi J, Jumpponen A, Körner C, Leadley PW, Loreau M, Minns A, Mulder CPH, O'Donovan G, Otway SJ, Pereira JS, Prinz A, Read DJ, Scherer-Lorenzen M, Schulze E-D, Siamantziouras A-SD, Spehn E, Terry AC, Troumbis AY, Woodward FI, Yachi S, Lawton JH (1999) Plant diversity and productivity experiments in European grasslands. Science 286:1123–1127

Hector A, Loreau M, Schmid B, project atB (2002) Biodiversity manipulation experiments: studies replicated at multiple sites. In: Loreau M, Naeem S, Inchausti P (eds) Biodiversity and ecosystem functioning. Synthesis and perspectives. Oxford Univ Press, Oxford, pp 36–46

Heywood VH, Watson RT (eds) (1995) Global biodiversity assessment. Cambridge Univ Press, Cambridge

Holden C (2003) Forest ecologists go mega. Science 300:1872

Hooper DU, Solan M, Symstad A, Díaz S, Gessner MO, Buchmann N, Degrange V, Grime P, Hulot F, Mermillod-Blondin F, Roy J, Spehn EM, van Peer L (2002) Species diversity, functional diversity, and ecosystem functioning. In: Loreau M, Naeem S, Inchausti P (eds) Biodiversity and ecosystem functioning. Synthesis and perspectives. Oxford Univ Press, Oxford, pp 195–208

Huston MA (1997) Hidden treatments in ecological experiments: re-evaluating the ecosystem function of biodiversity. Oecologia 110:449–460

Huston MA, McBride AC (2002) Evaluating the relative strengths of biotic versus abiotic controls on ecosystem processes. In: Loreau M, Naeem S, Inchausti P (eds) Biodiversity and ecosystem functioning: synthesis and perspectives. Oxford Univ Press, Oxford, pp 47–60

Kelty MJ, Cameron IR (1995) Plot design for the analysis of species interactions in mixed stands. Comm For Rev 74:322–332

Kempton RA Lockwood G (1984) Inter-plot competition in variety trials of field beans (*Vicia faba* L.). J Agric Sci 103:293–302

Kinzig AP, Pacala SW, Tilman D (2002) The functional consequences of biodiversity: empirical progress and theoretical extensions. Princeton Univ Press, Princeton

Kira T, Ogawa H, Shinozaki K (1953) Intraspecific competition among higher plants. 1. Competition-density-yield inter-relationships in regularly dispersed populations. J Inst Polytech Osaka Cy Univ D 4:1–16

Körner C (1993) Scaling from species to vegetation: the usefulness of functional groups. In: Schulze E-D, Mooney HA (eds) Biodiversity and ecosystem function. Springer, Berlin Heidelberg New York, pp 117–140

Kohyama T, Hotta M (1990) Significance of allometry in tropical saplings. Funct Ecol 4:515–521

Lamont BB (1995) Testing the effect of ecosystem composition/structure on its functioning. Oikos 74:283–295

Lawton JH, Naeem S, Thompson LJ, Hector A, Crawley MJ (1998) Biodiversity and ecosystem function: getting the Ecotron experiment in its correct context. Funct Ecol 12:848–852

Loreau M (2000) Biodiversity and ecosystem functioning: recent theoretical advances. Oikos 91:3–17

Loreau M, Hector A (2001) Partitioning selection and complementarity in biodiversity experiments. Nature 412:72–76

Loreau M, Naeem S, Inchausti P (eds) (2002) Biodiversity and ecosystem functioning: Synthesis and perspectives. Oxford Univ Press, Oxford

McCracken AR, Dawson WM (1998) Short rotation coppice willow in Northern Ireland since 1973: development of the use of mixtures in the control of foliar rust (*Melampsora* spp.). Eur J For Pathol 28:241–250

Mikola J, Salonen V, Setälä H (2002) Studying the effects of plant species richness on ecosystem functioning: does the choice of experimental design matter? Oecologia 133:594–598

Oleksyn J, Prus-Glowacki W, Giertych M, Reich PB (1994) Relation between genetic diversity and pollution impact in a 1912 experiment with east European *Pinus sylvestris* provenances. Can J For Res 24:2390–2394

Oliver CD, Larson BC (1996) Forest stand dynamics. Wiley, New York

Pacala SW, Deutschman DH (1995) Details that matter – the spatial distribution of individual trees maintains forest ecosystem function. Oikos 74:357–365

Palmer, MW, Clark DB, Clark DA (2000) Is the number of tree species in small tropical forest plots nonrandom? Comm Ecol 1:95–101

Petchey OL, Gaston KJ (2002) Functional diversity (FD), species richness and community composition. Ecol Lett 5:402–411

Peterken GF (2001) Ecological effects of introduced tree species in Britain. For Ecol Manage 141:31–42

Potvin C (2001) ANOVA: experiments layout and analysis. In: Scheiner S, Gurevitch J (eds) Design and analysis of ecological experiments. Oxford Univ Press, Oxford, pp 63–76

Raddi S, Stefanini FM, Camussi A, Giannini R (1994) Forest decline index and genetic variability in *Picea abies* (L.) Karst. For Genet 1:33–40

Rajora OP (1999) Genetic biodiversity impacts of silvicultural practices and phenotypic selection in white spruce. Theor Appl Genet 99:954–961

Rothe A, Binkley D (2001) Nutritional interactions in mixed species forests: a synthesis. Can J For Res 31:1855–1870

Schläpfer F, Schmid B (1999) Ecosystem effects of biodiversity: a classification of hypotheses and cross-system exploration of empirical results. Ecol Appl 9:893–912

Schmid B, Harper JL (1985) Clonal growth in grassland perennials. I. Density and pattern dependent competition between plants with different growth form. J Ecol 73:793–808

Schmid B (2002) The species richness-productivity controversy. Trends Ecol Evol 17:113–114

Schmid B, Pfisterer AB (2003) Species vs community perspectives in biodiversity experiments. Oikos 100:620–621

Schmid B, Hector A, Huston MA, Inchausti P, Nijs I, Leadley PW, Tilman D (2002) The design and analysis of biodiversity experiments. In: Loreau M, Naeem S, Inchausti P (eds) Biodiversity and ecosystem functioning: synthesis and perspectives. Oxford Univ Press, Oxford, pp 61–75

Schulze ED, Valentini R , Sanz M-J (2002) The long way from Kyoto to Marrakesh: implications of the Kyoto Protocol negotiations for global ecology. Global Change Biol 8:505–518

Smith TM, Shugart HH, Woodward FI (eds) (1997) Plant functional types – their relevance to ecosystem properties and global change. Cambridge Univ Press, Cambridge

Spaèkova I, Lepš J (2001) Procedure for separating the selection effect from other effects in diversity-productivity relationship. Ecol Lett 4:585–594

Stoll P, Prati D (2001) Intraspecific aggregation alters competitive interactions in experimental plant communities. Ecology 82:319–327

Stoll P, Weiner J, Schmid B (1994) Growth variation in a naturally established population of *Pinus sylvestris*. Ecology 75:660–670

Thorington RW, Tannenbaum B, Tarak A, Rudran R (1991) Distribución de los arboles en la ísla de Barro Colorado: una muestra de cinco hectareas. Smithsonian Tropical Research Institute, Panama

Tilman D, Reich PB, Knops J, Wedin D, Mielke T, Lehmann C (2001) Diversity and productivity in a long-term grassland experiment. Science 294:843–845

Vitousek PM, Hooper DU (1993) Biological diversity and terrestrial ecosystem biogeochemistry. In: Schulze ED, Mooney HA (eds) Biodiversity and ecosystem function. Springer, Berlin Heidelberg New York, pp 3–14

Walker B, Kinzig A, Langridge J (1999) Plant attribute diversity, resilience, and ecosystem function: the nature and significance of dominant and minor species. Ecosystems 2:95–113

WBGU – German Advisory Council on Global Change (2000) World in transition: conservation and sustainable use of the biosphere. Earthscan, London

Winer BJ (1971) Statistical principles in experimental design. McGraw-Hill, New York

Yoda K, Kira T, Ogawa H, Hozumi K (1963) Self thinning in overcrowded pure stands under cultivated and natural conditions. J Biol Osaka Cy Univ 14:107–129

17 The Functional Significance of Forest Diversity: a Synthesis

M. Scherer-Lorenzen, Ch. Körner, and E.-D. Schulze

17.1 A Lack of Functional Biodiversity Research in Forests?

Despite of the tremendous increase in knowledge about the relationship between biodiversity and ecosystem functioning during the last decade (Scherer-Lorenzen et al., Chap. 1, this Vol.), it should be noted that most of the studies were conducted with model systems, which – for very practical reasons – were small-statured, short-lived and even-aged, mainly herbaceous assemblages or microbial microcosms (e.g., Tilman et al. 1997b; Hector et al. 1999; Petchey et al. 2002; for an overview, see Schläpfer and Schmid 1999; Schmid et al. 2002). Experiments in forest ecosystems have been almost absent, with the exception of studies manipulating diversity of consumers or decomposers in the soil (e.g., Mikola and Setälä 1998; Laakso and Setälä 1999; see Scheu, Chap. 11, this Vol.). A manipulation of the producer level, i.e., trees, is obviously a difficult and long-lasting task and only recently attempts in this direction have been made (Scherer-Lorenzen et al., Chap. 16, this Vol.). The experiment by Ewel and colleagues in the tropics of Costa Rica (Berish and Ewel 1988; Ewel et al. 1991) has often been mentioned as the first manipulative diversity experiment indicating diversity effects on biogeochemistry (Vitousek and Hooper 1993). However, this experiment was designed to explore the possibilities of developing sustainable agroecosystems for the humid tropics, mimicking structural diversity of successional communities, and not to study the interaction of species richness and ecosystem functioning per se. Clear effects on soil chemistry were detectable between maize monocultures and highly-diverse (>100 species) treatments consisting of herbaceous and woody plants. Low and intermediate levels of diversity were lacking, which should be the part of the gradient where most effects are expected to occur, according to local deterministic processes involving species interactions (see below). Positive effects at such intermediate levels of tree species richness have been reported from afforestation experiments in Costa Rica, for example (Byard et al. 1996; Montagnini 2000). In contrast, mixture experiments from forestry

Ecological Studies, Vol. 176
M. Scherer-Lorenzen, Ch. Körner, and E.-D. Schulze (Eds.)
Forest Diversity and Function: Temperate and Boreal Systems

sciences lack mostly intermediate and high diversity treatments, because forestry is mainly interested in one- and two-species assemblages of economically important species (Pretzsch, Chap. 3, this Vol.; Jones et al., Chap. 6, this Vol.; Kelty 1992; Malcolm and Mason 1999). Therefore, in the search for biodiversity–ecosystem functioning relations in forest ecosystems, we still have to rely mostly on observational, comparative studies rather than on experimental, manipulative experiments. This bias is also reflected in the present book, and we have to keep in mind that, albeit such approaches cannot be used to determine causality due to covarying factors (Vilà et al., Chap. 4, this Vol.), they provide many valuable insights into the correlates of ecosystem functioning (Körner, Chap. 2, this Vol.).

17.2 Mechanisms of Mixture Effects, or: Are There Differences Between Grasslands and Forests?

There are no obvious biological reasons why the mechanisms responsible for diversity effects on ecosystem processes in grassland, agricultural systems, or microcosms should not work also in slow-growing forest communities. These mechanisms have been grouped into two classes, representing (1) local deterministic processes, such as niche differentiation or resource partitioning, and/or facilitation among different species, leading to increased yield of mixtures compared to that of the corresponding monocultures (complementary effect) (Loreau et al. 2002; Tilman and Lehman 2002); (2) local and regional stochastic processes involved in community assembly of experimental systems leading to a higher probability that a certain species with strong impacts on ecosystem processes is present at higher diversity ("sampling" or "selection-effect models") (Aarssen 1997; Huston 1997; Tilman et al. 1997a). In the following, we do not consider sampling effect models here, because they are mainly relevant for experimental systems with random sampling of species out of a fixed species pool and some sort of selection for species with extreme traits.

17.2.1 Niche Partitioning and Functional Traits

Niche partitioning among tree species is well known for certain mixed forest types, as shown by Pretzsch (Chap. 3, this Vol.), Jones (Chap. 6, this Vol.), and Kelty (1992). Following Harper (1977), such complementary species are termed species with "ecological combining abilities" (Harper 1977, p. 265 and p. 762 ff.). In general, this indicates that complementarity, with a more efficient resource use in mixtures compared to monocultures, occurs if the functional traits of species cause interspecific competition to be less than intraspecific competition (called "competitive production principle" by Vandermeer 1989). Examples from forestry are stratified mixtures of sun-adapted species in the overstory and shade-adapted species in the understory, the combination of early- and late-successional species, or ontogenetically early and late-culminating species (Assmann 1970; Kelty 1992; Körner, Chap. 2, this Vol.; Pretzsch, Chap. 3, this Vol.). The many citations from old German silvicultural books presented by Pretzsch (Chap. 3, this Vol.) show that this principle was well recognized by foresters more than 175 years ago. However, for a long time, foresters were also aware of negative effects of mixing species, which are of the same magnitude as potentially positive mixing effects, due to competitive interactions where the inferior competitor can only be sustained by silvicultural interference (Pretzsch, Chap. 3, this Vol.). Additionally, comparative studies have not found strong evidence for mixture effects on productivity in temperate forests of the northern hemisphere (Schulze et al. 1996), although Caspersen and Pacala (2001) reported an asymptotic increase in wood production with increasing tree species richness in North American forests.

Similarly, there seems to be no general relation between biodiversity and nutrient cycling in boreal and temperate forests (Schulze et al. 1996; Rothe and Binkley 2001), although niche complementarity models would lead one to expect so (Tilman 1999; Loreau 2000; Chesson et al. 2002). Again, certain mixtures do show enhanced nutrient uptake in comparison to the corresponding monospecific stands, whereas others do not. The combination of shallow- and deep-rooting species, such as Norway spruce with oak, European beech, or Scots pine provides an example for belowground niche partitioning, enhancing nutrient uptake under certain circumstances (Rothe and Binkley 2001; Thelin et al. 2002), as well as for stability against windthrow (Dhôte, Chap. 14, this Vol.). However, the distribution of roots within the soil profile must not necessarily coincide with nutrient uptake, and differentiations according to site properties are to be expected (Rothe and Binkley 2001). As shown by Hättenschwiler (Chap. 8, this Vol.), diversity effects on decomposition and nutrient mineralization are known for some mixtures due to inter- and intraspecific variations in litter quality. However, the relationship between litter species richness and process rate does not yet appear to be predictable, and the identity of species within a mixture, i.e., the functional traits of the species, has been shown to be more important than the number of species.

Resource partitioning and complementarity are discussed mostly in relation to productivity of the plant community or nutrient retention, but this concept is also applicable to other interactions between species. For example, Jactel (Chap. 12, this Vol.) has documented that complementary food sources in more diverse habitats can result in enhanced fecundity and longevity of specialized parasitoids, which increases the effectiveness of parasitism on forest pests, thus reducing damage in more diverse stands.

Whether mixtures will show complementarity or not depends on the ecological differences among species within a community, i.e., on the species' functional traits; hence, it should be possible to predict diversity effects on certain ecosystem functions if the species traits are known. Such traits are well known for tree species (at least for boreal and temperate species), although they can vary enormously depending on developmental stage or site factors (Körner, Chap. 2, this Vol.). The influence of tree species on ecosystem processes based on such traits is also well documented (Wirth et al., Chap. 15, this Vol.; Zinke 1962; Binkley and Valentine 1991; Binkley and Giardina 1998; Rothe and Binkley 2001; Augusto et al. 2002; Prescott 2002). The above-mentioned examples of positive mixture effects from forestry explicitly combine species with different traits, e.g., light-demanding and shade-tolerant species. Thus, the analysis of gradients in functional diversity in contrast to variations in species richness might be more promising for detecting complementarity effects. On the other hand, because functional groups are arbitrary divisions of a continuous niche space, and if several ecosystem functions are considered simultaneously, one may end up with the notion that each species represents its own functional group (Körner 1993; Körner, Chap. 2, this Vol.; Wirth, Chap. 15, this Vol.). This underlines the usefulness of species richness as one simple measure of ecological differences between species.

At this point, we want to emphasize that agroforestry systems explicitly make use of resource complementarity and facilitation to increase and/or stabilize yields by deliberately selecting species with differing functional traits, and that many aspects of the biodiversity–ecosystem functioning relation have been discussed in that area (Ewel 1986; Huxley 1999; Ashton 2000; Kelty 2000).

17.2.2 Differences Between Grassland and Forests

The strong evidence for diversity effects on ecosystem functioning through resource partitioning or facilitation mainly derives from experiments with fast-growing model ecosystems such as grasslands, where plant diversity was directly manipulated (Schläpfer and Schmid 1999; Schmid et al. 2002). In contrast, results from comparative studies are inconsistent, partly because both diversity and some measure of ecosystem functioning (in most cases, productivity) may be limited by the same site factors, leading to spurious correlations between the two (Wardle 2001), and because such "third variables" were often not adequately measured and incorporated into the statistical analysis (Schmid et al. 2002). An obvious difference between biodiversity–ecosystem functioning relationships found in grasslands and forests may thus be grounds for different types of scientific approach: comparisons of within-habitat diversity effects (experimental grasslands) versus across-habitat diversity effects (observational studies in forests).

What kind of biological differences between fast-growing grassland systems and forests are important when considering biodiversity–ecosystem functioning relations? While herbaceous/grassland communities rebuild most of their interacting aboveground structures year by year from close to zero, trees may take a hundred or more years to fill a large three-dimensional volume, which permits very small differences among individuals to accumulate in a compound-interest fashion. In herbaceous species such interest effects also exist, but are mainly limited to reproductive output and belowground structures (which are considered to interact "symmetrically"). Individuals of herbaceous systems reach maximum height year by year, whereas trees persist at gradually increasing height. This is not just a scaling issue in space and time, but a substantial qualitative difference in how species and their individuals interact. One consequence of this difference is that far more co-dominants tend to coexist in long established and non-fertilized grassland systems, whereas mature temperate or boreal forests commonly exhibit a dominance of few, mostly one to three, species. Such mature forest ecosystems are thus much more dependent on the characteristics of a small set of species than are grassland systems. In terms of plant life "strategies," grassland systems commonly retain a large r component and forests select for a large contribution of K components.

Additionally, the woody nature of the supporting structures of trees, i.e., stems, branches, and twigs, imply a "memory effect" in the crown architecture of trees that cannot be seen in grasslands. For example, under conditions of light competition in a closed stand, the crown may develop in an asymmetrical manner foraging for light. If a gap is formed later on the opposite side of the crown's main direction, the tree is not able to make use of this additional light in the short term, and the crown remains asymmetrical for at least several years or decades. In contrast, the rapid lateral growth of grassland species (especially clonal growing ones) and the greater flexibility of the non-lignified shoot system lead to a quick recovery of the open resource space created by the loss of neighboring individuals. Additionally, many grasses and herbs show photosynthetic activity in the shoot and can immediately make use of lateral incoming light without the need to regenerate or rearrange leaves. Thus, although the additional light in a forest gap might be used by tree seedlings or the herbaceous layer, the more flexible grassland system should have relatively more constant light use at the stand level than forest communities. This will have different consequences in both systems for light complementarity, as well as for other aspects of ecosystem functioning that depend on the aboveground structure, such as susceptibility to wind damage or the habitat function of the canopy.

Finally, most late successional grasslands, including natural ones, totally depend on disturbance (fire, grazers, mowing) and would convert to forests without these disturbances (Archibold 1995). Thus, grasslands (except for wet habitats) do not reach a "steady-state" situation (and the associated ecosystem function) unless they are regularly disturbed, with the degree of disturbance determining biodiversity and functioning. Forests may, at least theoretically, arrive at a "steady state" situation if there is no disturbance, and should have corresponding ecosystem functions. Although almost all boreal and temperate forests do also have natural disturbance regimes (fire, pest outbreaks), these disturbances generally do not permanently convert the forest into a totally different ecosystem type because regrowth of tree species is fast. Interestingly, in old-growth forests growing under favorable conditions, such as the Pacific Northwest in North America, or the coast of the South Island of New Zealand, certain ecosystem functions are, however, not associated with a "steady state" (in terms of species composition): these forests presumably continue to accumulate soil carbon, for example. Although nearly all modern temperate forests are heavily disturbed (managed), the evolutionary traits of trees have not been selected to the same extent as grassland species for operation under regular disturbance. These differences have to be accounted for when biodiversity theory, derived from regularly disturbed and fast growing systems, is applied to forests.

17.3 Research Needs

What do we need in order to gain more insight into the relationship between forest diversity and ecosystem functioning? Clearly, we need more comparative, observational studies to document patterns of forest biodiversity and correlate them to ecological processes within those ecosystems. There is a large variety of mixed species stands to be explored, with its diversity being determined by biogeographical and historical conditions, abiotic and biotic site factors, and human management interventions. However, unless site conditions are extremely similar, across-habitat or across-locality comparisons can be misleading because between site differences may obscure within-habitat effects of diversity on ecosystem processes (Vilá et al., Chap. 4, this Vol.; Lawton et al. 1998; Schmid 2002). As mentioned by Körner (Chap. 2, this Vol.), careful site characterization and large site numbers are needed to come to a reasonable signal-to-noise ratio. In addition, among-site abiotic variation has to be adequately accounted for by including these "third variables" as covariates in statistical analysis (Schmid et al. 2002). The exploration of forestry inventory data in combination with thinnings emphasizing biodiversity could also yield new insights. However, management may often confound diversity effects (Vilá et al., Chap. 4, this Vol.; Mund and Schulze, Chap. 10, this Vol.), so both aspects have to be clearly separated in such observational and inventory studies.

In order to detect causal mechanisms of diversity–functioning relationships, we certainly need more experimental work specifically designed for this task, as described by Scherer-Lorenzen et al. (Chap. 16, this Vol.). However, because life cycles of trees are up to two orders of magnitude longer than normally funded research projects and because results may only be gained after several years of observation (but see work with tropical systems: Scherer-Lorenzen et al., Chap. 16, this Vol.), we may also use model systems made up of tree seedlings or saplings (e.g., Körner and Arnore 1992). Although such model systems definitively will not reflect processes in natural, highly structured forests (not even to mention old-growth forests), they may provide initial insight into mechanisms and may thus be used to formulate more specific hypotheses. At this point, we have to note that the overwhelming number of studies on the response of forest ecosystems to increasing CO_2 concentrations also arise from studies with seedling or saplings (Körner 1995), and only in two cases was the whole canopy influence of CO_2 studied in diverse mature forest tree assemblages, a mixed Mediterranean evergreen/oak forest around a natural CO_2 spring in Tuscany (Tognetti et al. 1996) and a mature temperate forest in Switzerland (Pepin and Körner 2002).

Besides comparative studies in existing stands and experimental biodiversity plantations adopting the "synthetic-community approach" (Scherer-Lorenzen et al., Chap. 16, this Vol.), a third promising way for further research is the realization of removal or addition experiments. Such experiments, where single species or entire functional groups are removed from or added to existing communities, have some drawbacks (e.g., large disturbance effects, change of density, spatial segregation of species), but can be useful in certain circumstances (Freckleton and Watkinson 2000; Díaz et al. 2003). Certainly, it would be promising to re-analyze the magnitude of thinning experiments (removals) or conversions of conifer-dominated stands into more natural, multi-species communities (additions) from a biodiversity–ecosystem functioning perspective.

Besides the adoption of such complementary approaches, we propose the following points to be considered in future work, which should not only document patterns of biodiversity–ecosystem functioning relationships, but should also be mechanistically driven:

- One aspect that should receive more attention in analyzing diversity–functioning relationships in forests is the influence of confounding factors due to changing abiotic conditions (Vilá et al., Chap. 4, this Vol.) and management interventions (Mund and Schulze, Chap. 10, this Vol.; Cannell et al. 1992; Kelty et al. 1992; Olsthoorn et al. 1999). Because silvicultural treatments tend to standardize the basal area of the stands (Kramer 1988), potential diversity effects could simply be "thinned away" by extracting biomass that normally is not accounted for in inventories. For example, the analysis of carbon stocks in aboveground biomass in beech forests of Thuringia, Germany, shows no significant differences between pure and mixed stands at all ages on similar soil conditions, which is mainly a result of thinning interventions (C. Wirth, unpubl. data). At ages between 80 and

120 years, pure stands have even slightly higher carbon stocks than mixed stands. However, it may well be that the yield by thinning is higher in mixed than in monotypic stands. Thus, the wood extraction must be known when evaluating biodiversity effects in managed forests.

- Especially the question of spatial aggregation both horizontally (neighbor analyses) and vertically (canopy stratification, and the above-mentioned "memory effect" of woody structures) may be of importance in biodiversity studies of forests because the spatial distribution of species strongly affects ecosystem processes through alteration of competitive interactions (Pacala and Deutschman 1995; Stoll and Prati 2001). A major challenge for future work will thus be the integration of different spatial scales where biodiversity may have effects.
- A further important arena will be the study of the significance of intraspecific variation among forest tree taxa (Müller-Starck et al., Chap. 5, this Vol.). From the limited evidence we have, it seems effects of such differences can even exceed consequences of interspecific differences (e.g., in the case of susceptibility to certain diseases, Pautasso et al., Chap. 13, this Vol.).
- Totally overseen has been the importance of soil microbial biodiversity at this stage. Gleixner et al. (Chap. 9, this Vol.) show that carbon storage in soils depends to a large extent on the microbial biodiversity in the soil and not on the biodiversity in the stand. Nevertheless, Schulze et al. (2004) demonstrate that ecosystem diversity and the contribution of the soil to proteins identified in the soil water are very well coupled to the composition of the vegetation. However, the whole complex of interlinkages and feedback mechanisms between above- and belowground diversity and their influence on ecosystem functioning remains a large field for research (Jactel et al., Chap. 12, this Vol., Scheu et al., Chap. 11, this Vol.). In particular, it seems to be a safe prediction that the study of interactions between trees and their mycorrhizal partners will remain among the most promising focal points of forest biodiversity research (Smith and Read 1997).
- Given the longevity of forest ecosystems, long-term continuous studies are compulsory. For instance, disturbances such as fire (Wirth, Chap. 15, this Vol.) or large temporal fluctuations in the population size of forest insects have particular relevance for ecosystem dynamics and stability (Jactel et al., Chap. 12, this Vol.), but may have different consequences depending on forest biodiversity.

17.4 Conclusions

The contributions in this volume clearly show that forest diversity can have a variety of effects on ecosystem processes and characteristics, such as productivity and timber production (Pretzsch, Chap. 3, this Vol.; Vilá et al., Chap. 4, this Vol.; Müller-Starck et al., Chap. 5, this Vol.), biogeochemistry (Baldocchi, Chap. 7, this Vol.; Jones et al., Chap. 6, this Vol.; Hättenschwiler, Chap. 8, this Vol.; Gleixner et al., Chap. 9, this Vol.; Mund and Schulze, Chap. 10, this Vol.; Wirth, Chap. 15, this Vol.), associated fauna (Scheu, Chap. 11, this Vol.; Jactel et al., Chap. 12, this Vol.), and stability against disturbances (Pautasso et al., Chap. 13, this Vol.; Dhôte, Chap. 14, this Vol.). Although most presentations have both reviewed the existing literature and analyzed data of own research projects, evidence of consistent patterns of the relationship between forest diversity and ecological processes remained scarce.

Interestingly, the strongest positive tree diversity effects were related to some aspects of stand stability, such as resistance to pest insects (Jactel et al., Chap. 12, this Vol.) or susceptibility to fungal pathogens (Pautasso et al., Chap. 13, this Vol.), supporting the often expressed notion of higher stability in mixed stands. However, Dhôte (Chap. 14, this Vol.) could not support this rule of thumb concerning resistance to strong winds.

In most cases, a variety of possible answers to the central question "Does diversity matter?" emerged. For example, under certain site conditions, effects of tree species number on soil fauna diversity (Scheu, Chap. 11, this Vol.), litter decomposition (Hättenschwiler, Chap. 8, this Vol.), and resistance to pathogens (Pautasso et al., Chap. 13, this Vol.) could be detected, but effects differed at different sites and with the involvement of different tree species. This variety of possible answers has its roots in the distinction between the effects of species numbers or species identities. The notion that the identity of species within a mixture is more important than the number of species is the thread running through all chapters. Thus, it becomes clear that there is no "magic effect of numbers of species per se," and that any effect will arise instead from functional differences between species and from species interactions (Hector et al. 2000). There cannot be any relationship between species richness and ecosystem processes without these differences between species (Lawton et al. 1998). Therefore, the characteristic traits of species and thus the diversity of functions these species perform are important determinants for ecosystem processes (e.g., Baldocchi, Chap. 7, this Vol., Wirth et al., Chap. 15, this Vol.).

However, we have to recognize that our knowledge on patterns of diversity in forests and the implications this diversity may have for ecosystem processes or services are still very limited, largely because natural temperate forests hardly exist where most researchers are based (Europe, USA), and because the nature of trees prevents conventional short-term manipulative experimentation. The former limitation opens an arena for research in the last existing natural temperate forest ecosystems (e.g., in the Far East of Asia) and the latter urges better utilization of existing silviculture management results in our landscape (as exemplified by several authors in this volume).

It also appears that a priori definitions of functional tree types may be less practical and helpful in predicting forest functioning than post hoc assessments of the consequences of presence or absence of certain tree taxa (but see Wirth et al., Chap. 15, this Vol.). Furthermore, it is obvious that the significance of forest diversity (and the presence of certain taxa) depends on the forest functions considered. Windthrow risk, nitrate retention, catchment value for water yielding, or support of high wild ungulate diversity are subjects of concern that will always foster different answers with respect to whether and how forest diversity matters. We conclude that future explorations of the functional significance of forest species diversity needs to more clearly account for those differences in forest function. From what we have learned to date, it seems that for the most basic actual ecosystem functions, such as the direct contribution to biogeochemical cycles, diversity is of less significance than might be expected, and that more subtle and indirect and longer-term effects may become crucial. This and the needed time for such effects to materialize again suggest that the study of given forest assemblages will have to remain a focal point and will continue to cause forest biodiversity research to differ from grassland biodiversity research. It remains to be seen whether theory derived from grassland research will meet the given reality in forests of contrasting diversity.

References

Aarssen LW (1997) High productivity in grassland ecosystems: effects by species diversity or productive species? Oikos 80:183–184

Archibold OW (1995) Ecology of world vegetation. Chapman and Hall, London

Ashton PS (2000) Ecological theory of diversity and its application to mixed species plantation systems. In: Ashton MS, Montagnini F (eds) The silvicultural basis for agroforestry systems. CRC Press, Boca Raton, pp 61–77

Assmann E (1970) The principles of forest yield study. Pergamon Press, Oxford

Augusto L, Ranger J, Binkley D, Rothe A (2002) Impact of several common tree species of European temperate forests on soil fertility. Ann For Sci 59:233–253

Berish CW, Ewel JJ (1988) Root development in simple and complex tropical successional ecosystems. Plant Soil 106:73–84

Binkley D, Valentine D (1991) Fifty-year biogeochemical effects of green ash, white pine, and Norway spruce in a replicated experiment. For Ecol Manage 40:13–25

Binkley D, Giardina C (1998) Why do tree species affect soils – the warp and woof of tree-soil interactions. Biogeochemistry 42:89–106

Byard R, Lewis KC, Montagnini F (1996) Leaf litter decomposition and mulch performance from mixed and monospecific plantations of native tree species in Costa Rica. Agric Ecosyst Environ 58:145–155

Cannell MGR, Malcolm DC, Robertson PA (eds) (1992) The ecology of mixed-species stands of trees. Blackwell, Oxford

Caspersen JP, Pacala SW (2001) Successional diversity and forest ecosystem function. Ecol Res 16:895–903

Chesson P, Pacala S, Neuhauser C (2002) Environmental niches and ecosystem functioning. In: Kinzig AP, Pacala SW, Tilman D (eds) The functional consequences of biodiversity. Empirical progress and theoretical extensions. Princeton Univ Press, Princeton, pp 213–245

Díaz S, Symstad AJ, Chapin SF III, Wardle DA, Huenneke LF (2003) Functional diversity revealed by removal experiments. Trends Ecol Evol 18:140–146

Ewel J (1986) Designing agricultural ecosystems for the humid tropics. Annu Rev Ecol Syst 17:245–271

Ewel JJ, Mazzarino MJ, Berish CW (1991) Tropical soil fertility changes under monocultures and successional communities of different structure. Ecol Appl 1:289–302

Freckleton RP, Watkinson AR (2000) On detecting and measuring competition in spatially structured plant communities. Ecol Lett 3:423–432

Harper JL (1977) Population biology of plants. Academic Press, London

Hector A, Schmid B, Beierkuhnlein C, Caldeira MC, Diemer M, Dimitrakopoulos PG, Finn J, Freitas H, Giller PS, Good J, Harris R, Högberg P, Huss-Danell K, Joshi J, Jumpponen A, Körner C, Leadley PW, Loreau M, Minns A, Mulder CPH, O'Donovan G, Otway SJ, Pereira JS, Prinz A, Read DJ, Scherer-Lorenzen M, Schulze E-D, Siamantziouras A-SD, Spehn E, Terry AC, Troumbis AY, Woodward FI, Yachi S, Lawton JH (1999) Plant diversity and productivity experiments in European grasslands. Science 286:1123–1127

Hector A, Schmid B, Beierkuhnlein C, Caldeira MC, Diemer M, Dimitrakopoulos PG, Finn J, Freitas H, Giller PS, Good J, Harris R, Högberg P, Huss-Danell K, Joshi J, Jumpponen A, Körner C, Leadley PW, Loreau M, Minns A, Mulder CPH, O'Donovan G, Otway SJ, Pereira JS, Prinz A, Read DJ, Scherer-Lorenzen M, Schulze E-D, Siamantziouras A-SD, Spehn E, Terry AC, Troumbis AY, Woodward FI, Yachi S, Lawton JH (2000) No consistent effect of plant diversity on productivity. Response. Science 289:1255

Huston MA (1997) Hidden treatments in ecological experiments: re-evaluating the ecosystem function of biodiversity. Oecologia 110:449–460

Huxley P (1999) Tropical agroforestry. Blackwell, Oxford

Kelty MJ, Larson BC, Oliver CD (eds) (1992) The ecology and silviculture of mixed-species forests. Kluwer, Dordrecht

Kelty MJ (1992) Comparative productivity of monocultures and mixed-species stands. In: Kelty MJ, Larson BC, Oliver CD (eds) The ecology and silviculture of mixed-species forests. Kluwer, Dordrecht, pp 125–141

Kelty MJ (2000) Species interactions, stand structure, and productivity in agroforestry systems. In: Ashton MS, Montagnini F (eds) The silvicultural basis for agroforestry systems. CRC Press, Boca Raton, pp 183–205

Körner C (1993) Scaling from species to vegetation: the usefulness of functional groups. In: Schulze E-D, Mooney HA (eds) Biodiversity and ecosystem function. Springer, Berlin Heidelberg New York, pp 117–140

Körner C (1995) Towards a better experimental basis for upscaling plant responses to elevated CO_2 and climate warming. Plant Cell Environ 18:1101–1110

Körner C, Arnore JA III (1992) Responses to elevated carbon dioxide in artificial tropical ecosystems. Science 257:1672–1675

Kramer H (1988) Waldwachstumslehre. Parey, Hamburg

Laakso J, Setälä H (1999) Sensitivity of primary production to changes in the architecture of belowground food webs. Oikos 87:57–64

Lawton JH, Naeem S, Thompson LJ, Hector A, Crawley MJ (1998) Biodiversity and ecosystem function: getting the Ecotron experiment in its correct context. Funct Ecol 12:848–852

Loreau M (2000) Biodiversity and ecosystem functioning: recent theoretical advances. Oikos 91:3–17

Loreau M, Naeem S, Inchausti P (2002) Perspectives and challenges. In: Loreau M, Naeem S, Inchausti P (eds) Biodiversity and ecosystem functioning. Synthesis and perspectives. Oxford Univ Press, Oxford, pp 237–242

Malcolm DC, Mason WL (1999) Experimental mixtures of Scots pine and birch: 30 years effects on production, vegetation and soils. In: Olsthoorn AFM, Bartelink HH, Gardiner JJ, Pretzsch H, Hekhuis HJ, Franc A (eds) Management of mixed-species forest: silviculture and economics. IBN-DLO, Wageningen, pp 79–87

Mikola J, Setälä H (1998) Relating species diversity to ecosystem functioning: mechanistic background and experimental approach with a decomposer food web. Oikos 83:180–194

Montagnini F (2000) Accumulation in above ground biomass and soil storage of mineral nutrients in pure and mixed plantations in a humid tropical lowland. For Ecol Manage 134:257–270

Olsthoorn AFM, Bartelink HH, Gardiner JJ, Pretzsch H, Hekhuis HJ, Franc A (eds) (1999) Management of mixed-species forests: silviculture and economics. IBN-DLO, Wageningen

Pacala SW, Deutschman DH (1995) Details that matter – the spatial distribution of individual trees maintains forest ecosystem function. Oikos 74:357–365

Pepin S, Körner C (2002) Web-FACE: a new canopy free-air CO_2 enrichment system for tall trees in mature forests. Oecologia 133:1–9

Petchey OL, Casey T, Jiang L, McPhearson PT, Price J (2002) Species richness, environmental fluctuations, and temporal change in total community biomass. Oikos 99:231–240

Prescott CE (2002) The influence of the forest canopy on nutrient cycling. Tree Physiol 22:1193–1200

Rothe A, Binkley D (2001) Nutritional interactions in mixed species forests: a synthesis. Can J For Res 31:1855–1870

Schläpfer F, Schmid B (1999) Ecosystem effects of biodiversity: a classification of hypotheses and cross-system exploration of empirical results. Ecol Appl 9:893–912

Schmid B (2002) The species richness-productivity controversy. Trends Ecol Evol 17:113–114

Schmid B, Joshi J, Schläpfer F (2002) Empirical evidence for biodiversity-ecosystem functioning relationships. In: Kinzig A, Tilman D, Pacala S (eds) Functional consequences of biodiversity: experimental progress and theoretical extensions. Princeton Univ Press, Princeton, pp 120–150

Smith SE, Read DJ (1997) Mycorrhizal symbiosis. Academic Press, San Diego

Schulze E-D, Bazzaz FA, Nadelhoffer KJ, Koike T, Takatsuki S (1996) Biodiversity and ecosystem function of temperate deciduous broad-leaved forests. In: Mooney HA, Cushman JH, Medina E, Sala OE, Schulze E-D (eds) Functional roles of biodiversity: a global perspective. Wiley, Chichester, pp 71–98

Schulze W, Gleixner G, Mann M, Schulze E-D (2004) A protein fingerprint of biodiversity. Proc Natl Acad Sci USA (in press)

Stoll P, Prati D (2001) Intraspecific aggregation alters competitive interactions in experimental plant communities. Ecology 82:319–327

Thelin G, Sverdrup H, Holmquist J, Rosengren U, Linden M (2002) Sustainability in spruce and mixed-species stands. In: Sverdrup H, Stjernquist I (eds) Developing principles and models for sustainable forestry in Sweden. Kluwer, Dordrecht, pp 337–354

Tilman D (1999) The ecological consequences of changes in biodiversity: a search for general principles. Ecology 80:1455–1474

Tilman D, Lehman C (2002) Biodiversity, composition, and ecosystem processes: theory and concepts. In: Kinzig AP, Pacala SW, Tilman D (eds) The functional consequences of biodiversity: Empirical progress and theoretical extensions. Princeton Univ Press, Princeton, pp 9–41

Tilman D, Lehman C, Thomson K (1997a) Plant diversity and ecosystem productivity: theoretical considerations. Proc Natl Acad Sci USA 94:1857–1861

Tilman D, Knops J, Wedin D, Reich P, Ritchie M, Siemann E (1997b) The influence of functional diversity and composition on ecosystem processes. Science 277:1300–1302

Tognetti R, Giovanelli A, Longobucco A, Miglietta F, Raschi A (1996) Water relations of oak species growing in the natural CO_2 spring of Rapolano (central Italy). Ann Sci For 53:475–485

Vandermeer JH (1989) The ecology of intercropping. Cambridge Univ Press, Cambridge

Vitousek PM, Hooper DU (1993) Biological diversity and terrestrial ecosystem biogeochemistry. In: Schulze E-D, Mooney HA (eds) Biodiversity and ecosystem function. Springer, Berlin Heidelberg New York, pp 3–14

Wardle DA (2001) No observational evidence for diversity enhancing productivity in Mediterranean shrublands. Oecologia 129:620–621

Zinke PJ (1962) Pattern of influence of individual forest trees on soil properties. Ecology 43:130–133

Taxonomic Index (Genera)

Subject Index

Ecological Studies

Volumes published since 1998

Volume 136
Landscape Disturbance and Biodiversity in Mediterranean-Type Ecosystems (1998)
P.W. Rundel, G. Montenegro, and F.M. Jaksic (Eds.)

Volume 137
Ecology of Mediterranean Evergreen Oak Forests (1999)
F. Rodà et al. (Eds.)

Volume 138
Fire, Climate Change and Carbon Cycling in the North American Boreal Forest (2000)
E.S. Kasischke and B. Stocks (Eds.)

Volume 139
Responses of Northern U.S. Forests to Environmental Change (2000)
R. Mickler, R.A. Birdsey, and J. Hom (Eds.)

Volume 140
Rainforest Ecosystems of East Kalimantan: El Niño, Drought, Fire and Human Impacts (2000)
E. Guhardja et al. (Eds.)

Volume 141
Activity Patterns in Small Mammals: An Ecological Approach (2000)
S. Halle and N.C. Stenseth (Eds.)

Volume 142
Carbon and Nitrogen Cycling in European Forest Ecosystems (2000)
E.-D. Schulze (Ed.)

Volume 143
Global Climate Change and Human Impacts on Forest Ecosystems: Postglacial Development, Present Situation and Future Trends in Central Europe (2001)
J. Puhe and B. Ulrich

Volume 144
Coastal Marine Ecosystems of Latin America (2001)
U. Seeliger and B. Kjerfve (Eds.)

Volume 145
Ecology and Evolution of the Freshwater Mussels Unionoida (2001)
G. Bauer and K. Wächtler (Eds.)

Volume 146
Inselbergs: Biotic Diversity of Isolated Rock Outcrops in Tropical and Temperate Regions (2000)
S. Porembski and W. Barthlott (Eds.)

Volume 147
Ecosystem Approaches to Landscape Management in Central Europe (2001)
J.D. Tenhunen, R. Lenz, and R. Hantschel (Eds.)

Volume 148
A Systems Analysis of the Baltic Sea (2001)
F.V. Wulff, L.A. Rahm, and P. Larsson (Eds.)

Volume 149
Banded Vegetation Patterning in Arid and Semiarid Environments (2001)
D. Tongway and J. Seghieri (Eds.)

Volume 150
Biological Soil Crusts: Structure, Function, and Management (2001, 2003)
J. Belnap and O.L. Lange (Eds.)

Volume 151
Ecological Comparisons of Sedimentary Shores (2001)
K. Reise (Ed.)

Volume 152
Global Biodiversity in a Changing Environment: Scenarios for the 21st Century (2001)
F.S. Chapin, O. Sala, and E. Huber-Sannwald (Eds.)

Volume 153
UV Radiation and Arctic Ecosystems (2002)
D.O. Hessen (Ed.)

Volume 154
Geoecology of Antarctic Ice-Free Coastal Landscapes (2002)
L. Beyer and M. Bölter (Eds.)

Volume 155
Conserving Biological Diversity in East African Forests: A Study of the Eastern Arc Mountains (2002)
W.D. Newmark

Volume 156
Urban Air Pollution and Forests: Resources at Risk in the Mexico City Air Basin (2002)
M.E. Fenn, L. I. de Bauer, and T. Hernández-Tejeda (Eds.)

Volume 157
Mycorrhizal Ecology (2002, 2003)
M.G.A. van der Heijden and I.R. Sanders (Eds.)

Volume 158
Diversity and Interaction in a Temperate Forest Community: Ogawa Forest Reserve of Japan (2002)
T. Nakashizuka and Y. Matsumoto (Eds.)

Volume 159
Big-Leaf Mahogany: Genetic Resources, Ecology and Management (2003)
A. E. Lugo, J. C. Figueroa Colón, and M. Alayón (Eds.)

Volume 160
Fire and Climatic Change in Temperate Ecosystems of the Western Americas (2003)
T. T. Veblen et al. (Eds.)

Volume 161
Competition and Coexistence (2002)
U. Sommer and B. Worm (Eds.)

Volume 162
How Landscapes Change: Human Disturbance and Ecosystem Fragmentation in the Americas (2003)
G.A. Bradshaw and P.A. Marquet (Eds.)

Volume 163
Fluxes of Carbon, Water and Energy of European Forests (2003)
R. Valentini (Ed.)

Volume 164
Herbivory of Leaf-Cutting Ants: A Case Study on *Atta colombica* in the Tropical Rainforest of Panama (2003)
R. Wirth, H. Herz, R.J. Ryel, W. Beyschlag, B. Hölldobler

Volume 165
Population Viability in Plants: Conservation, Management, and Modeling of Rare Plants (2003)
C.A Brigham, M.W. Schwartz (Eds.)

Volume 166
North American Temperate Deciduous Forest Responses to Changing Precipitation Regimes (2003)
P. Hanson and S.D. Wullschleger (Eds.)

Volume 167
Alpine Biodiversity in Europe (2003)
L. Nagy, G. Grabherr, Ch. Körner, D. Thompson (Eds.)

Volume 168
Root Ecology (2003)
H. de Kroon and E.J.W. Visser (Eds.)

Volume 169
Fire in Tropical Savannas: The Kapalga Experiment (2003)
A.N. Andersen, G.D. Cook, and R.J. Williams (Eds.)

Volume 170
Molecular Ecotoxicology of Plants (2004)
H. Sandermann (Ed.)

Volume 171
Coastal Dunes: Ecology and Conservation (2004)
M.L. Martínez and N. Psuty (Eds.)

Volume 172
Biogeochemistry of Forested Catchments in a Changing Environment: A German Case Study (2004)
E. Matzner (Ed.)

Volume 173
Insects and Ecosystem Function (2004)
W. W. Weisser and E. Siemann (Eds.)

Volume 174
Pollination Ecology and the Rain Forest: Sarawak Studies (2005)
D. Roubik, S. Sakai, and A.A. Hamid (Eds.)

Volume 175
Antarctic Ecosystems: Environmental Contamination, Climate Change, and Human Impact (2005)
R. Bargagli

Volume 176
Forest Diversity and Function: Temperate and Boreal Systems (2005)
M. Scherer-Lorenzen, Ch. Körner, and E.-D. Schulze (Eds.)